Alternative Medicine Interventions for COVID-19

Muhammad Zia-Ul-Haq
May Nasser Bin-Jumah • Sarah I. Alothman
Hanan A. Henidi
Editors

Alternative Medicine Interventions for COVID-19

Editors
Muhammad Zia-Ul-Haq
Lahore College for Women University
Lahore, Pakistan

Sarah I. Alothman
Biology Department
Princess Nourah bint Abdulrahman University
Riyadh, Saudi Arabia

May Nasser Bin-Jumah
Biology Department
Princess Nourah bint Abdulrahman University
Riyadh, Saudi Arabia

Hanan A. Henidi
Health Sciences Research Centre
Princess Nourah bint Abdulrahman University
Riyadh, Saudi Arabia

ISBN 978-3-030-67988-0   ISBN 978-3-030-67989-7   (eBook)
https://doi.org/10.1007/978-3-030-67989-7

© The Editor(s) (if applicable) and The Author(s), under exclusive license to Springer Nature Switzerland AG 2021
This work is subject to copyright. All rights are solely and exclusively licensed by the Publisher, whether the whole or part of the material is concerned, specifically the rights of translation, reprinting, reuse of illustrations, recitation, broadcasting, reproduction on microfilms or in any other physical way, and transmission or information storage and retrieval, electronic adaptation, computer software, or by similar or dissimilar methodology now known or hereafter developed.
The use of general descriptive names, registered names, trademarks, service marks, etc. in this publication does not imply, even in the absence of a specific statement, that such names are exempt from the relevant protective laws and regulations and therefore free for general use.
The publisher, the authors, and the editors are safe to assume that the advice and information in this book are believed to be true and accurate at the date of publication. Neither the publisher nor the authors or the editors give a warranty, expressed or implied, with respect to the material contained herein or for any errors or omissions that may have been made. The publisher remains neutral with regard to jurisdictional claims in published maps and institutional affiliations.

This Springer imprint is published by the registered company Springer Nature Switzerland AG
The registered company address is: Gewerbestrasse 11, 6330 Cham, Switzerland

Qutb al-Aqtab, Shaykh al-Hadees, Barakat al-Asr, Hazrat Muhammad Zakariya Kandhlawi Saharanpuri Muhajir Madani Rahmatullah and Fakhar-ul-Mashaikh, Peer-e-Tariqat, Rahbar-e-Shareeyat, Wali-e-Kamil, Aarif Billah, Syed-ul-Aarfeen, Sanad ul-Wasleen, Sheikh-ul-Arab Walajam, Mujaddad-e-Dauran, Badr-ul-Auliya, Mureed Nawaz, Hazrat Dr. Shahid Awais Naqashbandi Mujaddidi Damat Barakatohum (Patron AAS academy).
Dr. Muhammad Zia-Ul-Haq

To my parents, husband, and children for supporting me in my journey in life and to Princess Nourah bint Abdulrahman University for all kind of support.
Dr. May Nasser Bin-Jumah

To my parents, husband, and children for their kindness and devotion and to Princess Nourah bint Abdulrahman University for its endless support.
Dr. Sarah I. Alothman

To my family and colleagues at the Health Sciences Research Center, Princess Nourah bint Abdulrahman University, for
their support and constant encouragement.
Dr. Hanan A. Henidi

Preface

The novel coronavirus has been the focus of the whole world since early 2020. Scientists and researchers are trying their best to find a cure for the disease caused by this virus, namely COVID-19. Slowing the COVID-19 pandemic depends on a combination of pharmacologic and nonpharmacologic interventions. Initial SARS-CoV-2 prevention includes social distancing, face masks, environmental hygiene, and hand washing. Although the most important pharmacologic interventions to prevent SARS-CoV-2 infection are likely to be vaccines, the repurposing of established drugs for short-term prophylaxis is another more immediate option. Similarly, many traditional systems of medicines are being practiced (with varying rates of success) in various parts of the world.

This book has a twin objective: (a) to provide a state-of-the-art and scientific review of alternative modes of treatment for COVID-19, and (b) to encourage research into this emerging topic. The book consists of nine chapters. The COVID-19 story, from the origin of this term to our current knowledge, is briefly reported in Chap. 1. The following eight chapters, written by highly qualified scientists in their fields, provide a detailed and thorough study of the various modes of COVID-19 treatment. Chapters 2 and 3 address the use of medicinal plants and traditional Chinese medicine against COVID-19, respectively. Chapter 4 details an overview of plant natural products as a remedy against COVID-19. The role of food in prevention and cure against this disease is described in Chap. 5. Globally, many researchers are engaged in carrying out research on existing drugs to develop remedies for COVID-19. This will considerably shorten the total duration of drug discovery and development, which extends from laboratory investigations to initial toxicity testing to animal studies to safety studies in humans and, finally, large-scale clinical trials. All this has been described in Chap. 6. Vaccines as most effective remedy against COVID-19 and recent developments in vaccine research are the subjects of Chaps. 7 and 8, respectively. Chapter 9 summarizes various modes of treatment and future prospects in COVID-19 research. It is expected that researchers in the medical profession and allied health professions will be interested in this book. We hope this book will also attract the attention of researchers in all fields as well as the layman as it presents existing literature in simple language.

We would like to thank each of the authors for their contributions and for their dedicated efforts in providing carefully prepared chapters. It is the aim of the editors that this book will be of benefit and a reference source to anyone researching the area on COVID-19 treatment options. We tried to include existing pertinent literature in the book; however, we might have missed some significant papers due to huge literature on this topic. Our apologies to the authors whose papers we could not include. We are obligated to the staff at Springer for their support in assembling this work and their efforts in keeping this book on schedule. Finally, we have a message for every reader of this book. These collaborative book projects of hundreds of thousands of words may always contain some errors or gaps. Therefore, instructive comments or even criticism are always welcome.

Lahore, Pakistan	Muhammad Zia-Ul-Haq
Riyadh, Saudi Arabia	May Nasser Bin-Jumah
Riyadh, Saudi Arabia	Sarah I. Alothman
Riyadh, Saudi Arabia	Hanan A. Henidi

Contents

1. **Introduction to COVID-19** 1
   Naheed Bano, Fatima Batool, and May Nasser Bin-Jumah

2. **Medicinal Plants as COVID-19 Remedy** 33
   Sara Zafar, Shagufta Perveen, Naeem Iqbal, M. Kamran Khan,
   Modhi O. Alotaibi, and Afrah E. Mohammed

3. **Traditional Chinese Medicines as Possible Remedy
   Against SARS-CoV-2** 63
   Saqib Mahmood, Tariq Mahmood, Naeem Iqbal, Samina Sabir,
   Sadia Javed, and Muhammad Zia-Ul-Haq

4. **Plant-Based Natural Products: Potential Anti-COVID-19 Agents** ... 111
   Sana Aslam, Matloob Ahmad, and Hanan A. Henidi

5. **Foods as First Defense Against COVID-19** 153
   Mahwish and Sarah I. Alothman

6. **Drugs for the Treatment of COVID-19** 193
   Sagheer Ahmed, Halimur Rehman, Rehan Salar,
   May Nasser Bin-Jumah, M. Tauseef Sultan, and Marius Moga

7. **COVID-19 Pandemic and Vaccines** 205
   Hina Qaiser, Roheena Abdullah, Tehreema Iftikhar,
   Hammad Majeed, and Imran Imran

8. **Updates in Vaccine Development Against COVID-19** 237
   Sagheer Ahmed, Rehan Salar, Halimur Rehman, Sarah I. Alothman,
   M. Riaz, and Muhammad Zia-Ul-Haq

9. **COVID-19: Recent Developments in Therapeutic Approaches** 249
   Umar Farooq Gohar, Irfana Iqbal, Zinnia Shah, Hamid Mukhtar,
   and Muhammad Zia-Ul-Haq

Index 275

About the Editors

Muhammad Zia-Ul-Haq holds a PhD from the University of Karachi, Pakistan, besides LLB from Punjab University, Lahore, and LLM-IP from Turin University, Italy. He is currently serving at the Office of Research, Innovation and Commercialization, Lahore College for Women University, as senior manager. Previously, he served as patent examiner in The Patent Office, IPO Pakistan (Ministry of Commerce), for 8 years. Dr. Zia-Ul-Haq received patents training from Japan, Korea, Malaysia, USA, and France. He has published 2 books with Springer and more than 120 research and review papers, with total IF of 120 and total Google Scholar Citations of 4000. He is peer reviewer of many journals published by Elsevier and Springer. Dr. Zia-Ul-Haq won RPA from Pakistan Council for Science and Technology (PCST), Ministry of Science and Technology (MOST), from 2010 to 2015.

May Nasser Bin-Jumah holds master's and doctoral degrees in biology- environment, pollution, and toxicity from King Saud University, Saudi Arabia. She teaches students different courses in biology at Princess Nourah bint Abdulrahman University. Dr. Bin-Jumah is interested in research in all subjects of biology and has published articles about the environment, pollution, anticancer, microorganisms, natural products, animal behavior, and neuroscience.

Sarah I. Alothman holds master's and doctoral degree in histology and cell biology (embryology) from Princess Nourah bint Abdulrahman University, Saudi Arabia. She teaches different courses in biology at Princess Nourah Bint Abdulrahman University. Dr. Alothman has published more than 40 research papers in embryology, toxicology, pollution, natural products, and many other fields.

Hanan A. Henidi is currently an associate researcher in the Health Sciences Research Center at Princess Nourah bint Abdulrahman University, Saudi Arabia. She holds a PhD in biochemistry from King Abdulaziz University, Saudi Arabia. Her current research interests are in the potential chemo modulatory effect of natural compounds against several diseases, and designing and screening of potential inhibitors against threatening diseases such as cancer.

Chapter 1
Introduction to COVID-19

Naheed Bano, Fatima Batool, and May Nasser Bin-Jumah

Introduction

SARS-CoV-2 is a zoonotic origin beta virus which has modified itself for man-to-man transfer (Fig. 1.1) and nowadays very famous for causing pandemic COVID-19 [1]. Till today reported cases of COVID -19 are in millions globally with thousands of reported deaths (https://coronavirus.jhu.edu). In response to the present challenges, people are facing travel bans and social distancing [2], with changes in anthropoid activity, followed by severe economic damage. Subsequently, each country is affected differentially with respect to social and economic losses by this COVID-19 pandemic. No doubt, China, which is considered as the origin land of this pandemic, is expected to recover faster than other countries like Europe and America. Other countries which will be more affected by long periods of social distancing are dependent on the tourism and automobile industries. It has also been reported that the economic growth will return to normal after three quarters 2023 (McKinsey report at https://t.co/iFC1k1A2WM). All aspects including conservation and health will be affected by this pandemic globally [3, 4].

If we look at the health aspects, the influence of this pandemic will be of three types: The first one is immediate impact which is due to rapid human limitation, inactivity, and confinement. The second medium impact includes crisis related to

---

N. Bano (✉)
Faculty of Veterinary and Animal Sciences (FVAS), MNS University of Agriculture, Multan, Pakistan

F. Batool
Department of Botany, Division of Science and Technology, University of Education Lahore, Punjab, Pakistan

M. N. Bin-Jumah
Biology Department, College of Sciences, Princess Nourah bint Abdulrahman University, Riyadh, Saudi Arabia

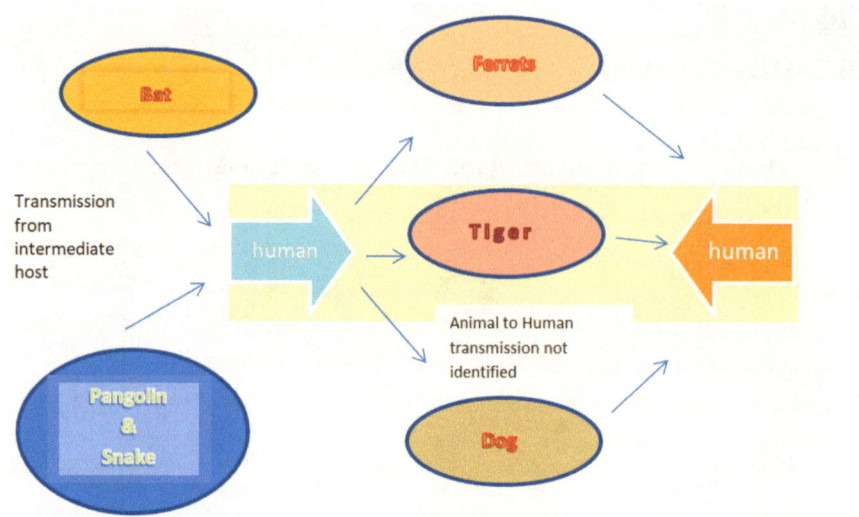

Fig. 1.1 Zoonotic link for coronavirus and flow path

animal farming and veterinary facilities which will ultimately in the long term affect the economy. The third impact is the improved consideration toward public well-being implications of this coronavirus contagion in pets and farm animals [5–7]. During the crisis of coronavirus, all countries should maintain necessary activities like veterinary inspections, food safety, food inspection, regional regulations, emergency conditions, and other preventive measures including research, economic impact, and vaccination against COVID-19 (https://www.oie.int/en/for-the-media/pressreleases/detail/article/covid-19-and-veterinary-activities-designatedas-essential/). Farmers play an important role in the economy of a country as animals help in improving the fiscal growth and quality of life [8].

Origin of COVID

The infectious pneumonia by coronavirus becomes a threat to humans globally. A massive pneumonia-like outbreak with strange etiology was reported in Wuhan, a city of China, in December 2019. Initial investigations reported that the patients have recently visited a market where seafood and other several animals are sold. Further results of isolated samples of pathogen showed that it is a new virus which was named as 2019-nCoV and is a member of corona family which is responsible for infecting humans. It was first identified in Wuhan. This epidemic outbreak remains different in different parts of the world. An association between the wholesale market of Wuhan and the coronavirus was reported in primary reported cases. And the possible source of this virus in that market was supposed to be the bat. Bat species are reported as the natural carrier (reservoir) of corona (SARS-CoV), as these bats which are supposed to carry alpha and beta corona are widely distributed in many areas of China.

These coronaviruses (beta viruses) can infect humans, farm animals, pets, laboratory experimental model animals, bats, whales, and other wild animals, whereas other gamma and delta viruses can cause infections in pigs, other mammals, aquatic animals like fish, and poultry birds. The researchers are working to find the truth, and till today SARS-CoV is assumed to be of animal origin. Recent report of COVID-19-infected dog who get it from his owner supports the hypothesis of transmission from animals to humans. The possibility of transmission by travelers also alerts all countries which resulted in the cancelation of events and tourist visa to all affected countries [9]. Some diagnostic methodologies were adopted like real-time PCR, and some repurposed drugs were used as vaccine is still not available. About the origin of COVID-19, there are many assumptions like some suggested snakes as outbreak source, but other studies reported bat and pangolin as source of emergence of SARS-CoV-2. The abovementioned assumptions are based on the similarity of genome sequence of COVID-19 with pangolin or bat origin coronavirus [10].

This new virus is identified as a family member of *Coronaviridae*, subfamily of *Orthocoronavirinae* and order *Nidovirales* [11], and named as SARS-CoV-2 or COVID-19. There are four genera of subfamily *Orthocoronavirinae* named as *Alphacoronavirus*, *Betacoronavirus*, *Gammacoronavirus*, and *Deltacoronavirus*. Among these four, two can infect mammals which are *Alphacoronavirus* and *Betacoronavirus*. Six strains of coronaviruses which can cause severe respiratory disorders in mammals including humans have been identified [12]. These viruses can be placed in two groups on the basis of phylogenetic analysis: HCOV-229E, TGEV (transmissible gastroenteritis virus), and PEDV (porcine epidemic diarrhea virus) are in group one, whereas SARS-CoV, bovine coronavirus, and hepatitis virus of animal origin are in group two. Recently known COVID-19 is *Betacoronavirus* and its subgenus is named *Sarbecovirus*. This virus is genetically different from SARS-CoV and MERS-CoV but all are *Betacoronavirus*. This presents 52% and 78% similarity with MERS-CoV and SARS-CoV, respectively [13].

In the past, coronavirus outbreaks were twice reported as host-specific: the first case, the 2002–2003 SARS, found in human, and the origin of the virus was reported as from bats, and the second case was from Middle East countries where it was MERS which showed linked with camels. But SARS-CoV-2 seems to be able to cross the host species barriers which make these jumping phenomena a regular feature of CoV species [14]. The reason behind these phenomena may be gene configuration, enzyme or gene stability associated with replication, and abundance of mutation rates. All these reasons may help in the maintenance of hosts and emergence of novel coronavirus. The expansion in host range is associated with glycoprotein spike which helps in the adaptation of virus to hosts which can be human or any other animal [15, 16].

CoV viruses are usually known for respiratory diseases with mild infections and very low mortality rate [17–19]. These coronaviruses (beta viruses) can infect humans, farm animals, pets, laboratory experimental model animals, bats, whales, and other wild animals, whereas other gamma and delta viruses can cause infections in pigs, other mammals, aquatic animals like fish, and poultry birds [20–22]. Several other groups of animals are reported as carriers (Fig. 1.2); these are porcine, bovine, camels, equine, canine, dogs and cats, lapin, avian, bats, rodents, mink, ferrets, snakes, marmots, frogs, pangolin, and hedgehog [23–28].

Coronavirus-infected laboratory animals include mice, rabbits, rats, ferrets, and guinea pigs. COVS were blamed for the respiratory infections, hepatitis, and enteritis in lab animals mentioned above. Among the different strains, MHV were found to be important and were observed in laboratory animals [29], the most infectious viral strain observed to affect the digestive system of mouse, with mortality rate observed as 100% [30]. In other cases, it was observed that MHV affected the

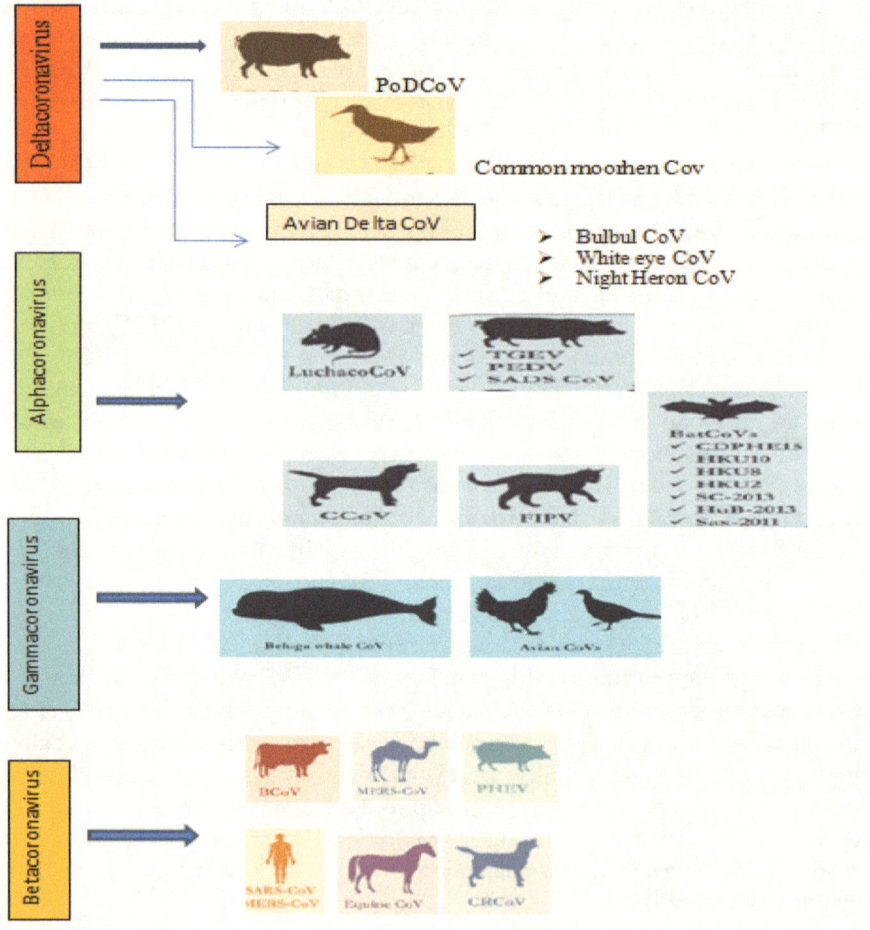

Fig. 1.2 Different animal host and carriers of coronaviruses belonging to different genus of *Coronaviridae* family

respiratory tract, endothelium, liver, hematopoietic tissues, lymph nodes, and central nervous system. Infection of rat by coronavirus starts from the respiratory epithelium which then spreads to the salivary gland [31, 32]. In ferrets, CoV signs were observed as the green slime [33]. Very little information about CoV is in record among guinea pig and rabbits, but it is observed to be responsible for diarrhea, atrophy, and malabsorption [34].

Bovine corona is reported as having zoonotic nature as it was isolated from humans. This can further be transmitted to other cattle with symptoms of bloody diarrhea [35]. This bovine corona may cause mortality by destroying the epithelium and villi of the intestinal tract (leading in bloody diarrhea) [36]. The carriers spread virus by oral or fecal secretions and stress conditions. This CoV was reported in domestic as well as wild animals including wood bison, water deer, reindeer, waterbuck, antelopes, sheep, goat, giraffe, camel, and white-tailed deer [37–39].

In foals, coronavirus has been reported in groups of less than 2 weeks, and symptoms include mild enteritis [21, 40]. In Japan, this equine coronavirus was also observed in 2011 in horse [41, 42], causing self-limiting enteritis [43]. The immune system plays an important role for the type of infection, and antigen-antibody reactions are responsible for lesions [44]. In 2008, coronavirus was reported from the liver of a captive whale [45]; similar virus was also reported in dolphin [46]. After that bats were reported as carriers of corona, and during flight in search of food, they can shed which may infect animals and humans whoever comes in direct or indirect contact [47]. In China, bats are used as food and also are of medicinal importance. So this animal (bats) is considered to be responsible for the spread of coronavirus [48, 49].

Emergence and Epidemiology

A massive pneumonia-like outbreak with strange etiology was reported in Wuhan, a city of China, in December 2019. Initial investigations reported that the patients have recently visited a market where seafood and other several animals are sold. Further results of isolated samples of pathogen showed that it is a new virus which was named as 2019-nCoV and is a member of corona family which are responsible for infecting humans [50–52]. Moreover, the World Health Organization named it COVID-19. A new name was also proposed by the International Committee on Taxonomy of Viruses which is SARS-CoV-2 which was not officially approved. The World Health Organization, after a massive increase in number of confirmed infected individuals, declared this pandemic on 30 January 2020 as international emergency for public health [53–55].

This family of virus is genetically diverse and is reported in a vast variety of animals belonging to different species of birds as well as mammals. In some cases corona-infected intestinal system while in others respiratory infections are confirmed in mammals and birds including humans. Corona got attention in 2002–2003, with reports of "strange pneumonia" in Guangdong Province which subsequently spread to Hong Kong. In Hong Kong, research separated the causative agent, and this pneumonia was later named as severe acute respiratory syndrome [52, 56]. This virus spread internationally and 8000 peoples were found infected in 26 countries with reported mortality rate of 10 percent (https://www.who.int/csr/sars/country/table2004_04_21/en/). This severe acute respiratory syndrome (SARS) at that time affected the economy and was globally a serious threat to health. After successive studies, it was found that the origin of SARS-CoV was bat and it came in humans via different intermediate animals (hosts) like dogs (*Nyctereutes procyonoides*) and civet (*Paguma larvata*). Another coronavirus which also infected human was MERS-CoV (Middle East respiratory syndrome corona); the mortality rate from this virus was higher than the SARS-CoV, but its transfer from human to human was not reported [57, 58].

Bat species are reported as the natural carrier (reservoir) of corona (SARS-CoV), as these bats which are supposed to carry alpha and beta corona are widely distributed in many areas of China. The molecular evolution and diversity of SARS-CoV have been intensively studied in bats [59, 60]. As our knowledge is very limited about the COVID-19, we may review the epidemiology of corona on the basis of the previous knowledge of MERS-CoV and SARS-CoV.

There are three phases of epidemiology curve of COVID-19. The first phase is the local outbreak which started in the visitors of seafood market. From the reported starting date of outbreak (December 2019) to 13 January 2020, confirmed reported cases were 41. Epidemiological studies confirmed that even during this first initial stage, close contact was the reason of transmission between humans [61–63]. After this, a second phase started from 13 January 2020; during this phase, there was an increase in the spread and expansion, and the outbreak and spread of this disease were confirmed outside of Wuhan. The transmission of this disease from man to man may be divided into two types: nosocomial infection and close contact. Nosocomial is spread within hospital, whereas the close contact is family transmission [64, 65]. First case outside China was reported in Thailand, and history of traveling showed that the person was a resident of Wuhan and traveled from there to Thailand. On 23 January, there were 846 confirmed cases from 29 provinces and 6 countries [66, 67]. This increase was 20-fold as compared to the first phase. Before the shutting down implementations in Wuhan, five million people had already left the city for home due to the upcoming event of Chinese New Year [68, 69].

From 26 January 2020, there was a start of third phase. It showed a quick growth of constellation cases. Among the total number of conformed cases, nearly 80% were in Shanghai, Beijing, Shandong, and Jiangsu on 10 February 2020. After 20 days, this number was reached to nearly 10,000 which showed 240-fold increase in confirmed cases. However, at that time, there were 441 confirmed patients outside China in 24 countries with low death rate out of China as only 1 patient died in the Philippines, whereas in China 1114 people died. The day 3rd Feb, 2020 can be written as a tip point of COVID-19, as from this time there was a decline in the confirmed patients and it showed a success of lockdown and the health care but if the transmission was reduced due to any other reason was still unknown [70, 71]. Furthermore, some specific infection transmission causes and failure of protection among medical and health care staff still showed further research and investigations.

If we compare the basic number (reproductive number), it is 1.4–3.9 for COVID-19, 2.3–3.7 for SARS-CoV, and 0.5–0.9 for MERS-CoV. As the reproductive number for SARS-CoV is smaller, it shows the possibility of easy control for this outbreak with rapid diagnosis and isolation of infected people [72–74]. However, this is very important to take in mind that only reproductive basic number is not important as it fluctuates with social behavior and environmental and biological factors. And the interpretation must be done with caution [75, 76].

Human-Animal Interaction as Risk Factor

Currently, we do not know about origin, detailed binding receptors of host, and natural host, as the theory of this virus as bat origin is still not confirmed [77, 78]. Some earlier studies reported SARS-CoV-2 as genetically analogous to corona of pangolin, whereas the coronavirus of bat origin did not show similarity and ancestral linkage. Recently one report confirmed this virus in pet dog which also showed some respiratory sickness and COVID-19 symptoms. Usually the dogs become affected by corona which causes severe respiratory symptoms [79–81].

Bovine corona and human corona are all related and are *Betacoronavirus* [80, 82], whereas the abovementioned viruses are different from enteric canine coronavirus which is *Alphacoronavirus* and may cause infection to dogs. Moreover, a bat origin virus was isolated, and phylogenetically it was found close relative of SARS-CoV-2, so it is considered as an intermediate host of the present outbreak strain [83–85].

If we look at the clinical picture of the COVID-19, the infection can be categorized as of four types: mild, modest, unambiguous, and acute. The symptoms and signs in patients are cough, fever, fatigue, diarrhea, and myalgia [86–88]. There was an abnormal ground glass bilateral opacity in patients when observed by chest computed tomography scan [89–91]. So the clinical and dialogical presentations of COVID-19 and SARS-CoV-2 are still further under research. There is similarity among the pathology of MERS-CoV, SARS, and COVID-19 [92–94]. Death due to COVID-19 is observed with alveolar damage, whereas the infection starts with epithelial cells which extend to the intestine, kidney, and brain [95–97].

A member of Chinese Health Commission raised an issue of evaluation of SARS-CoV-2 among animals. This issue was raised as the observations showed that the virus can move between mammals. If any pet animal come in contact with other animals or human, it can spread, so animals should be quarantined [98]. This is the first precautionary measure which must be taken to save loved ones [97, 98]. But as people came to know that corona may spread by pets, the response was severe and resulted in the killing of dogs and cats [99–101]. To change the attitude of the people, the World Health Organization gave a statement: it is not proved that cats or dogs can get infections from the virus [97, 98, 102]. Even after this statement, pet killing continued in China [103, 104]. A pet dog was reported to have infected by COVID-19 after RT-PCR testation [105]. A report from Hong Kong threatened humans as it was reported that a dog cannot be infected but can be a carrier who can spread the virus [106].

The French Food Agency requested to submit clear opinions about the role of pets and domestic animals in spreading coronavirus. A very less number of animals were experimentally proved to be infected with corona, among them a transgenic mouse was observed to be infected which expressed human ACE2 receptors for SARS-CoV-2 [105, 107]. We need further studies to observe interactions among COVID-19 and ACE2 in the tissues of other animals. Moreover, there is a need to study the dispersal of ACE2 in tissues, which is important for advance knowledge of infection and transmission to other animal species. Not only it is important to focus on the receptors' presence but also other cell factors necessary for replication of virus must be taken into consideration during intraspecific studies. To identify these factors, further studies and research are necessary [108, 109]. At the same time, there came a statement from the Ministry of Health of Singapore that pets are not a serious vector for transmission of COVID-19 but may spread this virus to human. They also not planned to isolate or quarantine animals [110].

After that, IDEXX Veterinary Laboratory tested pets for COVID-19 all over South Korea and the United States by using RT-qPCR, and there were no positive results. Although animals were from affected areas, it was confirmed from results that these pets were not living in homes with infected people [110, 111]. It was stated by the WHO that animal pets are safe from corona, but recently they admitted that they can be infected but it is not clear that they can spread the disease [112, 113]. A dog was found infected as confirmed by RT-PCR [114, 115]; after that, a cat was found COVID-19-positive [114, 116, 117]. Contrary to all previous reports, again virus was confirmed in cat as well as the presence of antigen, and it was a start of debate. But the Belgium scientist group does not recommend testing of pets as there is still unavailability of diagnostics [118]. Some scientists recommended pets as hosts of dead ending [119, 120]. It was also suggested that the risk of transmission of vector from animal to human is very less and human also must not rub their nose in front of animals [121, 122]. SARS-CoV-2-positive cats were found in New York, and their owners were also positive for this virus [123, 124].

In the respiratory and gut epithelium, antigen or RNA was present abundantly which was innervated by RT-PCR and immunohistobiochemistry assay. This was also reported that the virus in chicken, dog, pig, and duck cannot competently reproduce [125, 126]. In the studies related to then gene expression, it was noted that ACE2 was actively expressed in the lungs, retina, skin, and ear of cat, whereas in dogs it is actively expressed in the retina and skin. Moreover, cats and dogs are more affected by SARS-CoV-2 when compared with rodents [127, 128]. These results were also supported by the test positive results in lions and tigers, which become affected from their keeper.

Diagnosis of virus also faced difficulty due to the absence of clinical signs and symptoms. And radiography of chest also failed to test COVID-19 [129]. Other methods were adopted which include reverse transcription PCR, real-time PCR, computed tomography, X-ray, and enzyme-immune assay [123, 130–133]. For calorimetric detection, a loop-mediated amplification tool was designed [134]. And a web-based dash board was developed for tracking coronavirus [135, 136]. A quantitative detection (florescent-based) of N region by PCR assay has been developed [137].

Best Suitable Animal Model

The animal models are necessary to study and explore primates or in simple words nonhuman model (Table 1.1). Although presently we do not have a model for SARS-CoV-2, some researchers have explored primates like rhesus monkey as a model. In the past, many studies evaluated different vaccines against diseases like MERS-CoV, and primates were used as animal model [125, 137, 138]. During those studies, SARS-CoV-2 affected monkey, and the animal showed different signs and symptoms for nearly 16 days and are proved to be effective in pathogenic studies with next development of antiviral drugs and vaccines [139, 140].

Different animals have been reported susceptible for ferrets, dogs, cats, and other domestic animals. However, dog, chicken, duck, and pigs have been reported less susceptible, but cats and ferrets are efficient [141, 142]. Cats were reported to spread virus by spreading droplets. In the process of finding a best model, exploration is required for animals having ACE2 receptors as in human [143, 144].

Establishment of appropriate animal model will help us in understanding the process of disease transmission and will further help in therapeutic and prophylactic improvement [127, 145–147]. Primates are considered important as best models in human disease studies, whereas for pathogenic exploration and immune response studies other animals are recommended. These other animals which are used as model animals are hamsters, rabbits, mice, and mouse [12, 148]. Among these

Table 1.1 Animal models used to study different coronaviruses

Mouse model
Mouse-adapted viral strains
Humanized mouse
Receptor-transfected transgenic mice
Nonhuman primates
Old World monkeys
African green monkeys
Macaques
New World monkeys
Squirrel monkeys
Common marmosets
Animal models not usually selected
Hamsters
Rabbits
Ferrets

animals, mice and hamster have been used in the studies of MERS, SARS, and SARS-CoV-2 as animal model. Golden hamster has also been studied in a SARS-CoV study for testing vaccine [148, 149] and noted as a potential model for corona pathogenic studies and vaccine efficiency testing.

Transgenic animals (lab mice) may be suitable animal model in SARS-CoV-2 studies as the receptor ACE2 is different from human in different animals [150–152]. On the basis of similarity in their ACE2 receptors, different animals like bat, monkey, cat, orangutans, ferret, and pig may be used as model animals as they have similar affinity for coronavirus as human [153, 154]. For selecting animals as model, the important thing we have to take into consideration is that the selected animal model must be small and easily manipulated as mouse or rabbit, and the other point is the selected model must be cheaper and easily available [155–157]. In the beginning, mice were difficult to be accepted as animal models but transgenic mice are now believed to be good and are accepted as models of COVID-19 [41, 91, 158–161].

Many transgenic knockout mice like ACE2, Tmpress, Stat1, inbred, and HLA have been proved to be selected as model animals for COVID-19 or SARS-CoV-2 [162–164]. As it is proved that the present pandemic is caused by coronavirus, now scientists are trying to evaluate and find the most suitable animal model for this respiratory syndrome. Animal model will be selected on the basis of questions to be answered like the airborne transmission of coronavirus among two different species like between ferrets and cats. Among monkeys, the two age groups young and old were selected, and it was found to be affected by virus in the respiratory system, but no severe symptoms appear in both groups. So these animals can answer some other questions about induction and transmission of COVID-19 [164, 165], as the human showed a great range of clinical signs: from acute to severe pneumonia. If the data is explained on the basis of one animal model, it will be misleading, and disease outcome will be challenging for human [166, 167].

If we want to select an animal model for COVID-19 from the traditional models which have been selected previously, the main point for selection is those animals must not hide the pathology of pandemic especially when our focus is to check the viral medicines and vaccines [168]. If the animal model is able to mimic the pathology, researchers will be unable to understand the transmission and infection dynamics. Previously for the MERS-CoV and SARS-CoV-1 studies, some animal models were selected. These animal models include hamster, monkey, ferret, camel, and mice [169]. It is the economic truth that mice is best suited if we focus on the price and availability of this animal, but it has been suggested that this is not suitable for COVID-19 experiments because of the differences in receptors from human. But the transgenic mice have been reported good for selecting them as model animals as they can express the human angiotensin-converting enzyme 2 (HACE2). The receptors present in cells for the entry of SARS-CoV-2 are supposed to be important for examining COVID-19. In another study, transgenic mice showed partial ruthlessness [170, 171]. Shi et al. [5] inserted coronavirus in the nasal passage of different animals including cat, dog, ferret, pig, duck, and chicken. The aim was to check the pathogenesis and replication with transmission of disease. Animals shed virus but without clinical signs and symptoms. Similar results were reported in monkey

(*Rhesus macaques*) [91, 172]. Against viral protein, antibodies were produced by insertion of SARS-CoV-2 in inactive form in selected animal models (rats, mice, and monkey). After different studies, it was suggested that ferrets and cats are susceptible and can tolerate SARS-CoV-2 [173].

In animals like ferrets and cats, no signs of infections appeared, and there was no respiratory stress or weight loss. But no virus detected in digestive or respiratory tracts of other animals. However, in two dogs, viral RNA was detected in rectal secretions, and antibodies were also produced by these animals. From other reported data, it is also suggested that the birds (chicken and duck) and pig cannot tolerate viral infection [26, 174]. Vulnerability of any animal depends upon the presence of receptors in host or enzyme activity. These two factors are important factors for coronavirus to enter in cell. It was also observed that the virus is detected not only in the respiratory tract but also in the rectum of ferrets, monkey, and cat. This may be a trustworthy initial clinical sign in COVID patients and is proof for the theory of oral-fecal administration route [24].

It has also been proven by epidemiological studies that the fatality is higher in old aged people from corona. But our model animals are usually of young age. So it is also suggested that the old model animals must be chosen for experiments. But when young and old cats were selected, opposite was found true as young cats died on day 3 after infected as compared to older ones. In monkeys, these differences in severe or mild infections were not related to age as both young and old monkeys showed similar results [25, 159, 175]. The most important thing to be noted is that not any animal showed lethal effects of SARS-CoV-2 as compared to human. During the development of animal models, age and health factors must be fully given attentions, as these factors are important for evaluation of disease in human [176].

The virus transmission through air was observed in cases of pandemics SARS-CoV-2 and influenza. But the important thing was aerosol size as its wide range specified its range. It was also proved that this coronavirus can be transmitted through air as observed by Shi et al. [5], stating that 33 percent of the virus were transmitted by air in cats [177]. The airborne nature of pandemic was also proved by experiments where animals were separated by barriers and no physical contact was possible, but the air was shared by all containers. In case of influenza, it was also observed that virus replication in the upper region of respiratory tract is important for airborne transmission. The experiments on cats confirmed that the cats are vulnerable to viruses [178, 179]. Among other animals, SARS-CoV-2 was found to be replicated in the palate of ferret and cat. The airborne transmission among ferrets was reported to be 30 percent for SARS-CoV-2.

The transfer of coronavirus from human to cat was recently observed in big cats (tiger) in zoo. But the cats may not show symptoms of disease, and it can go unnoticed even if they are infected, which proved that cats are not good animal models for coronavirus studies [180, 181]. Additional serological and coronavirus-specific antibody studies are needed to find SARS-CoV-2 virus from animals. The animal models are very important for testing the potential of antiviral medicines, vaccines, and therapeutic strategy. Among mammals, cats, ferret, hamsters, and transgenic mice are important alternate of humans to study SARS-CoV-2. Because the transgenic mice and hamster showed prominent symptoms and signs, so they may be vigorous animal models for vaccine formation [182]. Cat and ferret are best suitable models for transmissible studies and effectiveness of vaccines. So the main focus is to choose a nonhuman primate animal as model. Therefore, a continuous experimental evaluation is needed to evaluate mice as model animal.

COVID and Animal Welfare

One factor that linked animals with COVID was that the first person who was found infected visited animal market. According to reports, that was wet market where live animals were sold [183]. This COVID has been sequenced and identified genetically as coronavirus that usually found in bats. So it is considered that bats are the natural reservoir of corona. As the coronavirus spread globally, the pets have also become victims. Some people left them home alone and some evacuated them. Although there is no evidence that pets can transmit virus, the fear of spreading the virus resulted in killing of pets [184]. So the current situation of coronavirus negatively impacted the animal welfare.

This is also very important to understand the relation between human, animal, and environment. For future and further prevention of coronavirus, one health collaborative approach of transdisciplinary help is necessary [185, 186]. As the coronavirus has resulted in shocking concerns for health and economic constancy all over the world and also appeared to be adverse, unexpected difficulties arise for animal welfare. Especially for the professionals dealing with veterinary and animal welfare, this is the time to fear how coronavirus is going to affect the pets. If we look at a picture of Wuhan (the place of origin of outbreak), many people who were evacuated from homes were forced to leave their pets behind alone. Although the owners left a sufficient amount of food and water at home as they were thinking to return soon, after more than 6 weeks, many of them still not returned [187]. It is estimated by the Chinese animal welfare organization that nearly 10,000 cats and dog were left behind in Hubei who faced starvation and death.

Concerns related to pets emerged among public and pet owners when a dog reported to be positive for corona. In another case, a dog was tested positive but apparently he was having no clinical signs. The swabs from oral and nasal cavity were weak positive [188, 189]. The dog was quarantined for 14 days, and without any clinical signs, after returning home, the dog was found dead. The cause of death remains unknown as the owner not allowed experts for postmortem. But it is believed that the death was not due to corona but age and age-related health issues may be the factor [190]. This case gives a spark of fear to public which resulted in the abuse of pets, as the public starts to believe that coronavirus may be spread among them by pets. After this case, an announcement was made from urban construction that if any cat or dog will be found on street, it will be killed to prevent the spread of coronavirus [191]. The same news about the killing of animals came from Hunan and Zhejiang also.

To stop the cruelty to animals, the Hong Kong Society makes a statement of information for public. This society informs general public that there is no need to fear from pets as there is no evidence about the transmission of virus from animals to human [192]. So people must cooperate in animal welfare. A large number of cases for uncontrolled movement of pets to become wild animals have been reported and spread widely on social media as people left them. Some people left their pets from the fear of transmission, and some left them because they become unemployed and now cannot afford a pet [193].

Another import side is that now people are isolating themselves during this outbreak and the rate for adoption of pets is rising. This is really good news that pet lovers and animals may enjoy this bond between animals and humans when this is a time of great need for all [194]. Some people who were not having pets before are now adopting animals, but some experts are also worried about the future of these pets as they think people will return these animals to shelters when life will return to normal [195].

Inclusive, the previous findings suggest that still there is no threat of transmission of SARS-CoV-2 from animals to humans. Nevertheless, the World Health Organization and the US Centers for Disease Control and Prevention tried to calm down the rising concern of people by issuing statements to not be afraid from pets, as they are not a source of spread of COVID-19 [1, 133]. According to the World Health Organization, our focus must be on the control of human-to-human transmission right now and we are not having justifications and proofs for taking any action against pets. So human activities must be restricted, and this requires reduction in community activities to reduce the spread of coronavirus [196]. However, when we critically look at the spread of pandemic, there is a complex relation between human, environment, and other living things like animals (pets or wild).

In the end, there is a need to develop one health approach strategies which will help in the prevention and control of COVID-19. One speculation that wild animals are associated with the spread of coronavirus must be elaborated and this needs some expertise and collaboration of wild life experts and forensic specialists [197]. Trading of wild animals in markets like in Wuhan must be regulated properly, as there exist poor regulation, illegal trends, and poor welfare standards. However, many experts hope for good wildlife trade globally after this crisis [198].

Molecular Differences Between Different Animal Species

It is needed to estimate the interaction residues of S receptor binding areas of coronavirus and human, as this estimation will help in the identification on how much it will affect other species. For this, researchers bring together all available amino acid sequences of ACE2. In an experiment, N-glycosylation treatment was used near binding site [199]. N near the binding area will negatively affect the binding of

S. During this, ACE2 was glycosylated at N (N53, N90, and N322). N53 was found conserved for all species, whereas N90 was not in glycosylation site of ACE2 in pig, mouse, civet, fox, chicken, ferret, and raccoon. N322 was not a site of glycosylation in ACE2 of cattle, rat, mouse, sheep, and pangolin [200]. However, some species have an additional N glycosylation area in ACE2. In chicken, L79 is the site of N glycosylation, and in *Rhinolophus sinicus*, rats and pangolin M82 are the sites of glycosylation. In rats, it was noted that glycosylation residue 82 may prevent S binding of SARS-CoV.

In ACE2, there are amino acids that may affect the binding of S which have been reported from some species [124]. Some detail of these is presented in Table 1.2. If we look at the binding site, it is entrusting for experimentation as ACE2 binding sites for monkey and chimpanzee are similar to man. And in other animal species, the differences in ACE2 amino acid from human to chicken are 11 and to rodents 9, while from cat to human, the differences are 3. Some attraction is in ACE2 protein of pet and farm animals as another possibility is that this protein may be reservoir of coronavirus. There are six differences in ACE2 of pig, and mostly are located at the border of site for binding. The important point here is N90 T as it affects the glycosylation site and the site will be loss. In SARS-CoV, E 329 is reported to form a bridge of salt with R 426, but in COVID-19, salt bridge is formed with D30 [124, 201]. Thus, N exchange with E 329 in ACE2 of porcine also affects the S binding in SARS-CoV; however, in COVID-19, S binding is not affected. The similar pattern of effects and changes has been observed between cattle, cat, and human ACE2. A few differences and changes have also been observed in the middle of binding site which gave confidence in decision that this exchange may be a big obstacle in infecting a cell with coronavirus from one species to another [202].

Table 1.2 Human ACE2 similarity with other animals

Animal	Similarity	Gene bank association number
Human	19/19	AAT45083.1
Chimpanzee	19/19	XP_016798468.1
Rat	10/19	AAW78017.1
Mouse	9/19	ABN80106.1
Cat	16/19	XP_023104564.1
Pig	13/19	XP_020935033.1
Macaque	19/19	XP_011733505.1
Horse	14/19	XP_001490241.1
Sheep	15/19	XP_011961657.1
Cattle	15/19	XP_005228485.1
Camel	14/19	XP_031301717.1
Ferret	11/19	BAE53380.1
Fox	14/19	XP_0258425131
Pangolin	13/19	XP_017505752.1
Chicken	8/19	XP_416822.2

Another term used for coronavirus is that this virus is RNA positive-sense virus. The newly identified virus (SARS-CoV-2) is from *Coronaviridae* family with four genera. SARS-CoV and MERS-CoV belong to one genus (*Betacoronavirus*), but SARS-CoV-2 is genetically different from both of the above [203, 204]. But this is important to understand that nearly 89% of SARS-CoV-2 is identical to the bat-SL-COVZC45 and bat-SL-COVZXC21 (SARS-CoV of bat origin). These were also named as ZXC21, and it was found to be nearly 82% identical to SARS-CoV Tor 2 (human), and ZC45 of bat origin was found to be identical to human SARS-CoV BJ01 2003 [15, 27, 28, 93, 205]. Among SARS-CoV-2 and MERS-CoV, it was found that these two are 50% identical, whereas SARS-CoV-2 and SARS-CoV were reported to be 79% identical. Phylogenetic analysis at the molecular level proved SARS-CoV-2 to be more closely related to bat SARS-CoV [206–209].

It was reported that there are 380 amino acids between SARS-CoV-2 sequences confirmed after genome analysis. This swap of amino acid may be helpful in functional as well as pathogenesis deviation of this pandemic coronavirus [180, 210]. The most important requirement for the recovery and prevention from coronavirus is immunity, which may be innate or adaptive. The humoral immune response is very important in corona outbreak especially antibodies of titer and subtype [211]. Although it is also of concern to the coronavirus studies that mouse and some other animals may hide the clinical signs or mimic COVID infection, the genetic diversity is also linked that may guide toward wrong interpretation between the results of primate and non- primate animals. Researchers suggest humanized animals may be good models for the study of organs and tissues of human in coronavirus infection studies [212]. In addition to infection studies, these will be helpful in the vaccine preparation and elevation of target agents in host.

In previous times, to study HIV, HCV, CMV, HBV, and VZV infections and other pathogens and zoonotic diseases, mice with grafted cells or tissues have been used extensively. As we all know, lung is the major target that is infected by coronavirus [213]. In one experiment, the human lung tissue was grafted under the kidney of mice which was immune-deficient [214]. The findings suggest the rapid tissue development which was associated with supporting CMV pathogenesis. A successful association was established by using by lung grafted mouse as animal model to study VZV infection. After this infection, pro-inflammatory cytokine and viral replication were observed in grafted lung.

In recent studies, researchers worked with humanized mouse (human lung grafted). This mouse named as BLT or LOM supported Zika virus, CMV, MERS-CoV, and RSV. Response of T cell suggested that dual chimera mouse (grafted with the lung and immune cell) may be best animal model for the study of SARS-CoV-2. Multiple organs may be failed due to SARS-CoV-2, as human ACE2 can be expressed in many organs like the liver, kidney, heart, bladder, intestine, and immune-related cells or tissues. Therefore, it is also suggested that different tissues may be grafted in mice to check the severity of SARS-CoV-2 on different human organs and tissues. So we can say that the humanized animal like mice is important in improving the understanding of mechanisms involved by coronavirus for infection and immune physiology [215, 216]. In the future, these human tissues containing mice will be proved as a matchless tool in obtaining the intuition for our strategies in developing vaccine, therapy, or intervention for coronavirus.

Vaccines for COVID from Animals

Till today, no license has been issued for any vaccine in the prevention of respiratory infection. Only vaccine for the prevention of respiratory infection from corona in chicken has been given license. Several proposals have been given for the development of vaccine for COVID-19 similar to MERS and SARS [217]. Vaccines must have to fill full the criteria of efficacy, immunity duration, and safety. However, the vaccine for COVID-19 also requires rapid production and large-scale development capacity [218]. Many traditional week and inactive vaccines have been used in experiments, but the high production of vaccines in large amount with biosafety culturing of cell is challenging. This culturing of cell with high level of biosafety is also limiting the rapid development during this emerging phase of coronavirus pandemic [219]. However, for the prevention of reversion (nsp16, exon, and accessory protein), live offset corona vaccine have been generated from clones of infected virus by deletion of many important virulence elements. But all this remains robust immunologically, inducing mucosal and humoral cell-mediated immunity with great protection. These offset vaccines remain effective in boosting immune response of host cells [220].

For the success of vaccine, it is important for a vaccine to be temperature-sensitive. The second remarks must include reverse genetics for further mutation which can be applied after generating mutant temperature-sensitive virus strain [126, 221]. A combined booster strain vaccine with S and N proteins like null (attenuated) vaccine may be potentially have cross-protection against heterogeneous mutant strain that have a potential to stumble beta COV, like bat coronavirus strain [222, 223]. It is important for vaccine technology that in this technology, subunit of viral particle is used.

New technologies for vaccine formulation also follow this principle and include viral protein, recombinant vector, and RNA. The use of viral protein and recombinant vector has advantage that these may provide a universal platform for introducing new antigen target from evolving virus [224]. Evolved viruses may mimic the week vaccines and infect the cells with the induction of antigen which will activate antibody formation and immune response from T cell. Underdeveloped vaccines for COVID-19 also have S protein and vector replicating or non-replicating with expressed S protein. There will be inconsistency between different strains so vaccines must be based on the S gene [225, 226]. Vaccines are made or classified on the basis of plasmid DNA, RBD protein, or RNA. Although DNA vaccines may be easy to produce, safe, and stable, the efficacy of these for human has not been checked. Corona DNA vaccines for SARS and MERS show great efficiency for inactive or recombinant virus and heterologous increase [227, 228].

There are still some issues related to vaccine administration, genome integration, and persistence. If we focus on the RNA vaccines, these are usually used as template which is used for protein production in animals which are vaccinated [229]. Supply of mRNA-based vaccine may be enhanced by using lipid particles for muscular administration. In the United States, a vaccine is in advance phase 1 and ready for trial on humans against COVID-19. One of the advantages of RNA vaccine is potential development in printing RNA, as it may produce massive amount of mRNA.

In recombinant vaccine scenario, vaccines are in different developmental stages for SARS including recombinant adenovirus also called Ad vector. This vector with chimpanzee's Ad63 has been used to overawe the preexisting extensive immunity in human Ad5. S or N protein of SARS-CoV or MERS-CoV, expressed from recombinant adenovirus vector, stimulated immunity or protection in animal models (ferret, mouse, and primates) [230]. Other famous vaccines for SARS include modified vaccinia Ankara (poxvirus vector), measles, and influenza, vesicular stomatitis and Newcastle disease virus [231]. These entire viruses express N and S proteins of SARS-CoV. These have been reported for protection and immunity in mouse which was used as model transgenic animal containing ACE2 and host receptor of MERS-CoV known as dipeptidyl peptidase. Intermediate hosts were also checked for the MERS vaccine. And vaccines like MVA and NCD were tested on camel which showed induction of antibodies, and MVA was reported as protected for nasal secretions and shedding [232]. The most important thing for scientists is the formulation of vaccine for livestock which may be intermediate hosts and pass on diseases to human. Similar strategy may be applied for influenza of swine and poultry; from results, we may be able to limit the risk of transmission of viruses to human [233].

During vaccination studies or formulation of vaccines, the most important point to be followed is biosafety. Even experimentation on animal studies also strictly follows the biosafety rules [234]. An experiment was designed to check the post-murine challenge of inactivated MERS-CoV vaccine or SARS-CoV vaccine with formalin in mice for effect on the lung. Addition of receptors to inactive SARS-CoV may reduce the pathogenesis of lungs. Another vaccine MVA-S is important for studying pathology of the liver in ferrets, but in other animals, it was not successful [235]. In hamsters, antibodies were enhanced by the use of S-based SARS protein vaccine. But in mice, S-based MERS vaccine was not effective. The main obstacle for the preparation of vaccine for COVID infection in cat was ADE. In this COVID infection, ADE was found to be triggered in cats as the virus-mediated antibody entered in macrophage by using the pathway of IG FC receptors. As some inconsistent events can be observed in animals, these are also important for the improvement of human immunological studies and in avoiding similar events in human [236].

The understanding with the pathogenesis, protection, and natural immunity of SARS-CoV-2 in human will be helpful in the preparation of vaccines for COVID-19 [237]. The route of pathogen infection to the specific organ is also of importance, as this knowledge will also positively help in the development of vaccines and prevention of disease from the target organ [238]. The other important point to be considered is the path to reach the target organ, e.g., lungs. There may be two routes for

SARS-CoV-2: one is by causing pneumonia by viremia and the other route is by infecting the upper respiratory tract [239]. If the other route is followed to infect, vaccines may consist of replicating vector or weak virus which may effectively encourage immunity to protect the respiratory system and may reduce nasal secretions.

On the other hand, if the infection path is viremia, parental vaccines that induce serum antibodies may successfully not only block viremia but also transduce blockage to infection [240]. This effect is the same as IM submission of slow influenza vaccine for the treatment of lung infection in humans. An annual insertion of vaccines or S subunit may be effective in humans who have been recovered from coronavirus [241]. This will be helpful in the prevention of corona infection by enhancing B- and T-cell response and immunity. Oral and nasal vaccines are reported to be effective in corona-affected people who may have diarrhea or fecal viral secretions. We can divide humans in three groups: first group includes people who have zero immunity, second group consists of immunity of different levels (these are those people who have been recovered), and third group of people has previously developed immunity for SARS. Therefore, the responses for vaccine in different groups of people will be different [242, 243]. So, the preexisting immunity levels must be studied for the effectiveness in all groups of population. But the focus must be given to the elder population, in which the death rate is reported high [244, 245].

There are different primates and non-primates who have been used for testing vaccines. These animals include African monkey, macaques, hamster, ferret, and mice. For prevention of CoV infections in animals, different vaccines to cure corona have been licensed. Not any vaccine was found effective in animals. In some animals, the digestive tract and intestine are the sites of corona infection, and this condition is severe in neonates [246]. To overcome these problems, oral vaccines have been developed which can be used for pregnant animal. These oral vaccines are helpful in preventing disease (intestinal infections) in mother and suckling neonate as antibodies will be transferred through milk. It was also observed in swine that milk antibodies were effective in passive immunity as compared to serum IgG. In bovine corona infections, diarrhea and lung infections are observed with nasal shedding. Although it has great impact on the economy, no vaccines have been developed for the prevention of bovine corona in cattle. In cattle, pneumonia and nasal shedding are associated with respiratory disease [154, 209]. This was also of great risk to use live weak coronavirus, but this type of vaccine based on bovine corona live virus has been licensed for oral administration in cattle. By using this vaccine, the risk for respiratory complex has been treated.

In camels, MERS-CoV is present in Saudi Arabia; a protein subunit (IM) or live weakened MERS-CoV may be a safe vaccine which may be helpful in controlling the transmission of virus to human and nasal shedding [177, 247]. If we observe the porcine corona, it resembles with COVID-19. As both may spread by nasal droplets, it infects the lungs. However, porcine corona infections are reported to be mild. But the dose of virus, itinerary of infection, and immunity levels are some factors which may enhance the severity of bovine and porcine coronavirus [248]. All the above-mentioned factors are also important in increasing the spread and severity of

COVID-19. During pandemic, first-generation vaccine development, production, and distribution are critical. Synthesis of nucleic acid-based vaccines with S and N protein may prove to be good vaccines. Another method for speeding up veterinary vaccination during epidemic is issuance of licenses. For present mutant coronavirus, this may be based on the data from humans [249]. After that, generating a second-generation vaccine for prevention of disease and death may also be helpful for future development. Coronavirus is likely to affect as second spill of SARS-CoV-2 via animal reservoir. For generating new generation of vaccine, strong immunity is needed against corona within *Betacoronavirus* ancestors [250]. The main focus must be on S and N proteins and conservation of apoptosis that may be effective in inducing cross-reactive immunity.

Any vaccine that is successful in inducing immunity is necessary at the present scenario. The need is passive immunization to treat people both therapeutically and prophylactically. In this situation, most emphasis is also in using plasma therapy [179]. This plasma can be collected from recovered people as their plasma may contain antibodies. The induction of double-blind trials is also of importance. It has been experimented, and a comparison was made in clinical treatment; short stay at hospital and reduction in mortality rate are associated with plasma therapy. But for plasma treatment, a plasma bank is necessary. These plasma banks can collect from donors (recovered people) but proper establishment is necessary.

To control the viral proteins in host, monoclonal antibodies may be rapidly produced and also important for mapping epitopes that act as protective guards and react against immune response as established by MAB management to SARS-CoV-2-defined animals. This may also help in guiding vaccine formation and design with positive immune prophylactic against COVID-19 [138]. Encoding of mRNA is among the recent strategies used for encoding virus via intra-tracheal aerosol. Nanoparticles against MERS may provide immunotherapy in model animals.

Since the first confirmed case of coronavirus, different chemicals, vaccines, or medicines and herbs have been tested for their activities [129]. Some genetically improved antiviral therapy, drug, vaccines, and antibodies were found to be useful in reducing the global threat [152, 247]. The world has faced many epidemic and pandemic threats of Ebola, Nipah, Zika, swine flu, MERS-CoV, and SARS-CoV-2. All these are important in developing effective vaccine, drugs, and therapeutic and advanced vaccines to combat COVID-19. Some initiatives in developing new vaccines include chimpanzee vectored adenovirus vaccine [109, 171].

Conclusion

Detection of most suitable animal models with antiviral drugs and SARS-CoV vaccines may be helpful in finding the most effective therapeutic against virus. Among animals, ferrets, cats, and hamsters are more attractive animals as models as compared to transgenic mouse and nonhuman primates. Different animals can be useful models for different parameters as for clinical symptom studies transgenic mice can be best model as they showed severe symptoms like reduction in weight. Hamsters will be proved as good models for testing the efficacy of vaccines. Cats may be suggested as model animals for viral transmission studies and for testing the bound blowout of virus and antivirus effectiveness. Nonhuman primate animals will be best suitable model animals for studying load of disease and testing vaccines and effectiveness of antiviral drugs before starting a speedy arrangement and production of vaccines for human. So we can say that continuous experiments on animal models from mice to cat, dogs, and primates will help in the establishment of data best suited for answering many problems about coronavirus (COVID-19).

References

1. Lin X, Gong Z (2020) Novel coronavirus pneumonia outbreak in 2019: computed tomographic findings in two cases. Korean J Radiol 21:365–368
2. Pung R, Chiew CJ, Young BE et al (2020) Investigation of three clusters of COVID-19 in Singapore: implications for surveillance and response measures. Lancet 395(10229):1039–1046. https://doi.org/10.1016/S0140-6736(20)30528-6
3. Propper C, Stoye G, Zaranko B (2020) The wider impacts of the coronavirus pandemic on the NHS. Institute for Fiscal Studies. https://doi.org/10.1920/BN.IFS.2020.BN0280. https://www.ifs.org.uk/publications/14798
4. Corlett RT, Primack RB, Devictor V, Maas B, Goswami VR, Bates AE, PinKoh L, Regan TJ, Loyola R, Pakeman RJ, Cumming GS, Pidgeon A, Johns D, Roth R (2020) Impacts of the coronavirus pandemic on biodiversity conservation. https://doi.org/10.1016/j.biocon.2020.108571.d
5. Shi J, Wen Z, Zhong G, Yang H, Wang C, Huang B, Liu R, He X, Shuai L, Sun Z et al (2020) Susceptibility of ferrets, cats, dogs, and other domesticated animals to SARS-coronavirus 2. Science 8:eabb7015
6. Tanner E, White A, Lurz PWW, Gortázar C, Díez-Delgado I, Boots M (2019) The critical role of infectious disease in compensatory population growth in response to culling. Am Nat 194(1):E1–E12. https://doi.org/10.1086/703437
7. Van der Waal K, Deen J (2018) Global trends in infectious diseases of swine. PNAS 115(45):11495–11500. https://doi.org/10.1073/pnas.1806068115
8. McKibbin WJ, Roshen F (2020) The global macroeconomic impacts of COVID-19: seven scenarios (March 2, 2020). CAMA Working Paper No. 19/2020. Available at SSRN. https://doi.org/10.2139/ssrn.3547729. https://ssrn.com/abstract=3547729
9. Gao Z (2020) Efficient management of novel coronavirus pneumonia by efficient prevention and control in scientific manner. Chin J Tuberc Respir Dis 43:E001
10. Gao ZC (2020) Efficient management of novel coronavirus pneumonia by efficient prevention and control in scientific manner. Zhonghua jie he he hu xi za zhi/Zhonghua jiehe he huxi zazhi/Chin J Tuberc Respir Dis 43(3):163–166. https://doi.org/10.3760/cma.issn.1001-0939.2020.03.002
11. Gorbalenya AE (2020) Severe acute respiratory syndrome-related coronavirus–the species and its viruses, a statement of the Coronavirus Study Group. BioRxiv 2020.02.07.937862:1–15. https://doi.org/10.1101/2020.02.07.937862
12. Gralinski LE, Menachery VD (2020) Return of the coronavirus: 2019-nCoV. Viruses 12(2):135
13. De Groot RJ, Baker SC, Baric R, Enjuanes L, Gorbalenya AE, Holmes KV, Perlman S, Poon L, Rottier PJM, Talbot PJ, Woo PCY, Ziebuhr J (2012) Family Coronaviridae. In: Virus taxon-

omy: ninth report of the International Committee on Taxonomy of Viruses. Academic Press, San Diego, CA. p 806–828
14. Su S, Wong G, Shi W, Liu J, Lai AC, Zhou J, Liu W, Bi Y, Gao GF (2016) Epidemiology, genetic recombination, and pathogenesis of coronaviruses. Trends Microbiol 24(6):490–502
15. Malik YS, Sircar S, Bhat S, Sharun K, Dhama K, Dadar M, Tiwari R, Chaicumpa W (2020) Emerging novel coronavirus (2019-nCoV)-current scenario, evolutionary perspective based on genome analysis and recent developments. Vet Q 40(1):1–12
16. Luo W, Yu H, Gou J, Li X, Sun Y, Li J, Liu L (2020) Clinical pathology of critical patient with novel coronavirus pneumonia (COVID-19). Pathol Pathobiol:2020020407. 104(11):e329–e331. https://doi.org/10.1097/TP.0000000000003412
17. Lv L, Li G, Chen J, Liang X, Li Y (2020) Comparative genomic analysis revealed specific mutation pattern between human coronavirus SARS-CoV-2 and Bat-SARSr-CoV RaTG13. BioRxiv 2020.02.27.969006:1–20. https://doi.org/10.1101/2020.02.27.969006
18. Chen J (2020) Pathogenicity and transmissibility of 2019-nCoV—a quick overview and comparison with other emerging viruses. Microbes Infect 22(2):69–71
19. Patel A, Jernigan DB (2020) Initial public health response and interim clinical guidance for the 2019 novel coronavirus outbreak—United States, December 31, 2019–February 4, 2020. Morb Mortal Wkly Rep 69(5):140
20. Salata C, Calistri A, Parolin C, Palù G (2019) Coronaviruses: a paradigm of new emerging zoonotic diseases. Pathog Dis 77(9):ftaa006
21. Ji W, Wang W, Zhao X, Zai J, Li X (2020) Cross-species transmission of the newly identified coronavirus 2019-nCoV. J Med Virol 92(4):433–440
22. Li Q, Guan X, Wu P, Wang X, Zhou L, Tong Y, Ren R, Leung KS, Lau EH, Wong JY (2020) Early transmission dynamics in Wuhan, China, of novel coronavirus–infected pneumonia. N Engl J Med. https://doi.org/10.1056/NEJMoa2001316
23. Li H, Wang YM, Xu JY, Cao B (2020) Potential antiviral therapeutics for 2019 novel coronavirus. Chin J Tuberc Respir Dis 43(3):170–172. https://doi.org/10.3760/cma.j.issn.1001-0939.2020.03.004
24. Woo PC, Lau SK, Lam CS, Lau CC, Tsang AK, Lau JH, Bai R, Teng JL, Tsang CC, Wang M (2012) Discovery of seven novel Mammalian and avian coronaviruses in the genus Deltacoronavirus supports bat coronaviruses as the gene source of Alphacoronavirus and Betacoronavirus and avian coronaviruses as the gene source of Gammacoronavirus and deltacoronavirus. J Virol 86(7):3995–4008
25. Woo PC, Lau SK, Lam CS, Tsang AK, Hui S-W, Fan RY, Martelli P, Yuen KY (2014) Discovery of a novel bottlenose dolphin coronavirus reveals a distinct species of marine mammal coronavirus in Gammacoronavirus. J Virol 88(2):1318–1331
26. Woo PC, Lau SK, Bai R, Teng JL, Lee P, Martelli P, Hui S-W, Yuen K-Y (2012) Complete genome sequence of a novel picobirnavirus, otarine picobirnavirus, discovered in California sea lions. J Virol 86(11):6377–6378
27. Hu B, Zeng L-P, Yang X-L, Ge X-Y, Zhang W, Li B, Xie J-Z, Shen X-R, Zhang Y-Z, Wang N (2017) Discovery of a rich gene pool of bat SARS-related coronaviruses provides new insights into the origin of SARS coronavirus. PLoS Pathog 13(11):e1006698
28. Hu D, Zhu C, Ai L, He T, Wang Y, Ye F, Yang L, Ding C, Zhu X, Lv R (2018) Genomic characterization and infectivity of a novel SARS-like coronavirus in Chinese bats. Emerg Microbes Infect 7(1):1–10
29. Hu B, Ge X, Wang L-F, Shi Z (2015) Bat origin of human coronaviruses. Virol J 12(1):221
30. Cui J, Li F, Shi Z-L (2019) Origin and evolution of pathogenic coronaviruses. Nat Rev Microbiol 17(3):181–192
31. FAO/WHO (2008) Microbiological hazards in fresh leafy vegetables and herbs: meeting report. Microbiol Risk Assess Ser 14:151
32. Gorbalenya AE et al (2020) Severe acute respiratory syndromerelated coronavirus: the species and its viruses – a statement of the Coronavirus Study Group. bioRxiv. Published online February 11, 2020. https://doi.org/10.1101/2020.02.07.937862

33. WHO (2003) Consensus document on the epidemiology of severe acute respiratory syndrome (SARS). In: World Health Organization
34. WHO (2020) Updated WHO recommendations for international traffic in relation to COVID-19 outbreak. In: World Health Organization
35. Dhama K (2013) Avian/bird flu virus: poultry pathogen having. J Med Sci 13(5):301–315
36. Monchatre-Leroy E, Boué F, Boucher J-M, Renault C, Moutou F, Ar Gouilh M, Umhang G (2017) Identification of alpha and beta coronavirus in wildlife species in France: bats, rodents, rabbits, and hedgehogs. Viruses 9(12):364
37. Dhama K, Chakraborty S, Kapoor S, Tiwari R, Kumar A, Deb R, Rajagunalan S, Singh R, Vora K, Natesan S (2013) One world, one health-veterinary perspectives. Adv Anim Vet Science 1(1):5–13
38. Dhama K, Karthik K, Khandia R, Chakraborty S, Munjal A, Latheef SK, Kumar D, Ramakrishnan MA, Malik YS, Singh R (2018) Advances in designing and developing vaccines, drugs, and therapies to counter Ebola virus. Front Immunol 9:1803
39. Dhama K, Khan S, Tiwari R, Dadar M, Malik Y, Singh K, Chaicumpa W (2020) COVID-19, an emerging coronavirus infection: advances and prospects in designing and developing vaccines, immunotherapeutics and therapeutics- a mini-review. Hum Vaccin Immunother 16:1232–1238. https://doi.org/10.1080/21645515.2020.1735227
40. Resta S et al (2018) Isolation and propagation of a human enteric coronavirus. Science 4717:978–981
41. Xu Y (2020) Genetic diversity and potential recombination between ferret coronaviruses from European and American lineages. J Infect 80(3):350–371
42. Compton SR, Ball-Goodrich LJ, Johnson LK, Johnson EA, Paturzo FX, Macy JD (2004) Pathogenesis of enterotropic mouse hepatitis virus in immunocompetent and immunodeficient mice. Comp Med 54(6):681–689
43. Funk CJ, Manzer R, Miura TA, Groshong SD, Ito Y, Travanty EA, Leete J, Holmes KV, Mason RJ (2009) Rat respiratory coronavirus infection: replication in airway and alveolar epithelial cells and the innate immune response. J Gen Virol 90:2956–2964
44. Doria-Torra G, Vidaña B, Ramis A, Amarilla S, Martínez J (2016) Coronavirus infection in ferrets: antigen distribution and inflammatory response. Vet Pathol 53(6):1180–1186
45. Lau SK, Woo PC, Yip CC, Fan RY, Huang Y, Wang M, Guo R, Lam CS, Tsang AK, Lai KK (2012) Isolation and characterization of a novel betacoronavirus subgroup A coronavirus, rabbit coronavirus HKU14, from domestic rabbits. J Virol 86(10):5481–5496
46. Suzuki T, Otake Y, Uchimoto S, Hasebe A, Goto Y (2020) Genomic characterization and phylogenetic classification of bovine coronaviruses through whole genome sequence analysis. Viruses 12(2):183
47. Torres-Medina A, Schlafer DH, Mebus CA (1985) Rotaviral and coronaviral diarrhea. Vet Clin North Am Food Anim Pract 1(3):471–493
48. Amer HM (2018) Bovine-like coronaviruses in domestic and wild ruminants. Anim Health Res Rev 19(2):113–124
49. Arabi YM et al (2017) Middle east respiratory syndrome. N Engl J Med 376:584–594
50. Kim JH, Jang JH, Yoon SW, Noh JY, Ahn MJ, Kim Y, Jeong DG, Kim HK (2018) Detection of bovine coronavirus in nasal swab of non-captive wild water deer, Korea. Transbound Emerg Dis 65(3):627–631
51. Zhong NS et al (2003) Epidemiology and cause of severe acute respiratory syndrome (SARS) in Guangdong, People's Republic of China, in February, 2003. Lancet 362:1353–1358
52. Liu Y-C et al (2020) A locally transmitted case of SARS-CoV-2 infection in Taiwan. N Engl J Med. Published online February 12, 2020. https://doi.org/10.1056/NEJMc2001573
53. Carman PS, Hazlett MJ (1992) Bovine coronavirus infection in Ontario 1990–1991. Can Vet J 33(12):812
54. Drosten C et al (2003) Identification of a novel coronavirus in patients with severe acute respiratory syndrome. N Engl J Med 348:1967–1976

55. Kang M et al (2020) Evidence and characteristics of human-to-human transmission of SARS-CoV-2. medRxiv. Published online February 17, 2020. https://doi.org/10.1101/2020.02.03.20019141
56. Coronavirus Disease 2019 (COVID-19) (2020) If you have animals Available at: https://www.cdc.gov/coronavirus/2019-ncov/daily-life-coping/animals.html
57. Dhama K, Pawaiya R, Chakraborty S, Tiwari R, Saminathan M, Verma AK (2014) Coronavirus infection in equines: a review. Asian J Anim Vet Adv 9(3):164–176
58. Yang Y et al (2020) Epidemiological and clinical features of the 2019 novel coronavirus outbreak in China. medRxiv. Published online February 21, 2020. https://doi.org/10.1101/2020.02.10.20021675
59. Guy JS, Breslin JJ, Breuhaus B, Vivrette S, Smith LG (2000) Characterization of a coronavirus isolated from a diarrheic foal. J Clin Microbiol 38(12):4523–4526
60. Guan Y et al (2003) Isolation and characterization of viruses related to the SARS coronavirus from animals in southern China. Science 302:276–278
61. Pusterla N, Mapes S, Wademan C, White A, Ball R, Sapp K, Burns P, Ormond C, Butterworth K, Bartol J (2013) Emerging outbreaks associated with equine coronavirus in adult horses. Vet Microbiol 162(1):228–231
62. Song HD et al (2005) Cross-host evolution of severe acute respiratory syndrome coronavirus in palm civet and human. Proc Natl Acad Sci U S A 102:2430–2435
63. Chan JF-W et al (2020) A familial cluster of pneumonia associated with the 2019 novel coronavirus indicating person-to-person transmission: a study of a family cluster. Lancet 395:514–523
64. Sanz MG, Kwon S, Pusterla N, Gold JR, Bain F, Evermann J (2019) Evaluation of equine coronavirus fecal shedding among hospitalized horses. J Vet Intern Med 33(2):918–922
65. Cohen J et al (2019) New SARS-like virus in China triggers alarm. Science 6475:234–235
66. Wang D et al (2020) Clinical characteristics of 138 hospitalized patients with 2019 novel coronavirus-infected pneumonia in Wuhan, China. JAMA. Published online February 7, 2020. https://doi.org/10.1001/jama.2020.1585
67. World Organisation for Animal Health (OIE) (2020) Questions and answers on the 2019 Coronavirus Disease (COVID-19) Available at: https://www.oie.int/en/scientific-expertise/specific-information-and-recommendations/questions-andanswers-on- 2019novel-coronavirus/. Accessed 3 Apr 2020
68. Jaimes JA, Millet JK, Stout AE, André NM, Whittaker GR (2020) A tale of two viruses: the distinct spike glycoproteins of feline coronaviruses. Viruses 12(1):83
69. Woo PC et al (2012) Discovery of seven novel mammalian and avian coronaviruses in the genus deltacoronavirus supports bat coronaviruses as the gene source of alphacoronavirus and betacoronavirus and avian coronaviruses as the gene source of gammacoronavirus and deltacoronavirus. J Virol 86:3995–4008
70. Mihindukulasuriya KA, Wu G, Leger JS, Nordhausen RW, Wang D (2008) Identification of a novel coronavirus from a beluga whale by using a panviral microarray. J Virol 82(10):5084–5088
71. Chinese SARS Molecular Epidemiology Consortium (2004) Molecular evolution of the SARS coronavirus during the course of the SARS epidemic in China. Science 303:1666–1669
72. Fan Y, Zhao K, Shi Z-L, Zhou P (2019) Bat coronaviruses in China. Viruses 11(3):210
73. He B et al (2014) Identification of diverse alphacoronaviruses and genomic characterization of a novel severe acute respiratory syndrome-like coronavirus from bats in China. J Virol 88:7070–7082
74. World Health Organization, March 8, 2020 (2020) Q&A on Coronaviruses (COVID-19). Available at: https://www.who.int/news-room/q-a-detail/q-a-coronaviruses
75. Wassenaar TM, Zou Y (2020) 2019_nCoV: rapid classification of betacoronaviruses and identification of traditional Chinese medicine as potential origin of zoonotic coronaviruses. Lett Appl Microbiol:1–7. https://doi.org/10.1111/lam.13285

76. Li Q et al (2020) Early transmission dynamics in Wuhan, China, of novel coronavirus–infected pneumonia. N Engl J Med. Published online January 29, 2020. https://doi.org/10.1056/NEJMoa2001316
77. Chang D et al (2020) Epidemiologic and clinical characteristics of novel coronavirus infections involving 13 patients outside Wuhan, China. JAMA. Published online February 7, 2020. https://doi.org/10.1001/jama.2020.1623
78. Li L et al (2020) An update on the epidemiological characteristics of novel coronavirus pneumonia (COVID-19). Zhonghua Liu Xing Bing Xue Za Zhi 41:139–144
79. Novel Coronavirus Pneumonia Emergency Response Epidemiology Team (2020) The epidemiological characteristics of an outbreak of 2019 novel coronavirus diseases (COVID-19) in China. Zhonghua Liu Xing Bing Xue Za Zhi 41:145–151
80. Subramanian MP, Meyers BF (2018)Bilateral versus single lung transplantation: are two lungs better than one? J Thorac Dis. 10(7):4588–4601. https://doi.org/10.21037/jtd.2018.06.56. PMID: 30174911
81. Szczepanski A, Owczarek K, Bzowska M, Gula K, Drebot I, Ochman M, Maksym B, Rajfur Z, Mitchell JA, Pyrc K (2019) Canine respiratory coronavirus, bovine coronavirus, and human coronavirus OC43: receptors and attachment factors. Viruses 11(4):328
82. Guan W-J et al (2020) Clinical characteristics of 2019 novel coronavirus infection in China. medRxiv. Published online February 9, 2020. https://doi.org/10.1101/2020.02.06.20020974
83. Lipsitch M et al (2003) Transmission dynamics and control of severe acute respiratory syndrome. Science 300:1966–1970
84. Riley S et al (2003) Transmission dynamics of the etiological agent of SARS in Hong Kong: impact of public health interventions. Science 300:1961–1966
85. Huang C, Wang Y, Li X, Ren L, Zhao J, Hu Y, Zhang L, Fan G, Xu J, Gu X (2020) Clinical features of patients infected with 2019 novel coronavirus in Wuhan, China. The Lancet 395(10223):497–506
86. Breban R et al (2013) Interhuman transmissibility of Middle East respiratory syndrome coronavirus: estimation of pandemic risk. Lancet 382:694–699
87. Chowell G et al (2004) Model parameters and outbreak control for SARS. Emerg Infect Dis 10:1258–1263
88. Tian S, Hu W, Niu L, Liu H, Xu H, Xiao S-Y (2020) Pulmonary pathology of early phase SARS-COV-2 pneumonia. Pathol Pathobiol 2020(2020020220). https://doi.org/10.20944/preprints202002.0220.v
89. Swerdlow DL, Finelli L (2020) Preparation for possible sustained transmission of 2019 novel coronavirus: lessons from previous epidemics. JAMA. Published online February 11, 2020. https://doi.org/10.1001/jama.2020.1960
90. Park JE et al (2018) MERS transmission and risk factors: a systematic review. BMC Public Health 18:574
91. Xu Z, Shi L, Wang Y, Zhang J, Huang L, Zhang C, Liu S, Zhao P, Liu H, Zhu L (2020) Pathological findings of COVID-19 associated with acute respiratory distress syndrome. Lancet Respir Med:1–3. https://doi.org/10.1016/S2213-2600(20)30076-X
92. Paul LD et al (2019) Complexity of the basic reproduction number (R0). Emerg Infect Dis 25:1–4
93. Chen N et al (2020) Epidemiological and clinical characteristics of 99 cases of 2019 novel coronavirus pneumonia in Wuhan, China: a descriptive study. Lancet 395:507–513
94. Xu Y-H, Dong J-H, An W-M, Lv X-Y, Yin X-P, Zhang J-Z, Dong L, Ma X, Zhang H-J, Gao B-L (2020) Clinical and computed tomographic imaging features of novel coronavirus pneumonia caused by SARS-CoV-2. J Infect. https://doi.org/10.1016/j.jinf.2020.02.017
95. Safoora (2020) Virus panic: owners throw away pets from high-rise buildings. Available at https://www.siasat.com/viruspanic-owners-throw-away-pets-highrise-buildings-1812619/
96. Wan Y, Shang J, Graham R, Baric RS et al (2020) Receptor recognition by the novel coronavirus from Wuhan: an analysis based on decade-long structural studies of SARS coronavirus. J Virol 94. https://doi.org/10.1128/JVI.00127-20

97. Williams S (2020) Coronavirus: rescuing China's animals during the outbreak. Available at https://www.bbc.com/news/world-asia-china-51614957
98. Daniels KF (2020) Animal humane experts estimate tens of thousands of pets have been abandoned, killed in China during coronavirus crisis. Available at https://www.nydailynews.com/coronavirus/ny-coronavirus-thousands-of-petsabandonedkilled-in-china-crisis-202003 16-3avi3u6osvdltnfttiqbfdttsi-story.html
99. McNamara T, Richt JA, Larry G (2020) A critical needs assessment for research in companion animals and livestock following the pandemic of COVID-19 in humans. Vector Borne Zoonotic Dis 20:393–405
100. Thomson B(2020) Chinese officials 'round up and execute villagers' pets to stop the spread of coronavirus' despite WHO experts saying the disease cannot be spread to animals. Available at https://www.dailymail.co.uk/news/article-8024787/Chineseofficials-continue-beat-dogs-death-fighting-coronavirus.html
101. Thomson B (2020) Community officers around China are filmed clubbing dogs to death in front of horrified residents 'in the name of curbing the coronavirus'. Available at https://www.dailymail.co.uk/news/article-7996217/Community-offciers-China-filmed-beatingdogs-death-prevent-coronavirus.html
102. Bao L, Deng W, Huang B, et al (2020) The pathogenicity of SARS-CoV-2 in hACE2 transgenic mice. Nature, 583: 830–833. https://doi.org/10.1038/s41586-020-2312-y. https://doi.org/10.1101/2020.02.07.939389
103. Simin W, Ge Y, Murphy F (2020) Update: a Hong Kong dog tested positive for coronavirus, and that's all anyone can agree on. Available at https://www.caixinglobal.com/2020-03-05/a-hong-kong-dog-tested-positive-for-coronavirus-and-thatsall-anyone-can-agree-on-101524525.html
104. Albawaba (2020) Chinese officials continue to kill pets to stop coronavirus spread. Available at https://www.albawaba.com/editors-choice/chinese-officials-continue-kill-pets-stopcoronavirus-spread-1340465
105. ANSES (2020) Opinion of the French Agency for Food, Environmental and Occupational Health & Safety on an urgent request to assess certain risks associated with COVID-19. Available at https://www.anses.fr/en/system/files/SABA2020 SA0037-1EN.pdf
106. Mahmud AH (2020) 'Theoretical possibility' that COID-19 can spread from animals to humans, but pets not a serious vector of transmission: MOH. Available at https://www.channelnewsasia.com/news/singapore/coronavirus-pets-spread-possibility-singapore-moh-covid-19-12508030
107. ScienceNews (2020) A cat appears to have caught the coronavirus, but it's complicated. Available at https://www.sciencenews.org/article/cats-animals-pets-coronavirus-covid19
108. Zhen L (2020) Cats are more susceptible to Covid-19 than dogs but pet lovers have 'no cause for alarm': researchers. Available at https://www.scmp.com/news/china/science/article/3078153/cats-more-susceptible-covid-19-dogs-no-pet-lovers-have-cause
109. World Small Animal Veterinary Association (WSAVA), April 3, 2020 (2020) COVID-19—advice and resources. Available at: https://wsava.org/news/highlightednews/the-new-coronavirus-and-companion-animals-advice-for-wsava-members/. Accessed 3 Apr 2020
110. Zhou Y (2020) WHO says pets are generally safe from being infected with coronavirus. Available at https://qz.com/1816853/your-dogs-and-cats-cannot-spread-the-coronavirus/
111. AVMA (2020) COVID-19: FAQs for pet owners. Available at https://www.avma.org/sites/default/files/2020-03/covid-19-faqpet-owners_031620.pdf
112. AVMA (2020) COVID-19: what veterinarians need to know. Available at https://www.avma.org/sites/default/files/2020-03/COVID-19-What-veterinarians-need-to-know_031520.pdf
113. Systematic Reviews for Animals & Food (2020) A rapid review of evidence of infection of pets and livestock with human-associated coronavirus diseases, SARS, MERS, and COVID-19, and evidence of fomite potential of pets and livestock. Available at http://www.syreaf.org/wp-content/uploads/2020/04/Rapid-Review-of-pets-as-fomites_3.pdf

114. Bryner J (2020) Cat infected with COID-19 from owner in Belgium. Available at https://www.livescience.com/cat-infectedcovid-19-from-owner.html
115. Tiwari PM et al (2018) Engineered mRNA-expressed antibodies prevent respiratory syncytial virus infection. Nat Commun 9(1):3999
116. Brown L (2020) First known cat infected with coronavirus reported in Belgium. Available at https://nypost.com/2020/03/27/firstknown-cat-infected-with-coronavirus-reported-in-belgium/
117. Hellewell J, Abbott S, Gimma A, Bosse NI, Jarvis CI, Russell TW, Munday JD, Kucharski AJ, Edmunds WJ, Funk S et al (2020) Feasibility of controlling COVID-19 outbreaks by isolation of cases and contacts. Lancet Glob Health 8(4):e488–e496
118. AFSCA (2020) COVID-19 in a cat in Belgium (in Dutch). Available at http://www.afsca.be/newsletter-da-vt/newsletter285_nl.asp
119. Stone J (2020) How much should you worry about your pets? A cat was confirmed to be infected with COVID19 by its owner. Available at https://www.forbes.com/sites/judystone
120. Stalin RV et al (2018) Chimeric camel/human heavy-chain antibodies protect against MERS-CoV infection. Sci Adv 4(8):eaas9667
121. Sun K, Gu L, Ma L, Duan Y (2020) Atlas of ACE2 gene expression in mammals reveals novel insights in transmission of SARSCov-2. bioRxiv. https://doi.org/10.1101/2020.03.30.015644
122. Jarvis C (2020) Mar 16, 2020 which species transmit Covid-19 to humans? we're still not sure. Accessed 1 Apr 2020. https://www.the-scientist.com/news-opinion/which-species-transmit-covid-19-to-humans-were-stillnot-sure-67272
123. Corman V, Eckerle I, Bleicker T, Zaki A, Landt O, Eschbach-Bludau M, van Boheemen S, Gopal R, Ballhause M, Bestebroer T (2012) Detection of a novel human coronavirus by real-time reverse-transcription polymerase chain reaction. Eurosurveillance 17(39):20285
124. Dhama K, Sharun K, Tiwari R, Sircar S, Bhat S, Malik YS, Singh KP, Chaicumpa W, Bonilla-Aldana DK, Rodriguez-Morales AJ (2020) Coronavirus disease 2019 – COVID-19. Preprints (202003000). https://doi.org/10.20944/preprints202003.0001.v1
125. Dong E, Du H, Gardner L (2020) An interactive web-based dashboard to track COVID-19 in real time. Lancet Infect Dis. https://doi.org/10.1016/S14733099(20)30120-1
126. Jalava K (2020) First respiratory transmitted food borne outbreak? Int J Hyg Environ Health 226:113490
127. Munster VJ, Koopmans M, van Doremalen N, van Riel D, de Wit E (2020) A novel coronavirus emerging in China—key questions for impact assessment. N Engl J Med 382(8):692–694
128. Liu Z, Xiao X, Wei X, Li J, Yang J, Tan H, Zhu J, Zhang Q, Wu J, Liu L (2020) Composition and divergence of coronavirus spike proteins and host ACE2 receptors predict potential intermediate hosts of SARS-CoV-2. J Med Virol 92(6):595–601. https://doi.org/10.1002/jmv.25726
129. Cheng Y et al (2005) Use of convalescent plasma therapy in SARS patients in Hong Kong. Eur J Clin Microbiol Infect Dis 24(1):44–46
130. de Wit E, van Doremalen N, Falzarano D et al (2016) SARS and MERS: recent insights into emerging coronaviruses. Nat Rev Microbiol 14(8):523–534
131. Daszak P (2020) A qualitative study of zoonotic risk factors among rural communities in southern China. Int Health 12(2):77–85
132. Daszak P, Olival K, Li H (2020) A strategy to prevent future epidemics similar to the 2019-nCoV outbreak. Biosaf Health 2(1):6–8. https://doi.org/10.1016/j.bsheal.2020.01.003
133. Wei M et al (2020) Novel coronavirus infection in hospitalized infants under 1 year of age in China. JAMA. Published online February 14, 2020. https://doi.org/10.1001/jama.2020.2131
134. De Wit E, Feldmann F, Cronin J, Jordan R, Okumura A, Thomas T, Scott D, Cihlar T, Feldmann H (2020) Prophylactic and therapeutic remdesivir (GS-5734) treatment in the rhesus macaque model of MERS-CoV infection. Proc Natl Acad Sci U S A 117(12):6771–6776
135. Munster VJ, Feldmann F, Williamson BN, van Doremalen N, Pérez-Pérez L, Schulz J, Meade-White K, Okumura A, Callison J, Brumbaugh B et al (2020) Respiratory disease and virus shedding in rhesus macaques inoculated with SARS-CoV-2. Nature. https://doi.org/10.1038/s41586-020-2324-7

136. Liu J, Zheng X, Tong Q, Li W, Wang B, Sutter K, Trilling M, Lu M, Dittmer U, Yang D (2020) Overlapping and discrete aspects of the pathology and pathogenesis of the emerging human pathogenic coronaviruses SARS-CoV, MERS-CoV, and 2019-nCoV. J Med Virol 92(5):491–494
137. Murdoch DR, French NP (2020) COVID-19: another infectious disease emerging at the animal-human interface. N Z Med J 133(1510):12–15
138. Wang C et al (2020) A human monoclonal antibody blocking SARS-CoV-2 infection. bioRxiv, preprint. https://doi.org/10.1101/2020.03.11.987958
139. Nishiura H, Linton NM, Akhmetzhanov AR (2020) Initial cluster of novel coronavirus (2019-nCoV) infections in Wuhan, China is consistent with substantial human-to-human transmission. JCM 9(2):488
140. Roberts A, Lamirande EW, Vogel L, Jackson JP, Paddock CD, Guarner J, Zaki SR, Sheahan T, Baric R, Subbarao K (2008) Animal models and vaccines for SARS-CoV infection. Virus Res 133(1):20–32
141. OIE (2020) Animal and environmental investigation to identify the zoonotic source of the COVID-19 virus. Accessed 3 Mar 2020. COVID19_21Feb.pdf
142. Parry NMA (2020) COVID-19 and pets: when pandemic meets panic. Forensic Sci Int Rep 2:100090
143. Andersen KG, Rambaut A, Lipkin WI, Holmes EC, Garry RF (2020) The proximal origin of SARS-CoV-2. Nat Med 26(4):450–452
144. Gu J, Han B, Wang J (2020) COVID-19: gastrointestinal manifestations and potential fecal-oral transmission. Gastroenterology 158(6):1518–1519
145. Ar Gouilh M, Puechmaille SJ, Diancourt L, Vandenbogaert M, Serra-Cobo J, Lopez Roig M, Brown P, Moutou F, Caro V, Vabret A, EPICOREM Consortium et al (2018) SARS-CoV related betacoronavirus and diverse alphacoronavirus members found in western old-world. Virology 517:88–97
146. Au S (2020) Revisiting the role of telemedicine under the 2019 novel coronavirus outbreak. EJGG 2(1):26–27
147. Gretebeck LM, Subbarao K (2015) Animal models for SARS and MERS coronaviruses. Curr Opin Virol 13:123–129
148. TBRI (2020) Texas Biomedical Research Institute. Texas Biomed accelerates multi-species study to identify COVID-19 animal model. Accessed 1 Apr 2020. https://eurekalert.org/pub_releases/2020-03/tbritba032620.php
149. Dhama K, Sharun K, Tiwari R, Dadar M, Malik YS, Singh KP, Chaicumpa W (2020) COVID-19, an emerging coronavirus infection: advances and prospects in designing and developing vaccines, immunotherapeutics, and therapeutics. Hum Vaccin Immunother 18:1–7
150. Dhama K, Patel SK, Sharun K, Pathak M, Tiwari R, Yatoo MI (2020) SARS-CoV-2: jumping the species barrier, lessons from SARS and MERS, its zoonotic spillover, transmission to humans, preventive and control measures and recent developments to counter this pandemic virus. Preprints. https://doi.org/10.20944/preprints202004.0011.v1
151. Ge XY, Li JL, Yang XL, Chmura AA, Zhu G, Epstein JH, Mazet JK, Hu B, Zhang W, Peng C et al (2013) Isolation and characterization of a bat SARS-like coronavirus that uses the ACE2 receptor. Nature 503(7477):535–538
152. Vlasova AN et al (2016) Gnotobiotic neonatal pig model of rotavirus infection and disease. In: Svensson L et al (eds) Viral gastroenteritis: molecular epidemiology and pathogenesis. Elsevier, Amsterdam, pp 219–241
153. Gorbalenya AE, Baker SC, Baric RS, de Groot RJ, Drosten C, Gulyaeva AA, Haagmans BL, Lauber C, Leontovich AM, Neuman BW et al (2020) The species severe acute respiratory syndrome-related coronavirus: classifying 2019-nCoV and naming it SARS-CoV-2. Nat Microbiol 5(4):536–544
154. Taconic Biosciences (2020) Taconic biosciences' coronavirus (COVID-19) toolkit. Accessed 1 Apr 2020. https://www.taconic.com/taconic-biosciences-coronavirus-covid-19-toolkit.html#comm

155. Hoffmann M, Kleine-Weber H, Schroeder S, Krüger N, Herrler T, Erichsen S, Schiergens TS, Herrler G, Wu NH, Nitsche A, Müller MA, Drosten C, Pöhlmann S (2020) SARS-CoV-2 cell entry depends on ACE2 and TMPRSS2 and is blocked by a clinically proven protease inhibitor. Cell 181(2):271–280.e8
156. Rockx B et al (2020) Comparative pathogenesis of COVID-19, MERS, and SARS in a non-human primate model. Science 368:1012
157. Rudraraju R et al (2019) How live attenuated vaccines can inform the development of broadly cross-protective influenza vaccines. J Infect Dis 219:S81–S87
158. Lakdawala SS, Menachery VD (2020) The search for a COVID-19 animal model. A comparison of SARS-CoV-2 replication, transmission, and disease in mice to monkeys. sciencemag.org. Science 368(6494):942–943
159. Gao Q et al (2020) Science. https://doi.org/10.1126/science.abc1932
160. Yu L, Wu S, Hao X, Li X, Liu X, Ye S, Han H, Dong X, Li X, Li J et al (2020) Rapid colorimetric detection of COVID-19 coronavirus using a reverse transcriptional loop-mediated isothermal amplification (RT-LAMP) diagnostic plat-form: iLACO. medRxiv, 2020.02.20.20025874. https://doi.org/10.1101/2020.02.20.20025874
161. Zhang N, Wang L, Deng X, Liang R, Su M, He C, Hu L, Su Y, Ren J, Yu F et al (2020) Recent advances in the detection of respiratory virus infection in humans. J Med Virol 92(4):408–417. https://doi.org/10.1002/jmv.25674
162. Zhang X, Deng T, Lu J, Zhao P, Chen L, Qian M, Guo Y, Qiao H, Xu Y, Wang Y (2020) Molecular characterization of variant infectious bronchitis virus in China, 2019: implications for control programs. Transbound Emerg Dis:1–7
163. Zhao S, Lin Q, Ran J, Musa SS, Yang G, Wang W, Lou Y, Gao D, Yang L, He D (2020) Preliminary estimation of the basic reproduction number of novel coronavirus (2019-nCoV) in China, from 2019 to 2020: a data-driven analysis in the early phase of the outbreak. Int J Infect Dis 92:214–217
164. Zhou P, Fan H, Lan T, Yang X-L, Shi W-F, Zhang W, Zhu Y, Zhang Y-W, Xie Q-M, Mani S (2018) Fatal swine acute diarrhoea syndrome caused by an HKU2-related coronavirus of bat origin. Nature 556(7700):255–258
165. Zhou P, Yang X-L, Wang X-G, Hu B, Zhang L, Zhang W, Si H-R, Zhu Y, Li B, Huang, C-L (2020) Discovery of a novel coronavirus associated with the recent pneumonia outbreak in humans and its potential bat origin. bioRxiv, 2020.01.22.914952: 1–18. https://doi.org/10.1101/2020.01.22.914952
166. Zhou P, Yang X, Wang X, Hu B, Zhang L, Zhang W, Si H, Zhu Y, Li B, Huang C (2020) A pneumonia outbreak associated with a new coronavirus of probable bat origin. Nature 579:270–273
167. Wölfel R et al (2019) Virological assessment of hospitalized cases of coronavirus disease 2019. medRxiv, preprint. https://doi.org/10.1101/2020.03.05.20030502
168. Wilder-Smith A, Chiew CJ, Lee VJ (2020) Can we contain the COVID-19 outbreak with the same measures as for SARS? Lancet Infect Dis, S1473-3099(20)30129-8. https://doi.org/10.1016/S1473-3099(20)30129-8
169. Wilson ME, Chen LH (2020) Travellers give wings to novel coronavirus (2019-nCoV). J Travel Med. https://doi.org/10.1093/jtm/taaa015
170. Wong AC, Li X, Lau SK, Woo PC (2019) Global epidemiology of bat coronaviruses. Viruses 11(2):174
171. Vergara-Alert J et al (2017) Livestock susceptibility to infection with middle east respiratory syndrome coronavirus. Emerg Infect Dis 23(2):232–240
172. Saif LJ (2017) Coronaviruses of domestic livestock and poultry: interspecies transmission, pathogenesis and immunity. In: Perlman S et al (eds) Nidovirales. American Society of Microbiology, Washington, DC, pp 279–298
173. Wong MC, Cregeen SJJ, Ajami NJ, Petrosino JF (2020) Evidence of recombination in coronaviruses implicating pangolin origins of nCoV-2019. bioRxiv, 2020.02.07.939207:1–9. https://doi.org/10.1101/2020.02.07.939207

174. Jordan B (2017) Vaccination against infectious bronchitis virus: a continuous challenge. Vet Microbiol 206:137–143
175. Schindewolf C, Menachery VD. (2019) Middle east respiratory syndrome vaccine candidates: cautious optimism. Viruses 11(1): pii: E74
176. Wood C (2020) Infections without borders: a new coronavirus in Wuhan, China. Br J Nurs 29(3):166–167
177. Wu A, Yu B, Zhang K, Xu Z, Wu D, He J, Luo J, Luo Y, Yu J, Zheng P (2020) Transmissible gastroenteritis virus targets Paneth cells to inhibit the self-renewal and differentiation of Lgr5 intestinal stem cells via Notch signaling. Cell Death Dis 11(1):1–16
178. Xiao K, Zhai J, Feng Y, Zhou N, Zhang X, Zou J-J, Li N, Guo Y, Li X, Shen X (2020) Isolation and characterization of 2019-nCoV-like coronavirus from Malayan Pangolins. bioRxiv, 2020.02.17.951335:1–31. https://doi.org/10.1101/2020.02.17.951335
179. Saif LJ (2020) Vaccines for COVID-19: perspectives. Prospects, and challenges based on candidate SARS, MERS and animal coronavirus vaccines. Allergy Immunol EMJ. 2020. https://doi.org/10.33590/emj/200324
180. Su S, Wong G, Shi W et al (2016) Epidemiology, genetic recombination, and pathogenesis of coronaviruses. Trends Microbiol 24:490–502
181. Rembaut A (2020) Phylogenetic analysis of nCoV-2019 genomes. http://virological.org/t/phylodynamic-analysis-44-genomes-29-jan-2020/356. Accessed 3 Feb 2020
182. Zhou P, Yang X-L, Wang X-G et al (2020) Discovery of a novel coronavirus associated with the recent pneumonia outbreak in humans and its potential bat origin. bioRxiv, published online Jan 23. https://doi.org/10.1101/2020.01.22.914952
183. Briese T, Mishra N, Jain K et al (2014) Middle East respiratory syndrome coronavirus quasispecies that include homologues of human isolates revealed through whole-genome analysis and virus cultured from dromedary camels in Saudi Arabia. MBio 5:e01146–e01114
184. Lahm SA, Kombila M, Swanepoel R, Barnes RF (2007) Morbidity and mortality of wild animals in relation to outbreaks of Ebola haemorrhagic fever in Gabon, 1994–2003. Trans R Soc Trop Med Hyg 101:64–78
185. Ng OW, Tan YJ (2017) Understanding bat SARS-like coronaviruses for the preparation of future coronavirus outbreaks—implications for coronavirus vaccine development. Hum Vaccin Immunother 13:186–189
186. Guan Y, Zheng BJ, He YQ et al (2003) Isolation and characterization of viruses related to the SARS coronavirus from animals in southern China. Science 302:276–278
187. Kan B, Wang M, Jing H et al (2005) Molecular evolution analysis and geographic investigation of severe acute respiratory syndrome coronavirus-like virus in palm civets at an animal market and on farms. J Virol 79:11892–11900
188. Zumla A, Dar O, Kock R et al (2016) Taking forward a 'One Health' approach for turning the tide against the Middle East respiratory syndrome coronavirus and other zoonotic pathogens with epidemic potential. Int J Infect Dis 47:5–9
189. Karesh WB, Dobson A, Lloyd-Smith J et al (2012) Ecology of zoonoses: natural and unnatural histories. Lancet 380:1936–1945
190. Pongpirul WA et al (2020) Journey of a Thai taxi driver and novel coronavirus. N Engl J Med. Published online February 12, 2020. https://doi.org/10.1056/NEJMc2001621
191. Bastola A et al (2020) The first 2019 novel coronavirus case in Nepal. Lancet Infect Dis 20:279–280
192. Holshue ML et al (2020) First case of 2019 novel coronavirus in the United States. N Engl J Med. Published online January 31, 2020. https://doi.org/10.1016/S1473-3099(20)30067-0
193. Huang C et al (2020) Clinical features of patients infected with 2019 novel coronavirus in Wuhan, China. Lancet 395:497–506
194. Chen H et al (2020) Clinical characteristics and intrauterine vertical transmission potential of COVID-19 infection in nine pregnant women: a retrospective review of medical records. Lancet 20:30360–30363

195. Xu X et al (2020) Imaging features of 2019 novel coronavirus pneumonia. Eur J Nucl Med Mol Imaging. https://doi.org/10.1007/s00259-020-04720-2
196. Hoehl S et al (2020) Evidence of SARS-CoV-2 infection in returning travelers from Wuhan, China. N Engl J Med. Published online February 18, 2020. https://doi.org/10.1056/NEJMc2001899
197. Zhu H et al (2020) Clinical analysis of 10 neonates born to mothers with 2019-nCoV pneumonia. Transl Pediatr 9:51–60
198. Anderson RM et al (2004) Epidemiology, transmission dynamics and control of SARS: the 2002–2003 epidemic. Philos Trans R Soc Lond B Biol Sci 359:1091–1105
199. Chowell G et al (2015) Transmission characteristics of MERS and SARS in the healthcare setting: a comparative study. BMC Med 13:210
200. Zhou P et al (2020) A pneumonia outbreak associated with a new coronavirus of probable bat origin. Nature. Published online February 3, 2020. https://doi.org/10.1038/s41586-020-2012-7
201. Wu F et al (2020) A new coronavirus associated with human respiratory disease in China. Nature. Published online February 3, 2020. https://doi.org/10.1038/s41586-020-2008-3
202. Li G et al (2018) Origin, genetic diversity, and evolutionary dynamics of novel porcine circovirus 3. Adv Sci (Weinh) 5:1800275
203. Lu G et al (2015) Bat-to-human: spike features determining 'host jump' of coronaviruses SARS-CoV, MERS-CoV, and beyond. Trends Microbiol 23:468–478
204. Millet JK, Whittaker GR (2015) Host cell proteases: critical determinants of coronavirus tropism and pathogenesis. Virus Res 202:120–134
205. Hoffmann M et al (2020) The novel coronavirus 2019 (2019-nCoV) uses the SARS-coronavirus receptor ACE2 and the cellular protease TMPRSS2 for entry into target cells. bioRxiv. Published online January 31, 2020. https://doi.org/10.1101/2020.01.31.929042
206. Coutard B et al (2020) The spike glycoprotein of the new coronavirus 2019-nCoV contains a furin-like cleavage site absent in CoV of the same clade. Antiviral Res 176:104742
207. Walls AC et al (2020) Structure, function and antigenicity of the SARS-CoV-2 spike glycoprotein. bioRxiv. Published online February 20, 2020. https://doi.org/10.1101/2020.02.19.956581
208. Deng Y et al (2006) Structures and polymorphic interactions of two heptad-repeat regions of the SARS virus S2 protein. Structure 14:889–899
209. Barr S (2020) Independent. March 26, Coronavirus pandemic sees huge increase in cat and dog adoptions. Available at: https://www.independent.co.uk/life-style/coronavirus-dog-cat-pet-adoption-battersea-rehome-covid-19-a9426741.html
210. Heald-Sargent T, Gallagher T (2012) Ready, set, fuse! The coronavirus spike protein and acquisition of fusion competence. Viruses 4:557–580
211. Lan J et al (2020) Crystal structure of the 2019-nCoV spike receptor-binding domain bound with the ACE2 receptor. bioRxiv. Published online February 20, 2020. https://doi.org/10.1101/2020.02.19.956235
212. Li F et al (2005) Structure of SARS coronavirus spike receptorbinding domain complexed with receptor. Science 309:1864–1868
213. Zumla A et al (2016) Coronaviruses - drug discovery and therapeutic options. Nat Rev Drug Discov 15:327–347
214. Xia S, Yan L (2019) A pan-coronavirus fusion inhibitor targeting the HR1 domain of human coronavirus spike. Sci Adv 5:eaav4580
215. Gadalla MR, Veit M (2020) Toward the identification of ZDHHC enzymes required for palmitoylation of viral protein as potential drug targets. Expert Opin Drug Discovery 15:59–177
216. Wang N et al (2019) Structural definition of a neutralization-sensitive epitope on the MERS-CoV S1-NTD. Cell Rep 28:3395–3405
217. Letko M et al (2018) Adaptive evolution of MERS-CoV to species variation in DPP4. Cell Rep 24:1730–1737
218. Zhang S et al (2018) Structural definition of a unique neutralization epitope on the receptor-binding domain of MERS-CoV spike glycoprotein. Cell Rep 24:441–452

219. Su S et al (2016) Epidemiology, genetic recombination, and pathogenesis of coronaviruses. Trends Microbiol 24:490–502
220. Zhou P et al (2018) Fatal swine acute diarrhoea syndrome caused by an HKU2-related coronavirus of bat origin. Nature 556:255–258
221. Hu B et al (2017) Discovery of a rich gene pool of bat SARS related coronaviruses provides new insights into the origin of SARS coronavirus. PLoS Pathog 13:e1006698
222. Katoh K et al (2002) MAFFT: a novel method for rapid multiple sequence alignment based on fast Fourier transform. Nucleic Acids Res 30:3059–3066
223. Kumar S et al (2016) MEGA7: molecular evolutionary genetics analysis version 7.0 for bigger datasets. Mol Biol Evol 33:1870–1874
224. Stamatakis A (2014) RAxML version 8: a tool for phylogenetic analysis and post-analysis of large phylogenies. Bioinformatics 30:1312–1313
225. Huang Y-W et al (2013) Origin, evolution, and genotyping of emergent porcine epidemic diarrhea virus strains in the United States. MBio 4:e00737–e00713
226. Tao Y et al (2017) Surveillance of bat coronaviruses in Kenya identifies relatives of human coronaviruses NL63 and 229E and their recombination history. J Virol 91:e01953–e01916
227. Yang L et al (2019) Broad cross-species infection of cultured cells by bat HKU2-related swine acute diarrhea syndrome coronavirus and identification of its replication in murine dendritic cells in vivo highlight its potential for diverse interspecies transmission. J Virol 93:e01448–e01419
228. Karakus U et al (2019) MHC class II proteins mediate cross species entry of bat influenza viruses. Nature 7746:109–112
229. He W et al (2019) Genetic analysis and evolutionary changes of porcine circovirus 2. Mol Phylogenet Evol 139:106520
230. Chen W et al (2005) SARS-associated coronavirus transmitted from human to pig. Emerg Infect Dis 11:446–448
231. Hu D et al (2018) Genomic characterization and infectivity of a novel SARS-like coronavirus in Chinese bats. Emerg Microbes Infect 7:154
232. Lam TT-Y et al (2020) Identification of 2019-nCoV related coronaviruses in Malayan pangolins in southern China. bioRxiv. Published online February 18, 2020. https://doi.org/10.1101/2020.02.13.945485
233. Menachery VD et al (2015) A SARS-like cluster of circulating bat coronaviruses shows potential for human emergence. NatMed 21:1508–1513
234. Paraskevis D et al (2020) Full-genome evolutionary analysis of the novel corona virus (2019-nCoV) rejects the hypothesis of emergence as a result of a recent recombination event. Infect Genet Evol 79:104212
235. Luk HKH et al (2019) Molecular epidemiology, evolution and phylogeny of SARS coronavirus. Infect Genet Evol 71:21–30
236. Wong MC et al (2020) Evidence of recombination in coronaviruses implicating pangolin origins of nCoV-2019. bioRxiv. Published online February 13, 2020. https://doi.org/10.1101/2020.02.07.939207
237. Xiao K et al (2020) Isolation and characterization of 2019-nCoV-like coronavirus from Malayan pangolins. bioRxiv. Published online February 20, 2020. https://doi.org/10.1101/2020.02.17.951335
238. Liu P et al (2020) Are pangolins the intermediate host of the 2019 novel coronavirus (2019-nCoV)? bioRxiv. Published online February 20, 2020. https://doi.org/10.1101/2020.02.18.954628
239. Lu R et al (2020) Genomic characterisation and epidemiology of 2019 novel coronavirus: implications for virus origins and receptor binding. Lancet 395:565–574
240. Hui KP et al (2017) Tropism and innate host responses of influenza A/H5N6 virus: an analysis of ex vivo and in vitro cultures of the human respiratory tract. Eur Respir J 49:1601710
241. Wrapp D et al (2020) Cryo-EM structure of the 2019-nCoV spike in the prefusion conformation. Science. Published online February 19, 2020. https://doi.org/10.1126/science.abb2507

242. Sun C et al (2020) SARS-CoV-2 and SARS-CoV spike-RBD structure and receptor binding comparison and potential implications on neutralizing antibody and vaccine development. bioRxiv. Published online February 20, 2020. https://doi.org/10.1101/2020.02.16.951723
243. Li F (2016) Structure, function, and evolution of coronavirus spike proteins. Annu Rev Virol 3:237–261
244. Cui J et al (2019) Origin and evolution of pathogenic coronaviruses. Nat Rev Microbiol 17:181–192
245. Huang C, Wang Y, Li X, Ren L et al (2020) The continuing 2019-nCoV epidemic threat of novel coronaviruses to global health - the latest 2019 novel coronavirus outbreak in Wuhan, China. Lancet 395:497–506. https://doi.org/10.1016/S0140-6736(20)30183-5
246. Hui DS, Azhar EI, Madani TA, Ntoumi F et al (2020) The continuing 2019-nCoV epidemic threat of novel coronaviruses to global health - the latest 2019 novel coronavirus outbreak in Wuhan, China. Int J Infect Dis 91:264–266
247. Teenan T (2002) March 26, Daily beast. Animal shelters have welcomed the coronavirus-related boom in pet adoption and fostering. but some are also planning for large numbers of pets being returned, and financial peril. Available at: https://www.thedailybeast.com/coronavirus-sparks-a-pet-adoption-and-fostering-boombut-animal-shelters-worry-it-may-go-bust. Accessed 3 Apr 2020
248. Lu R, Zhao X, Li J et al (2020) Genomic characterization and epidemiology of 2019 novel coronavirus: implications for virus origins and receptor binding. Lancet 395:565–574. https://doi.org/10.1016/S0140-6736(20)30251-8
249. Li H, Mendelsohn E, Zong C et al (2019) Human-animal interactions and bat coronavirus spillover potential among rural residents in Southern China. J Biosaf Health Educ 1:84–90. https://doi.org/10.1016/j.bsheal.2019.10.004
250. Kruse H, Kirkemo AM, Handeland K (2004) Wildlife as source of zoonotic infections. Emerg Infect Dis 10:2067–2072. Centers for Disease Control and Prevention, March 27, 2020

Chapter 2
Medicinal Plants as COVID-19 Remedy

Sara Zafar, Shagufta Perveen, Naeem Iqbal, M. Kamran Khan, Modhi O. Alotaibi, and Afrah E. Mohammed

Introduction

The pandemic (COVID-19) first appeared in Wuhan (Hubei Province) city of China in late December [1]. The World Health Organization (WHO) reported the disease as pandemic and declared a health emergency globally [2]. SARS-CoV-2 is causative agent of novel coronavirus (COVID-19) infection. It is fatal and spread more rapidly as compared with SARS and MERS [3]. Coronavirus is a rapidly dividing virus; it injects its genome into host genes where it multiplies and uses hosts machinery for its growth. SARS-CoV-2 (COVID-19) infection is in focus over all human infections; it is destroying the social life of people and the world economy. No specific medicine is available so far; however, the use of medicinal plants might be a successful therapeutic strategy against the disease. Coronavirus (family *Coronaviridae*) infects humans and a broad range of animals including bats, cats, cattle, and camels, resulting in respiratory, gastrointestinal, neurological, and hepatic infections [4]. The family of coronavirus, i.e., *Coronaviridae*, is divided into subfamily *Orthocoronavirinae*, which infects human respiratory track; the subfamily is further divided into genera: *Alphacoronavirus*, *Betacoronavirus*, *Gammacoronavirus*, and *Deltacoronavirus* [5]. Seven members of CoV affect human health, and they are known as human CoVs (HCoVs) which include HCoV-229E and HCoV-NL63 from *Alphacoronavirus* genus and HCoV-OC43, SARS-CoV, HCoV-HKU1, SARS-CoV-2, and MERS-CoV from *Betacoronavirus* genus [6]. The size of coronavirus (HCoVs) range between 120 and 160 nm,

S. Zafar (✉) · S. Perveen · N. Iqbal · M. K. Khan
Department of Botany, Government College University Faisalabad, Faisalabad, Punjab, Pakistan

M. O. Alotaibi · A. E. Mohammed
Department of Biology, College of Science, Princess Nourah bint Abdulrahman University, Riyadh, Saudi Arabia

enveloped as a single-stranded RNA which is responsible for the infection in the upper respiratory tract. The infectious respiratory tract shows symptoms such as bronchitis, pneumonia, common cold, etc. A suitable environment for the establishment of coronavirus infection is winter in the documented cases [7]. Coronavirus infection is fatal for the immunosuppressed patients, newborn, and aged people, for the mortality rate of such people are higher than normal and healthy people [8]. The use of natural plant-derived compounds or herbal treatment can be effective against this disease especially TCM (traditional Chinese medicine) [9]. TCM helps to cure and prevent against coronavirus disease.

Some drugs are also used for the cure of coronavirus infection like antiviral drugs [10], antimalarial drugs [11], anti-HIV drugs [12], anti-inflammatory drugs [13], monoclonal ribavirin [14], sofosbuvir [15], lopinavir [16], and monoclonal antibodies [17], but they are not recommended by the WHO for COVID-19 treatment. Due to the absence of proper specific treatment of coronavirus, natural products are best alternative drug therapies [18]. The leaf, root, and stem extract of mulberry tree (*Morus* spp.) possess antiviral properties; it can be used as therapeutic agent for non-enveloped and enveloped viral pathogens. Mulberry tree *Morus alba* var. *alba*, *Morus alba* var. *rosa*, and *Morus rubra* root, stem bark, and leaf extract in water and alcohol helps to recover viral infections [19]. Plants like jackfruit and coffee are mostly studied in this respect because these plants have potential to fight against viral infections [20]. Asian populations use *Morus* spp. against sore throat, fever, rheumatism, and blood pressure to protect the liver and enhance eyesight [21]. *Morus* genus belongs to family Moraceae and has antibacterial and antifungal properties [22].

Herbal medication plays a crucial role to control the infectious diseases, while in the past herbal medication also proved significant during the SARS coronavirus (SARS-CoV) [23]. The Indian herbal plants used for prevention and treatment of various respiratory diseases help improve the immune system [3]. Some Indian plants effective against coronavirus are *Vitex trifolia*, *Indigofera tinctoria* (AO), *Cassia alata*, *Leucas aspera*, *Abutilon indicum*, *Gymnema sylvestre*, *Clerodendrum inerme* Gaertn., *Evolvulus alsinoides*, *Clitoria ternatea*, and *Pergularia daemia* [24]. Among them, *Sphaeranthus indicus* and *Vitex trifolia* are using NF-kB pathway; it is a pathway that decreases inflammatory cytokines in SARS-CoV [25, 26]. Another plant which inhibits viral replication of SARS-CoV and SARS-CoV-2 is *Glycyrrhiza glabra* [27]. A key herb is *Clerodendrum inerme* Gaertn. which inactivate the viral ribosome, while *Strobilanthes Cusia* is an herb which blocks the synthesis of viral genome [28]. Himalayan forests have many medicinal plant species effective in bronchitis [29, 30]. *Verbascum thapsus*, *Justicia adhatoda*, and *Hyoscyamus niger* are effective against influenza viruses; another useful herb *Hyoscyamus niger* acts as bronchodilator with inhibitory role on Ca^{2+} channel [30]. The plant species effective against coronavirus family are *Platycodon grandiflorum* [31], *Reseda luteola* [32], *Glycyrrhiza glabra* [33], *Peganum harmala* L. [34], *Mentha piperita* [35], *Anthemis hyalina* [36], *Camellia japonica* [37], and *Zingiber officinale* [38] that contain active compounds platycodin D, luteolin, glycyrrhizin, harmine, menthol, essential oils oleanane triterpenes, 6-gingerol, sinigrin and hesperetin, respectively. The drugs used as a curative treatment of COVID-19 infection include chloroquine, Arbidol, favipiravir, remdesivir, and interferon therapy [39], but they are not recommended by regulatory authorities. Plant metabolites can inhibit the activity of enzymes in the replication cycle of Coronaviruses including papain-like protease and 3CL protease, inhibit the fusion of the S protein of CoVs and ACE2 of the host, and can inhibit cell signaling pathways [40].

Urtica dioica (stinging nettle) is a plant whose lectins are effective for the inhibition of virus attachment to active compound present in *Utrica dioica* Agglutinin (UDA) [41]. *Glycyrrhiza glabra* (licorice) is another plant; its roots are used as a medicine because it contains effective compound glycyrrhizin which is helpful in the treatment of jaundice, gastritis, and bronchitis due to its anti-inflammatory activity by stimulating the interferon in the body. Glycyrrhizin is composed of glycyrrhetinic acid, flavonoids, β-sitosterol, glycyrrhetinic acid, and hydroxyl coumarins [42]. A secondary metabolite quercetin is a flavonoid discovered in garlic and onion species [43]. Quercetin-3-β-galactoside showed inhibitory effects for SARS-CoV infection by inhibiting the 3CL protargeted for the treatment of the COVID-19 [44]. The extract of plant *Lycoris radiata* in chloroform and ethanol has anti-SARS-CoV characteristics (about 2.1 to 2.4 ug/ml). Active compound present in this plant is lycorine having antiviral properties [45]. In vitro terpenoids and lignoids extracted from *Cryptomeria japonica*, *Juniperus formosana*, and *Chamaecyparis* have inhibitory effect for 3CL protease of SARS-CoV [45].

Existing Plants with Potential Therapeutic Applications for Coronavirus Family (SARS-CoV)

Herbal treatment showed positive results against viral infections of SARS-CoV [23]. Baicalin compound found in *Scutellaria* plant genus have the ability to inhibit the activity of SARS-CoV [46]. *Morus* plant species studied many times for the treatment against diseases in many parts of the world and showed potential against bacterial and fungal infections [22, 47, 48]. Extracts from *Morus alba* minimized the viral load and viral cytopathogenic effects [19]. *Morus alba* leaf extract showed antiviral properties against human coronaviruses [19]. *Clerodendrum inerme* L. proved effective against the SARS-CoV-2 through inactivating the ribosome and protein translation of virus [49]. Organosulfur compounds found in *Allium* genus especially *Allium porrum* can potentially inhibit the attachment of SARS-CoV with cells [50]. A flavonoid compound quercetin found in onion and garlic has potential against SARS-CoV activity [51]. Extract from four different herbs and lycorine compound showed positive results for the treatment of SARS-CoV [45]. Molecules like crocin, digitoxigenin, and β-eudesmol naturally occurring in plants have potential to inhibit the coronavirus activity [52]. Medicinal plants such as *Caesalpinia sappan* L. and citrus species have active compounds that have inhibitory effect against the action of coronavirus [53]. *Cullen corylifolium* L. seed extract hampered the activity of SARS-CoV [54]. Different flavonoid compounds found in *Paulownia tomentosa* fruit significantly inhibited the protein of SARS-CoV [55]. Nine chalcone compounds and four coumarin compounds extracted from *Angelica keiskei* showed inhibitory capability against protein of SARS-CoV [56]. SARS-CoV protein showed inhibited activity with treatment of nine phlorotannins extracted from *Ecklonia cava* [57]. *Torreya nucifera* L. extract consist of four bioflavonoids showed potential against SARS-CoV action [58]. Fruits and vegetables contain a flavonoid named quercetin having antiviral potential against SARS-CoV [59]. Plant extract obtained from *Houttuynia cordata* boosted the cell immunity against SARS-CoV [60]. *Andrographis paniculata* have potential against viral proteins [61]. An active compound luteolin found in *Reseda luteola* inhibited the entrance of SARS-CoV into the cell [62]. *Anthemis hyalina* potentially decreased the receptor potential protein gene expression against coronavirus [36]. Phytochemicals found in *Cedrela sinensis* inhibited the viral replication of SARS-CoV [63]. *Clerodendrum inerme* have potential to inhibit the activation of SARS-CoV-2 [64]. Flavonoids and alkaloids present in *Hyoscyamus niger* played a role in inhibition of calcium channels and dilation of bronchi during infection of SARS-CoV-2 [65]. An active compound named 1, 8-cineoli present in *Vitex trifolia* proved effective against SARS-CoV-2 [66].

Plants as Specific Inhibitors of HCoV Target Proteins

A lot of medicinal plants show inhibitory effect against SARS-CoV and other viral infections; these plants show therapeutic potency against coronavirus infection and examined with ethnobotanical evidence against viral infections (influenza, etc.). Coronavirus mostly use ACE-2 (angiotensin-converting enzyme-2) receptor, RdRp (RNA-dependent RNA polymerase), 3CLpro (3-chymotrypsin-like protease), PLpro (papain-like protease) enzyme, and other factors and gain entry into human cell to complete life cycle. The selected plants are tested by various scientists globally to act on the abovementioned specific receptors and proteins and inhibit RNA replication. SARS-CoV-2 uses the ACE-2 receptor to enter into human cells. Several medicinal plants act on ACE-2 receptor and block the entry of CoVs. Literature studied exhibit many plants acting on ACE-2 receptor; these plants may become promising antiviral agents and can help control COVID-19 infection [4, 67].

Bupleurum Species

Bupleurum species are widely dispersed in Northern Hemisphere and are oldest herbal medicine in China used for the treatment of viral infections [68]. *Radix bupleuri* is used as antiviral, anti-inflammatory, and antitumor agent [69]. Triterpene, saponin, and glycosides are the main components of this medicinal plant with vigorous effects.

Artemisia annua *L.*

The Chinese medicinal plant is effective for its antimalarial, antiviral, and anticancer effects [70, 71]. The plant has been known to possess antiviral activity for the treatment of poliovirus, HIV, hepatitis C, and type 2 dengue virus. The extract contains quercetin, flavonoid, polyphenols, triterpenes, sterols, saponins, and other molecules [72]. The extract has been used to treat SARS-CoV and MERS infection [45] and is currently used against SARS-CoV-2 infection [73]. The plant extract inhibits the activity of 3CLpro the virus main proteinase and controls the activity of replication complex, also produced by SARS-CoV-2 to increase the production of pro-inflammatory cytokines prostaglandin E2 (PGE2) TNF-a, IFN-ϒ, and enhance CD4+ and CD8+ T-cell populations [4].

Isatis indigotica *Fortune ex Lindl.*

The Chinese herbal plant belongs to Cruciferae family. The extract inhibits SARS-CoV-1 entry and replication in host [72]. However, *Radix isatidis* (dried root) of the plant is used by researchers for extraction of compounds against SARS-CoV-1 infection. Its root contains bioactive compounds such as indirubin, indigo, β-sitosterol, and many other bioactive compounds [74, 75] against SARS-CoV-1 infection. Indigo, sinigrin, and hesperetin inhibit virus entry and replication by inhibiting SARS-CoV1 3CLpro [76]. Viral replication is mediated by proteolytic processing of polypeptides into functional proteins by 3CLpro [77]. The plant *Isatis indigotica* can be used as therapeutic cure against SARS-CoV-2.

Alcea digitata *(Boiss.) Alef.*

The plant belongs to Malvaceae family with potent antiviral, antimicrobial, and laxative therapeutic properties [78, 79]. The flower extract of the medicinal plant is used to cure respiratory infections and head and neck cancer [79]. The medicinal plant has strong potential to act on ACE-2 receptor. The extract can be effective against SARS-CoV-2 infection (Fig. 2.1).

Lycoris radiata *(L'Hér.) Herb.*

Lycoris radiata origin was found in China, Japan, Korea, and Nepal [80]. It shows antiviral activity against SARS-CoV [81, 82], poliovirus, measles, and herpes simplex virus [82, 83]. The active compound lycorine is extracted from the flower and stem cortex of *L. radiata* medicinal plant and is used for the treatment of various

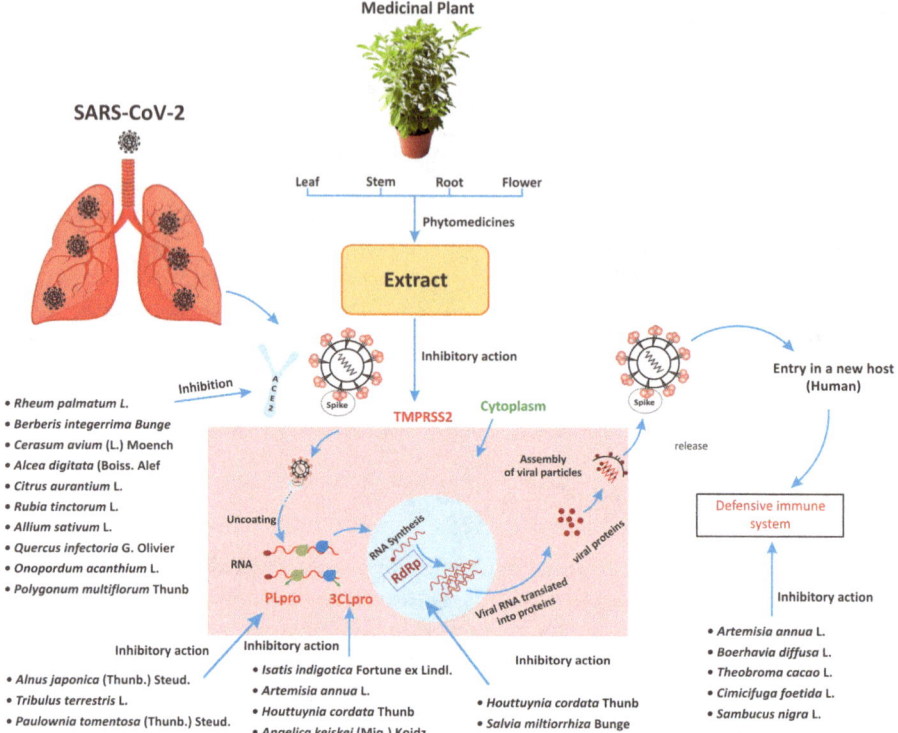

Fig. 2.1 Some medicinal plants' possible specific inhibitory sites to act against SARS-CoV2 action

diseases. It is recommended for the development of drugs against SARS-CoV infection [83, 84]. The antiviral activity of the plant is due to the inhibiting action against virus replication by inhibiting autophagy [83]; the plant extract exhibits anti-SARS-CoV effect with 50% effective concentration (EC_{50}) in the range of 2.4–88.2 μg/mL. The active component lycorine is extracted from the flower and stem cortex of L. radiata medicinal plant is used for the treatment of various diseases [85]. The plant extract inhibits autophagy by reduced JNK/MAPK phosphorylation due to viral replication [83, 84].

Pyrrosia lingua *(Thunb.) Farw.*

The plant belongs to family Polypodiaceae and occurs in China, Korea, Japan, and other regions of Asia [86]. It has antiviral, antibacterial, and anticancer activities and contains flavonoids, chlorogenic acid, and other bioactive compounds [87]. The leaf extract is used for the treatment of HIV, SARS, and other infections caused by viruses [86]. The plant inhibits SARS-CoV-1 infection, but mechanism of action needs to be studied in detail [4].

Houttuynia cordata *Thunb.*

The Chinese herbal plant belongs to family Saururaceae and is mostly found in Southeast Asia. It possesses anti-allergic, antioxidant, and anticancer properties [88]. The plant is well known for the treatment of pneumonia and SARS-CoV-1 infection [4, 89]. The bioactive compounds present in the plant extract are rutin, hyperin, quercetin, linoleic acid, aspartic acid, amino acids, vitamins, potassium, copper, zinc, iron, and others [4, 90]. The leaf extract was effective in SARS-CoV-1 patients [76]. It inhibits 3CLpro activity and RdRp activity of SARS-CoV-1 [60]. Hence, it blocks the entry of virus and interferes with viral replication [60], and the plant can be used against SARS-CoV-2 infections also.

Torreya nucifera *L.*

The traditional medicinal plant is mostly found in snowy regions of Korea. The leaf extract is used mostly to cure stomachache and rheumatoid arthritis [88, 91]. Ryu et al. [58] isolated 12 bioactive compounds from ethanolic extract of the plant leaves and observed inhibitory effect of the biflavonoid amentoflavone against SARS-CoV-1 by blocking the activity of 3CLpro of coronavirus [58].

Lindera aggregata *(Sims) Kosterm.*

A traditional Chinese medicinal plant belongs to the family Lauraceae and is native to China and Japan [90]. The root extract is used to cure chest pain, inflammation, and other diseases. The bioactive compounds, i.e., flavonoids, alkaloids, tannins, and sesquiterpene lactones, exhibit antiviral, antitumor, and antimicrobial activities [92, 93]. The extract is used as tea for its protective effect against oxidative stress [93]. The root extract inhibits SARS-CoV-1 possibly by inhibiting viral replication and blocking viral entry into host [45].

Rheum palmatum *L.*

The *R. palmatum* belongs to family Polygonaceae and is native to mountainous regions of China [94]. It is traditionally used as a laxative and to cure stomachache, liver bile disease, and gastroenteritis. It contains many phytoactive compounds like emodin, chrysophanol, and aloe emodin [23, 62] and possesses antiviral, antipyretic, and antibacterial activity [62]. The root tubers are effective against SARS-CoV-1 S protein in binding ACE-2 receptor, thus blocking entry and replication of CoVs into cells [71, 95]. The bioactive compound emodin extracted from root tubers can be effective against COVID-19. It may provide us new insight for the treatment of SARS-CoV-2.

Cerasus avium *(L.) Moench.*

The medicinal plant extract has potent antiviral and antimicrobial characteristics [4]. Its stem contains many bioactive compounds [96]. The plant extracts block the ACE-2 receptor and prevent viral processing [97]. The plant extract can be used against SARS-CoV-2 infection due to its great potential to act on ACE-2 [67]. Ziai and Heidari [97] studied 20 plant extracts and found that largest inhibitory effect on ACE was due to *Cerasus avium* (L.) Moench. It was suggested that herbal products could be used for the treatment of the COVID-19 infection following the essential in vitro and in vivo evaluation.

Polygonum multiflorum *Thunb.*

The plant belongs to Polygonaceae family and is native of China, Japan, and Korea [98]. Root tuber is mostly used for the treatment of the waist, knee pain, paralysis, malaria and other diseases due to its anti-inflammation, anticancer, and anti-CoV functions [99]. The most effective bioactive compound present in PMT is emodin against SARS-CoV-1 infection by acting on ACE-2 receptor [100]. Thus due to the potent bioactive compound emodin, it is a highly recommended therapeutic agent for the treatment of SARS-CoV-2 infection.

Citrus aurantium *L.*

The bitter orange plant belongs to family Rutaceae [101]. The plant extracts contain essential bioactive components like phenolics, vitamin C, and carotenes [102]. The medicinal plant fruit extract is a potent remedy for anxiety, gastrointestinal problems, and prostate cancer [101, 103] but has strong efficacy [67] to block ACE-2 receptor.

Rubia tinctorum *L.*

The plant belongs to family Rubiaceae, a native to North Africa, Southern Europe, and Western Asia [104]. The extract is effective to treat kidney stones and urinary disorders. The root contains red coloring matter due to anthraquinone and pseudo-purpurin used for dyeing purposes [105]. The studies show the effective use of the plant extract to act on ACE-2 receptor [4, 106].

Onopordum acanthium *L.*

The plant family is Asteraceae and commonly known as Scotch thistle. It is present all over the world. The extract shows biological activities with antiviral and anti-inflammatory effects. The plant extract is used as cardiotonic agent. The plant extract contains many bioactive compounds like flavonoids, triterpenoids, phenylpropanoids, and sterols [4]. The plant extract is potent inhibitor of ACE-2 due to the presence of bioactive compound tannin [107]. This proves to be an active therapeutic agent against SARS-CoV-2 infection.

Quercus infectoria *G. Olivier*

Quercus tinctoria or gall oak belongs to Fagaceae family. The plant extract is used to treat diarrhea, tonsillitis, and internal hemorrhages [4]. The plant extract contains phenolics and flavonoids having antiviral, antifungal, and antidiabetic characteristics. The plant extract has strong potential to block ACE-2 receptor due to the presence of many bioactive compounds [4]. Similarly, the extract can be used as a therapeutic agent for treating COVID-19 infection.

Crataegus microphylla *C. Koch*

The medicinal plant family is Rosaceae. All the parts can be used as therapeutic agents [108]. The plant extract is used as antioxidant and antidiabetic agent and is used to cure coronary dilation and heart muscle activation [4]. The plant extract contains bioactive compounds like phenols, triterpenes, and catecholamines which help to control many diseases [108, 109]. Plant extracts inhibit ACE-2 receptor and inhibit viral entry in the cell [107]. A 330 µG/mL concentrated plant extract inhibits viral binding to ACE-2 receptor; the IC_{50} was examined at 80.9 ± 1.3 [107]. The occurrence of some potent bioactive constituents and their effectiveness against SARS-CoV-2 should be tested as therapeutic agent against COVID-19 infection.

Berberis integerrima Bunge

The plant family is Berberidaceae. It is native to Iran and contains different kinds of alkaloids [110]. The root extract contains bioactive compounds like palmatine, ascorbic acid, ursolic acid, beta-carotene, coumarin, and tannin [110]. The plant extract has antiviral, antioxidant, anti-hyper-glycemic, and anticancer therapeutic characteristics and a liver protection agent. The plant was tested in vitro. It was shown that 330 µg/mL of the extract inhibits ACE-2 receptor 88.2 ± 1.7 IC_{50} [107]. The plant *B. integerrima* needs to be studied in further detail for its therapeutic values and may show considerable results against SARS-CoV-2.

Alnus japonica *(Thunb.) Steud.*

The plant is native to Japan, China, Korea, and Russia [111]. The plant has antiviral, anti-inflammatory, and anticancer effects. The plant extract is used to treat fever, lymphatic cancer, and gastroenteric problems [112]. It has antiviral effects and can be used as therapeutic agent to cure SARS-CoV infection. The bioactive compounds extracted from plant bark are hirsutenone, rubranoside, and rubranol [56]. The bioactive compounds are potent therapeutic agents against SARS-CoV PLpro infection. The efficacy of plant bark against SARS-CoV was established by Park et al. [56]. The medicinal plant is the potent inhibitor of SARS-CoV PLpro and IC_{50} value between 3 and 44.5 µM [56]. The plant bark can inhibit SARS-CoV PLpro activity.

Paulownia tomentosa *(Thunb.) Steud.*

The Chinese plant family is Scrophulariaceae. The plant is native to Korea, Taiwan, and central and western China. The plant has antibacterial, antiviral, and antioxidant characteristics [113]. The plant extract is used to cure respiratory tract infection, tonsillitis, traumatic bleeding, bacteriological diarrhea, conjunctivitis, swelling, and hemorrhoids [114]. The plant extract has potent antiviral effect against SARS-CoV PLpro. The fruit contains bioactive compounds like tomentin, geranylated flavonoids, etc. [55]. The bioactive compounds from fruit extract were found effective against SARS-CoV. The plant extract inhibits SARS-CoV PLpro activity with IC50 value between 5.0 and 14.4 µM [55]. The bioactive compound tomentin E exhibited highest inhibition to SARS-CoV infection with lowest value of IC_{50} 5.0 ± 0.06 µM [55].

Psoralea corylifolia *L.*

The plant is native to Asia and belongs to family Leguminosae. The plant has antibacterial, antiviral, antidepressant, and antioxidant activities [115]. It contains various bioactive constituents like neobavaisoflavone, corylifol A, bavachinin, and psoralidin [90]. The seed extract shows inhibition of SARS-CoV PLpro, and the IC50 was 15 µg/mL [116]. The IC50 of the tested bioactive compounds against SARS-CoV PLpro infection were in range of 4.2–38.4 µM.

Tribulus terrestris *L.*

The medicinal plant belongs to family Zygophyllaceae is native to Pakistan, South America, Spain, Mexico, Bulgaria, China, and India. The plant has antiviral, antidiabetic, anti-inflammatory, antioxidant, and anti-urolithic characteristics [117]. The plant contains alkaloids and flavonoids as main bioactive constituents [118, 119]. The fruit extract showed inhibition against SARS-CoV PLpro [118]. The IC50 of all the bioactive compounds studied were in range of 15.8 and 70.1 μM [118].

Medicinal Plants as Potential Source of Natural Antiviral Agents Against COVID -19

Medicinal plants show their antiviral potential by inhibiting viral entry into the host cell as a part of their mechanism of action [72]. Flavonoids, like apigenin and quercetin, exhibited antiviral activity against SARS-CoV through the inhibition of the activity of Mpro enzymes having IC50 of 38.4 ± 2.4 μM and 23.8 μM, respectively [120] (Table 2.1).

Table 2.1 Antiviral potential of medicinal plants against coronavirus (CoV) strains

Medicinal plants	Family	Mode of action	References
Agrimonia pilosa	Rosaceae	Terpenoids having antiviral activity against influenza viruses (H1N1 and H3N2)	[121]
Allium cepa L.	Amaryllidaceae	Crushed and mixed with honey against influenza virus	[86, 122]
Allium sativum L.	Amaryllidaceae	Essential oil effective against SARS-CoV2	[50]
Alnus japonica	Betulaceae	Polyphenols having antiviral activity against SARS-CoV	[56]
Andrographis paniculata	Acanthaceae	Antiviral activity against SARS-CoV	[123]
Artemisia annua L.	Asteraceae	Antiviral activity against SARS-CoV	[45]
Astragalus mongholicus	Fabaceae	Inhibited avian infectious bronchitis virus (IBV) replication	[124]
Azadirachta indica	Meliaceae	Antiviral activity against poliovirus, bovine herpes virus type 1, duck plague virus, and herpes simplex virus type I	[125]
Acacia nilotica	Fabaceae	Antiviral activity against HIV-1 protease viruses	[126]
Andrographis paniculata	Acanthaceae	Used in control of SARS-CoV and likely SARS-CoV-2	[127]
Angelica keiskei (Miq.) Koidz	Apiaceae	Alkylated chalcones from ethanolic extract inhibited SARS-CoV PLpro	[128]

(continued)

Table 2.1 (continued)

Medicinal plants	Family	Mode of action	References
Berberis vulgaris L.	Berberidaceae	Boiled extract and poultice Antiviral activity and likely against SARS-CoV-2	[86, 129]
Betula papyrifera	Betulaceae	Methanolic plant extract exhibited antiviral activity against coronavirus	[130]
Camellia sinensis (L.) Kuntze	Theaceae	Boiled and drunk exhibited antiviral activity	[86]
Carica papaya L.	Caricaceae	Leaf juice; fruit can be eaten Antiviral activity against dengue virus	[131]
Cassia tora	Fabaceae	Inhibition activity of SARS-CoV-3CL	[85]
Cedrela sinensis	Meliaceae	Inhibits viral replication SARS-CoV	[63]
Chrysanthemi flos	Compositae	SARS-CoV-2	[132]
Citrus sinensis	Rutaceae	Polyphenols having antiviral activity against SARS-CoV	[37]
Coriandrum sativum	Apiaceae	Antiviral activity against coronavirus	[123]
Cullen corylifolium (L.) Medik.	Leguminosae	Ethanolic extract of seeds inhibits SARS-CoV PLpro (papain-like protease nonstructural protein encoded by SARS-CoV-2 genome)	[116]
Dioscorea batatas	Dioscoreaceae	Inhibition activity of SARS-CoV-3CL	[85]
Epimrdii herba	Berberidaceae	SARS-CoV-2	[132]
Eucalyptus polybractea	Myrtaceae	Aerosol and vapor of eucalyptus oil inhibit avian influenza virus	[133]
Eugenia jambolana	Myrtaceae	Inhibition of protease activity avian influenza	[134]
Euphorbia helioscopia herb	Euphorbiaceae	SARS-CoV-2	[132]
Forsythiae fructus	Oleaceae	SARS-CoV-2	[132]
Fritillaria thunbergii	Liliaceae	Inhibits H1N1 replication	[98]
Gentiana scabra	Gentianaceae	Terpenoids having antiviral activity against SARS-CoV	[135]
Glycyrrhiza uralensis Fisch.	Fabaceae	Antiviral activities against SARS coronavirus	[136]
Gentiana scabra	Gentianaceae	Inhibition activity of SARS-CoV-3CL	[85]
Helianthus annuus L.	Asteraceae	Flower and seed extract effective against human infectious diseases	[137]
Houttuynia cordata aq.	Saururaceae	Antiviral activity against SARS-CoV	[45, 138]
Hyoscyamus niger	Solanaceae	Inhibition of Ca^{2+} channels and bronchodilator Antiviral activity against SARS-CoV-2	[65]
Isatis indigotica	Brassicaceae	Extract effective against SARS-CoV	[139, 140]
Juniperus oxycedrus L. subsp. *oxycedrus* L.	Cupressaceae	Antiviral activity against SARS-CoV	[141]

(continued)

2 Medicinal Plants as COVID-19 Remedy

Table 2.1 (continued)

Medicinal plants	Family	Mode of action	References
Laurus nobilis L.	Lauraceae	Antiviral activity against SARS-CoV	[142]
Lepidium meyenii	Brassicaceae	Antiviral action against flu A and flu B infection	[143]
Lindera aggregata	Lauraceae	Antiviral activity against SARS-CoV	[45]
Lonicerae japonicae Flos	Caprifoliaceae	Antiviral activity against SARS-CoV-2	[132]
Lycoris radiata	Asparagales	Antiviral activity against SARS-CoV	[45]
Mori cortex	Moraceae	Antiviral activity against SARS-CoV-2	[132]
Mentha piperita	Lamiaceae	Essential oils exhibit antiviral activity against IBV	[123]
Moringa oleifera	Moringaceae	Blocks initial stages of viral replication FMDV	[90]
Nigella sativa	Ranunculaceae	Potential inhibitor of SARS-CoV-2	[144]
Panax ginseng	Araliaceae	Carbohydrates having antiviral activity against human rotavirus	[145]
Paulownia tomentosa (Thunb.) Steud.	Paulowniaceae	Ethanolic extract of fruits inhibits SARS-CoV PLpro (papain-like proteinase)	[55]
Peucedani radix	Apiaceae	Obtained from dried roots of *Peucedanum praeruptorum* Dunn inhibits SARS-CoV-2	[132]
Punica granatum	Lythraceae	Antiviral activity against SARS-CoV	[123]
Portulaca oleracea L	Portulacaceae	Water extract exhibits good virucidal activity	[146]
Polygonum multiflorum	Polygonaceae	Polyphenols having antiviral activity against SARS-CoV	[100]
Psoralea corylifolia	Fabaceae	Polyphenols having antiviral activity against SARS-CoV	[116]
Pyrrosia lingua	Polypodiaceae	Antiviral activity against SARS-CoV	[45]
Rheum officinale	Polygonaceae	Polyphenols having antiviral activity against SARS-CoV	[100]
Rheum palmatum	Polygonaceae	Anti-SARS-CoV 3CLpro activity	[147]
Rhizoma fagopyri cymosi	Polygonaceae	Rhizome of *Fagopyrum dibotrys* Antiviral activity against SARS-CoV-2	[132]
Rhizoma cibotii	Dicksoniaceae	Root of *Cibotium barometz* (L.) J. Sm. inhibits SARS-CoV-1	[135]
Rosmarinus officinalis	Lamiaceae	Blocks viral replication Human respiratory syncytial virus	[76]
Salvia miltiorrhiza Bunge	Lamiaceae	Ethanolic extract caused inhibition of SARS-CoV PLpro	[56]
Sambucus nigra	Adoxaceae	Inhibited avian infectious bronchitis (IBV) virus replication	[148]
Senna angustifolia	Fabaceae	Polyphenols having antiviral activity against dengue virus serotype −2(DENV-2)	[149]
Taxillus chinensis	Loranthaceae	Inhibition activity of SARS-CoV-3CL	[85]

(continued)

Table 2.1 (continued)

Medicinal plants	Family	Mode of action	References
Thuja orientalis L.	Cupressaceae	Inhibits SARS-CoV	[141]
Toona sinensis Roem (Chinese mahogany)	Meliaceae	Inhibits SARS-CoV-1 replication	[140]
Torreya nucifera	Taxaceae	Inhibits SARS-CoV-3CLpro activity	[58]
Tridex procumbens	Asteraceae	Polyphenols having antiviral activity against dengue virus serotype −2(DENV-2)	[149]
Vitex trifolia	Lamiaceae	Antiviral activity against SARS-CoV-2	[66]
Zanthoxylum piperitum	Rutaceae	Leaf extract showed antiviral activity against influenza virus	[150]
Zingiber officinalis	Zingiberaceae	Antiviral activity against avian influenza virus	[38]

Allium Sativum (Garlic)

Garlic is alleged to have antimicrobial, antifungal, and antiviral properties. Garlic is emphasized as a remedy for viral infection since centuries [151]. Garlic extracts boost the immune system. Its antiviral effects are due to sulfur-containing compounds such as allicin, diallyl disulfide, and diallyl trisulfide reacting with thiol groups of enzymes: alcohol dehydrogenase, thioredoxin reductase, and disulfide bonds critical for microorganism survival [64]. It is supposed to lower cholesterol, reduce blood pressure, prevent atherosclerosis, and act as an anticoagulant [152]. It is also reported to have anticarcinogenic and immunomodulatory effects [153]. Allicin is produced on crushing raw garlic; it possesses antibacterial properties, but it is an unstable compound and cannot be detected after intake [154]. There are insufficient clinical trials regarding the effects of garlic in preventing or treating the common cold.

The safety of consuming small quantities of raw garlic is evident in its worldwide use as a culinary spice. It is the most popular herbal remedy. Fresh garlic crushed and mixed with honey possess antiviral and other biological properties and strengthen the immune system due to the presence of several bioactive sulfur-containing compounds, i.e., sulfoxide, proteins, and polyphenols [155, 156]. Its adverse effects include bad breath and body odor and allergic reactions, manifesting in minor respiratory or skin symptoms [157]. Allicin and its derivatives, ajoene, allitridin, and garlicin, are the most effective organic sulfur compounds (OSCs) used to prevent viral infections. *Allium sativum* may be an acceptable preventive measure against COVID-19 infection to boost immune system cells and to repress the production and secretion of proinflammatory cytokines as well as an adipose tissue-derived hormone leptin having the proinflammatory nature [151].

Azadirachta indica *(Neem)*

Every part of the tree such as root, leaves, bark, seeds, and oil has antibacterial and antiviral properties. The main bioactive compounds are limonoids and terpene. Neem tree purifies air and possesses antimalarial, antifungal, wound healing, antidiabetic, and anticancerous properties. Phytochemicals obtained from neem seed oil possess antimalarial, antifungal, and antipyretic activity. However, azadirachtin in seed oil possess insecticidal, antifeedant, and antimalarial activity and prevent development of parasites. The extract of leaf is effective against eczema. Leaf aqueous extract shows antiviral properties against chikungunya, measles, and vaccinia virus. Leaves are taken orally to kill harmful worms and are used in bathing to prevent chicken pox infection. However, excess use may have harmful effects and cause sterility [158].

Curcuma domestica *(Turmeric, Haldi)*

A perennial herb, the rhizome portion, is used for medicinal purpose and possesses anti-inflammatory properties. It is used to cure flatulence, jaundice, and hemorrhage and also used as ointment for skin problems. It has antiviral properties and is used as a cleansing agent of the respiratory tract and helps to fight against cold and flu. Bioactive compound curcumin boosts immunity and reduces inflammation, pain, and sinusitis. It has antiseptic, antiviral, antifungal, and antibacterial properties. Curcumin reduces RNA replication and enhances the immune system [158].

Echinacea purpurea L. *(Echinacea)*

Echinacea extract is used as herbal medicine in the treatment of cold and influenza and possesses potent antiviral activities against human and avian influenza viruses, as well as herpes simplex virus, respiratory syncytial virus, and rhinoviruses [183].

Foeniculum vulgare *Mill. (Fennel)*

Fennel herb is a therapeutic plant (*Foeniculum vulgare* Mill) belongs to the family Apiaceae. Fennel seeds effectively control numerous infectious disorders of bacterial, fungal, viral, and protozoal origin. It is used in a wide range against gastrointestinal and endocrinal reproduction and respiratory problems and as a lactation improving agent as well as used as diuretic, anti-inflammatory, and antioxidant remedy. Fennel seed essential oil is used as flavoring agent in food products and an ingredient of cosmetic and pharmaceutical products [159]. Fennel herbal tea containing fennel seed extract is a common household remedy traditionally used for the treatment of a variety of symptoms of the gastrointestinal and respiratory tract in some areas of Europe and Asia [160]. Fennel has many biological activities due to its volatile and nonvolatile compounds, and it has been used in traditional Chinese medicine for treating various conditions, particularly rheumatism, cold pain, and stomach disorder [160]. The current study has proved that the nutritional syndromes can be corrected by healthful intakes. The functional and nutraceutical foods are effective in the management of human health, on the basis of their compositional profiling. It is recommended that fennel is an extraordinary source of major and minor nutrients. It acts as an antioxidant, preventing the deteriorating health conditions such as hyperglycemia [161]. Its bioactive composition makes it a potential candidate for inclusion in nutraceutical industry. *Foeniculum vulgare* essential oil exhibit antiviral effects against herpes viruses boost immune system and decrease inflammation [162].

Glycyrrhiza glabra *(Licorice)*

Licorice has been used in traditional Chinese medicine and other natural practices for centuries. Glycyrrhizin, liquiritigenin, and glabridin are just some of the active substances in licorice that have powerful antiviral properties [170]. The dried roots of licorice are crushed and boiled to prepare an extract and dried to a dark paste or powder to be taken orally for the treatment of different types of chronic infections [171]. The active compounds triterpene, saponins, and particularly glycyrrhizin are responsible for immunomodulatory, antiviral, and other biological activities [172]. Also animal studies have revealed its efficacy against the influenza virus by stopping the virus replication [173]. Glycyrrhizic acid present in the plant inhibits virus growth and inactivates virus particles [174]. Licorice root extract is effective against HIV, RSV, herpes viruses, and severe acute respiratory syndrome-related coronavirus (SARS-CoV), which causes a serious type of pneumonia [34]. Licorice is used in a variety of ways for medicinal purposes; mixing the herb with *Aloe vera* gel helps to treat eczema. The herb is extracted in hot water to make herbal tea for a sore throat. The licorice extract is used as a treatment for ulcers or stomach problems [175].

Melissa officinalis *L. (Lemon Balm)*

Lemon balm is used as herbal medicine in herbal teas, often in combination with other herbs such as spearmint [176]. Lemon balm has been widely used as a topical antiviral treatment for genital and oral herpes. In vitro studies also exhibited antiviral effects against avian influenza (bird flu) and HIV-1 [177, 178].

Mentha piperita *L. (Peppermint)*

Menthol and rosmarinic acid in the essential oil in peppermint contain antiviral and anti-inflammatory activity. Peppermint tea is used to treat morning sickness. The extracts possess antiviral activity against respiratory syncytial virus (RSV) and reduce inflammation [179]. Herbal extracts must be used to treat the viral infection and spread, but the recommended use and side effects must also be taken into account.

Nigella sativa *L. (Black Seeds)*

Nigella sativa, a folk medicine, referred to by prophet Mohammed (PBUH), contains all kinds of cure except death [163]. It is also identified as the curative black cumin in the Holy Bible [164]. The seeds of *Nigella sativa* and its main active constituent, thymoquinone, are reported to be medicinally very effective against various illnesses including different chronic illness, neurological and mental illness, cardiovascular disorders, cancer, diabetes, inflammatory conditions, and infertility, as well as various infectious diseases due to bacterial, fungal, parasitic, and viral infections. The medicinal use of black cumin seeds in various traditional herbal systems is known for a wide range of ailments which include neurologic disorders, hypertension, diabetes, inflammatory disorders, cancer, paralysis, and digestive tract-related problems administered in different kinds of preparations, a weapon to the drug scientists [165].

Origanum vulgare *L. (Oregano)*

Oregano has therapeutic properties; it has slightly bitter taste. Oregano tea as herbal extract is used to soothe sore throat, cough, and digestive disorders. It shows antiviral characteristics against equine arteritis virus and canine distemper virus [189].

Ocimum bacilicum *L. (Basil)*

The extracts of *Ocimum bacilicum*, sweet basil, exhibited antiviral activities against herpes viruses (HSV), adenoviruses (ADV), hepatitis B virus, coxsackievirus B1 (CVB1), and enterovirus 71 (EV71) [184].

Rosmarinus officinalis *L. (Rosemary)*

Rosemary extract has great therapeutic potential due to bioactive molecule oleanolic acid. Rosemary extract exhibited antiviral activity against herpes viruses and hepatitis A [180, 181]. Carnosic acid is a bioactive compound in rosemary and exhibits strong antiviral activities against human respiratory syncytial virus (hRSV) [182].

Salvia officinalis *L. (Sage)*

Sage tea is used traditionally for the cure of digestive disturbances, bronchitis, cough, asthma, inflammations, depression, excessive sweating, and skin disease. The essential oil has been used in the treatment of respiratory system and metabolic disturbances [185]. *Salvia officinalis* extract inhibits HIV activity by preventing the virus from entering the target cells [186]. It also shows antiviral activity against HSV-1 and Indiana *Vesiculovirus* that show disease symptoms in cows, pigs, and horses. The crude extracts of S. officinalis contain diterpenes safficinolid and sageon. Safficinolid showed antiviral properties against vesicular stomatitis virus (VSV) infection, while sageon exhibited virus inactivation activity against VSV and HSV-1 (herpes simplex virus type 1) [187, 188].

Senna alexandrina *Mill. (Senna Makki)*

Also known as *Senna* leaves – an Asian herb used in medicines to boost immunity – it is usually prescribed for short-term use to people suffering from constipation; "cleansing" herb flushes out toxins from your body if used properly [190]. *Senna alexandrina* is traditionally used for its antioxidant and anti-inflammatory properties [191]. The leaves were used for treating anemia, bronchosis, burns, cancer, constipation, herpes, leukemia, mycosis, nausea, neural disorders, viral diseases, antihelmenthic, and wound. Pharmacological investigations show that sennosides A and B account for the entire activity of the *Senna* leaves and pods. In folk medicine, *Senna* was applied for curing different ailments as it shows anthelmintic, antiherpetic, antiviral, and antibacterial (*Staphylococci* and *Bacillus coli*) properties. The methanolic and ethyl acetate extract of *Senna alexandrina* leaves predominantly contain Alkaloids and Flavonoids. These phytochemicals could contribute to the various medicinal applications of the plant [192].

Other plants including pomegranate (*P. granatum*), long pepper (*Piper longum* L.), black pepper (*P. nigrum*), barberry (*Berberis vulgaris* L.), black caraway (*Nigella sativa* L.), papaya (*Carica papaya* L.), white mulberry (*Morus alba* L.), Chinese boxthorn (*Lycium barbarum* L.), orange (*Citrus aurantium* L.), European plum (*Prunus domestica* L.), and soybean (*Glycine max* [L.] Merr.) possess immunomodulatory activity, and these food plants boost the immune response of the host and could be used as a preventive measure against COVID-19 [193]. In SARS cases, a number of coinfections are reported. Moreover, most of the SARS-CoV-2 patients also suffered from other health-related problems. The food plants discussed above are easily accessible and may serve as a cheap way of maintaining healthy immune system and general good health. It is recommended to include these medicinal food plants in the daily diet as a preventive measure for effective health management and to cure respiratory tract infections.

Zingiber officinale *Rosc. (Ginger)*

Ginger products, such as teas and lozenges, are popular natural cure. Ginger has been used as a spice as well as medicine in India and China since ancient times. Ginger is thought to act directly on the gastrointestinal system used to treat morning sickness, bloating, flatulence, and dyspepsia [166]. The plant extract of *Z. officinale* showed excellent antiviral properties against avian influenza, RSV, and *Feline calicivirus* (FCV), which is comparable to human norovirus [167] as well as antichikungunya activity [168]. Additionally, specific compounds in ginger, such as gingerols and zingerone, have been found to inhibit viral replication and prevent viruses from entering host cells [169].

The chief active consitutents of ginger are volatile oil (zingiberol, zingiberene, farnesene, curcumene and D-camphor, Shogaols, Gingerols, Paradol), Terpenoids and flavonoids [158]. The volatile essential oils and non-volatile compound oleoresin give ginger pungency and a hot sensation.

Limitations

Some food plants may have toxic effects if taken in an excessive amount. The prolonged and high consumption level of licorice and glycyrrhizin may lead to certain harmful impacts on human health like hypertension- and hypokalemic-induced secondary disorders. However, licorice taken orally produces less side effects; however, longer oral administration can cause toxic effects [194]. Similarly, garlic ingested in large amounts may cause diarrhea, dizziness, nausea, vomiting, headache, and flatulence especially on an empty stomach.

The recommended dose of the garlic is one tablet of dried garlic powder twice or thrice per day for adults [195]. Curcumin dose of 0.9–3.6 g per day for 1–4 months may cause nausea and diarrhea [196]. Higher amount of green tea consumption may cause gastrointestinal problems taken on an empty stomach [197]. Herbal extracts provide a rich resource for novel antiviral drug development. Their antiviral therapeutic mechanisms of interaction with the viral life cycle, viral entry, replication, assembly, and release, as well as on the targeting of virus-host-specific interactions, must be identified. Natural products serve as an excellent source of novel antivirals, to develop effective protective/therapeutic strategies against viral infections. The viral outbreaks, which remain without vaccines for long, can be controlled by herbal treatment as they possess robust antiviral activity. The most efficient way to strengthen our immune system is to use nutrient-rich natural food [72].

Hippocrates, the Greek Father of Medicine, always said health starts from the gut, so it is important to get our gastrointestinal health in check. Phytochemicals have been used throughout history to fight against ailments, since the inception of pharmaceutical drugs. Research has shown that many natural compounds suggest to have pharmacological effects and have antiviral activity and inhibitory effect on viral attachment and penetration: herbal tea consists of compounds proven to be effective as natural remedies throughout use over the ages [198].

It is important that before using herbs, healthcare professional should be consulted to avoid allergic reaction. Most importantly, herbal dosages will depend on people's health and age, so be sure to follow the instructions on the labels and ask your healthcare professional [199].

The only real protection against a future COVID-19 pandemic or any other viral pandemic is the one that was demonstrated to work in the SARS, MERS, COVID-19, and annual influenza pandemics [200]. Senna-induced dermatitis is rare but may occur when patients need a higher dose. Currently, the major reported side effects of Senna include abdominal cramping, electrolyte and fluid deficiencies, and malabsorption. The pathogenesis of the skin reaction to Senna overdose is unclear. It may be that the high concentration of digestive enzymes in the diarrheal stool due to the increased gut transit time may cause a chemical contact burn to the skin [201].

The plants presented in the table are currently used to treat symptoms of SARS-CoV-2 such as fever, inflammation, or cardiovascular disorders. The efficiency of medicinal plants and herbal extracts should be further studied and validated in COVID-19 patients. It will clarify the mechanism and pathways targeted by them to improve their clinical usefulness.

Conclusions

Recent studies revealed that combined therapy of herbal medicine with Western medicine helps to improve the immunity and minimize the coronavirus threats. This study also emphasizes the potential role of herbal medicine for the treatment of COVID-19 in a better way. Herbal medicines could improve the functioning of lungs. However, it is important to have an updated knowledge on the risk benefits of food plants.

References

1. Wang M et al (2020) Remdesivir and chloroquine effectively inhibit the recently emerged novel coronavirus (2019-nCoV) in vitro. Cell Res 30(3):269–271
2. W. H. Organization (2020) Laboratory testing for coronavirus disease 2019 (COVID-19) in suspected human cases: interim guidance, 2 Mar 2020. World Health Organization
3. Vellingiri B et al (2020) COVID-19: a promising cure for the global panic. Sci Total Environ 725:138277
4. Siddiqui AJ et al (2020) Plants-derived biomolecules as potent antiviral phytomedicines: new insights on ethnobotanical evidences against coronaviruses. Plan Theory 9(9):1244
5. Ye Z-W, Yuan S, Yuen K-S, Fung S-Y, Chan C-P, Jin D-Y (2020) Zoonotic origins of human coronaviruses. Int J Biol Sci 16(10):1686
6. Kannan S, Ali PSS, Sheeza A, Hemalatha K (2020) COVID-19 (Novel Coronavirus 2019)-recent trends. Eur Rev Med Pharmacol Sci 24(4):2006–2011
7. Ramshaw RE et al (2019) A database of geopositioned Middle East Respiratory Syndrome Coronavirus occurrences. Scientific Data 6(1):1–13
8. Shen L et al (2019) High-throughput screening and identification of potent broad-spectrum inhibitors of coronaviruses. J Virol 12:93
9. Dudani T, Saraogi A (2020) Use of herbal medicines on coronavirus. Acta Scientific Pharmaceut Sci 4:61–63
10. Razonable RR (2011) Antiviral drugs for viruses other than human immunodeficiency virus. Mayo Clinic Proc 86(10):1009–1026: Elsevier
11. Edwin GT, Korsik M, Todd MH (2019) The past, present and future of anti-malarial medicines. Malar J 18(1):1–21
12. De Clercq E (2009) Anti-HIV drugs: 25 compounds approved within 25 years after the discovery of HIV. Int J Antimicrob Agents 33(4):307–320
13. Stebbing J et al (2020) COVID-19: combining antiviral and anti-inflammatory treatments. Lancet Infect Dis 20(4):400–402
14. Elfiky AA (2020) Anti-HCV, nucleotide inhibitors, repurposing against COVID-19. Life Sci 248:117477
15. Cheema SUR, Rehman MS, Hussain G, Cheema SS, Gilani N (2019) Efficacy and tolerability of sofosbuvir and daclatasvir for treatment of hepatitis C genotype 1 & 3 in patients undergoing hemodialysis-a prospective interventional clinical trial. BMC Nephrol 20(1):438
16. Wu Y et al (2020) Nervous system involvement after infection with COVID-19 and other coronaviruses. Brain Behav Immun 87:18
17. Zheng M et al (2020) Functional exhaustion of antiviral lymphocytes in COVID-19 patients. Cell Mol Immunol 17(5):533–535
18. Naithani R, Mehta RG, Shukla D, Chandersekera SN, Moriarty RM (2010) Antiviral activity of phytochemicals: a current perspective. In: Dietary components and immune function. Springer, Totowa, pp 421–468
19. Thabti I et al (2020) Advances on antiviral activity of Morus spp. Plant extracts: human coronavirus and virus-related respiratory tract infections in the spotlight. Molecules 25(8):1876
20. Patharakorn T, Talawat S, Promboon A, Wetprasit N, Ratanapo S (2010) Antimutagenicity and anti-HSV-2 activity of mulberry tea (Morus rotunbiloba Koidz). Kasetsart J.(Nat. Sci.) 5:44
21. Dhiman S, Kumar V, Mehta C, Gat Y, Kaur S (2020) Bioactive compounds, health benefits and utilisation of Morus spp. – a comprehensive review. J Hortic Sci Biotechnol 95(1):8–18
22. Wei H, Zhu J-J, Liu X-Q, Feng W-H, Wang Z-M, Yan L-H (2016) Review of bioactive compounds from root barks of Morus plants (Sang-Bai-Pi) and their pharmacological effects. Cogent Chem 2(1):1212320
23. Yang X et al (2020) Clinical course and outcomes of critically ill patients with SARS-CoV-2 pneumonia in Wuhan, China: a single-centered, retrospective, observational study. Lancet Respir Med 8:475

24. Vimalanathan S, Ignacimuthu S, Hudson J (2009) Medicinal plants of Tamil Nadu (Southern India) are a rich source of antiviral activities. Pharm Biol 47(5):422–429
25. Alam G, Wahyuono S, Ganjar IG, Hakim L, Timmerman H, Verpoorte R (2002) Tracheospasmolytic activity of viteosin-A and vitexicarpin isolated from Vitex trifolia. Planta Med 68(11):1047–1049
26. Srivastava RAK, Mistry S, Sharma S (2015) A novel anti-inflammatory natural product from Sphaeranthus indicus inhibits expression of VCAM1 and ICAM1, and slows atherosclerosis progression independent of lipid changes. Nutrition Metabol 12(1):20
27. Nourazarian SM, Nourazarian A, Majidinia M, Roshaniasl E (2016) Effect of root extracts of medicinal herb Glycyrrhiza glabra on HSP90 gene expression and apoptosis in the HT-29 colon cancer cell line. Asian Pac J Cancer Prev 16(18):8563–8566
28. Tsai J, Wilson M (2020) COVID-19: a potential public health problem for homeless populations. Lancet Public Health 5(4):e186–e187
29. Amber R, Adnan M, Tariq A, Mussarat S (2017) A review on antiviral activity of the Himalayan medicinal plants traditionally used to treat bronchitis and related symptoms. J Pharm Pharmacol 69(2):109–122
30. Gilani AH et al (2008) Gastrointestinal, selective airways and urinary bladder relaxant effects of Hyoscyamus niger are mediated through dual blockade of muscarinic receptors and Ca2+ channels. Fundam Clin Pharmacol 22(1):87–99
31. Nair R (2012) HIV-1 reverse transcriptase inhibition by Vitex negundo L. leaf extract and quantification of flavonoids in relation to anti-HIV activity. J Cell Mol Biol 10(2):53–59
32. Zhang Y, Buckles E, Whittaker GR (2012) Expression of the C-type lectins DC-SIGN or L-SIGN alters host cell susceptibility for the avian coronavirus, infectious bronchitis virus. Vet Microbiol 157(3–4):285–293
33. Yi L et al (2004) Small molecules blocking the entry of severe acute respiratory syndrome coronavirus into host cells. J Virol 78(20):11334–11339
34. Cinatl J, Morgenstern B, Bauer G, Chandra P, Rabenau H, Doerr H (2003) Glycyrrhizin, an active component of liquorice roots, and replication of SARS-associated coronavirus. Lancet 361(9374):2045–2046
35. Moradi M-T, Karimi A, Fotouhi F, Kheiri S, Torabi A (2017) In vitro and in vivo effects of Peganum harmala L. seeds extract against influenza A virus. Avicenna J Phytomed 7(6):519
36. Lelešius R et al (2019) In vitro antiviral activity of fifteen plant extracts against avian infectious bronchitis virus. BMC Vet Res 15(1):178
37. Ulasli M et al (2014) The effects of Nigella sativa (Ns), Anthemis hyalina (Ah) and Citrus sinensis (Cs) extracts on the replication of coronavirus and the expression of TRP genes family. Mol Biol Rep 41(3):1703–1711
38. Yang J-L, Ha TKQ, Oh WK (2016) Discovery of inhibitory materials against PEDV corona virus from medicinal plants. Japanese J Veterinary Res 64(1 Suppl):S53–S63
39. Sanders JM, Monogue ML, Jodlowski TZ, Cutrell JB (2020) Pharmacologic treatments for coronavirus disease 2019 (COVID-19): a review. JAMA 323(18):1824–1836
40. Mohammadi N, Shaghaghi N (2020) Inhibitory effect of eight secondary metabolites from conventional medicinal plants on COVID_19 virus protease by molecular docking analysis. Preprint 11987475:1. https://doi.org/10.26434/chemrxiv
41. Van der Meer F et al (2007) The carbohydrate-binding plant lectins and the non-peptidic antibiotic pradimicin A target the glycans of the coronavirus envelope glycoproteins. J Antimicrob Chemother 60(4):741–749
42. Ramos-Tovar E, Muriel P (2019) Phytotherapy for the liver. In: Dietary interventions in liver disease. Elsevier, Boston, pp 101–121
43. Massi A et al (2017) Research progress in the modification of quercetin leading to anticancer agents. Molecules 22(8):1270
44. Russo M, Moccia S, Spagnuolo C, Tedesco I, Russo GL (2020) Roles of flavonoids against coronavirus infection. Chem Biol Interact 328:109211

45. Li S-y et al (2005) Identification of natural compounds with antiviral activities against SARS-associated coronavirus. Antivir Res 67(1):18–23
46. Yuen KY et al (2009) Baicalin as a treatment for SARS infection, ed: Google Patents
47. Du J, He Z-D, Jiang R-W, Ye W-C, Xu H-X, But PP-H (2003) Antiviral flavonoids from the root bark of Morus alba L. Phytochemistry 62(8):1235–1238
48. Lee HY et al (2007) Inhibition of HCV replicon cell growth by 2-arylbenzofuran derivatives isolated from Mori Cortex Radicis. Planta Med 73(14):1481
49. Keyaerts E et al (2007) Plant lectins are potent inhibitors of coronaviruses by interfering with two targets in the viral replication cycle. Antivir Res 75(3):179–187
50. Thuy BTP et al (2020) Investigation into SARS-CoV-2 resistance of compounds in garlic essential oil. ACS Omega 5(14):8312–8320
51. Nguyen TTH et al (2012) Flavonoid-mediated inhibition of SARS coronavirus 3C-like protease expressed in Pichia pastoris. Biotechnol Lett 34(5):831–838
52. Aanouz I, Belhassan A, El-Khatabi K, Lakhlifi T, El-Ldrissi M, Bouachrine M (2020) Moroccan medicinal plants as inhibitors against SARS-CoV-2 main protease: computational investigations. J Biomol Struct Dyn:1–9. https://doi.org/10.1080/07391102.2020.1758790
53. Laksmiani NPL, Larasanty LPF, Santika AAGJ, Prayoga PAA, Dewi AAIK, Dewi NPAK (2020) Active compounds activity from the medicinal plants against SARS-CoV-2 using in silico assay. Biomed Pharmacol J 13(2):873–881
54. Jo S, Kim S, Shin DH, Kim M-S (2020) Inhibition of SARS-CoV 3CL protease by flavonoids. J Enzyme Inhib Med Chem 35(1):145–151
55. Cho JK et al (2013) Geranylated flavonoids displaying SARS-CoV papain-like protease inhibition from the fruits of Paulownia tomentosa. Bioorg Med Chem 21(11):3051–3057
56. Park J-Y et al (2012) Diarylheptanoids from Alnus japonica inhibit papain-like protease of severe acute respiratory syndrome coronavirus. Biol Pharm Bull 35:2036
57. Park J-Y et al (2013) Dieckol, a SARS-CoV 3CLpro inhibitor, isolated from the edible brown algae Ecklonia cava. Bioorg Med Chem 21(13):3730
58. Ryu YB et al (2010) Biflavonoids from Torreya nucifera displaying SARS-CoV 3CLpro inhibition. Bioorg Med Chem 18(22):7940–7947
59. Chiang L, Chiang W, Liu M, Lin C (2003) In vitro antiviral activities of Caesalpinia pulcherrima and its related flavonoids. J Antimicrob Chemother 52(2):194–198
60. Lau K-M et al (2008) Immunomodulatory and anti-SARS activities of Houttuynia cordata. J Ethnopharmacol 118(1):79–85
61. Prathapan A, Vineetha V, Abhilash P, Raghu K (2013) Boerhaaviadiffusa L. attenuates angiotensin II-induced hypertrophy in H9c2 cardiac myoblast cells via modulating oxidative stress and down-regulating NF-κβ and transforming growth factor β1. Br J Nutr 110(7):1201–1210
62. Zheng L et al (2018) Combination of comprehensive two-dimensional prostate cancer cell membrane chromatographic system and network pharmacology for characterizing membrane binding active components from radix et rhizoma rhei and their targets. J Chromatogr A 1564:145–154
63. Younus I, Siddiq A, Ishaq H, Anwer L, Badar S, Ashraf M (2016) Evaluation of antiviral activity of plant extracts against foot and mouth disease virus in vitro. Pak J Pharm Sci 29(4):1263–1268
64. Mehrbod P, Amini E, Tavassoti-Kheiri M (2009) Antiviral activity of garlic extract on influenza virus. Iranian J Virol 3(1):19–23
65. Sood R et al (2012) Antiviral activity of crude extracts of Eugenia jambolana Lam. against highly pathogenic avian influenza (H5N1) virus. Indian J Exp Biol 50:179
66. Liou C-J, Cheng C-Y, Yeh K-W, Wu Y-H, Huang W-C (2018) Protective effects of casticin from Vitex trifolia alleviate eosinophilic airway inflammation and oxidative stress in a murine asthma model. Front Pharmacol 9:635
67. Heidary F, Varnaseri M, Gharebaghi R (2020) The potential use of persian herbal medicines against COVID-19 through angiotensin-converting enzyme 2. Arch Clin Infect Dis 15(COVID-19)

68. Yao R-y, Zou Y-F, Chen X-F (2013) Traditional use, pharmacology, toxicology, and quality control of species in Genus Bupleurum L. Chinese Herbal Med 5(4):245–255
69. Yang F, Dong X, Yin X, Wang W, You L, Ni J (2017) Radix bupleuri: a review of traditional uses, botany, phytochemistry, pharmacology, and toxicology. BioMed Res Int 2017: 7597596. https://doi.org/10.1155/2017/7597596
70. Efferth T, Romero MR, Wolf DG, Stamminger T, Marin JJ, Marschall M (2008) The antiviral activities of artemisinin and artesunate. Clin Infect Dis 47(6):804–811
71. Ho WE, Peh HY, Chan TK, Wong WF (2014) Artemisinins: pharmacological actions beyond anti-malarial. Pharmacol Ther 142(1):126–139
72. Lin L-T, Hsu W-C, Lin C-C (2014) Antiviral natural products and herbal medicines. J Tradit Complement Med 4(1):24–35
73. Law S, Leung AW, Xu C (2020) Is the traditional Chinese herb "Artemisia annua" possible to fight against COVID-19? Integr Med Res 9(3):100474. https://doi.org/10.1016/j.imr.2020.100474
74. Chen Y, Fan C-L, Wang Y, Zhang X-Q, Huang X-J, Ye W-C (2018) Chemical constituents from roots of Isatis indigotica. China J Chinese Materia Medica 43(10):2091–2096
75. Zhang D et al (2019) Alkaloid enantiomers from the roots of Isatis indigotica. Molecules 24(17):3140
76. Lin C-W et al (2005) Anti-SARS coronavirus 3C-like protease effects of Isatis indigotica root and plant-derived phenolic compounds. Antivir Res 68(1):36–42
77. Chang S-J, Chang Y-C, Lu K-Z, Tsou Y-Y, Lin C-W (2012) Antiviral activity of Isatis indigotica extract and its derived indirubin against Japanese encephalitis virus. Evid-Based Complement Alternat Med 2012:925830
78. Ameri A, Heydarirad G, Rezaeizadeh H, Choopani R, Ghobadi A, Gachkar L (2016) Evaluation of efficacy of an herbal compound on dry mouth in patients with head and neck cancers: a randomized clinical trial. J Evid-Based Complement Alternat Med 21(1):30–33
79. Rezaeipour N et al (2017) Efficacy of a Persian medicine herbal compound (Alcea digitata Alef and Malva sylvestris L.) on prevention of radiation induced acute mucositis in patients with head and neck cancer: a pilot study. Int J Cancer Manag 10(9):e8642. https://doi.org/10.5812/ijcm.8642
80. Lamoral-Theys D et al (2010) Lycorine and its derivatives for anticancer drug design. Mini-Rev Med Chem 10(1):41–50
81. Ieven M, Van den Berghe D, Vlietinck A (1983) Plant antiviral agents. Planta Med 49(10):109–114
82. Liu J, Yang Y, Xu Y, Ma C, Qin C, Zhang L (2011) Lycorine reduces mortality of human enterovirus 71-infected mice by inhibiting virus replication. Virol J 8(1):483
83. Wang H, Guo T, Yang Y, Yu L, Pan X, Li Y (2019) Lycorine derivative LY-55 inhibits EV71 and CVA16 replication through downregulating autophagy. Front Cell Infect Microbiol 9:277. https://doi.org/10.3389/fcimb.2019.00277
84. Mukhtar M, Arshad M, Ahmad M, Pomerantz RJ, Wigdahl B, Parveen Z (2008) Antiviral potentials of medicinal plants. Virus Res 131(2):111–120
85. Islam MT et al (2020) Natural products and their derivatives against coronavirus: a review of the non-clinical and pre-clinical data. Phytother Res 34(10):2471–2492
86. Fan Y, Feng H, Liu L, Zhang Y, Xin X, Gao D (2020) Chemical components and antibacterial activity of the essential oil of six pyrrosia species. Chem Biodivers 17(10):e2000526. https://doi.org/10.1002/cbdv.202000526
87. Xiao W et al (2017) Comparative evaluation of chemical profiles of pyrrosiae folium originating from three pyrrosia species by HPLC-DAD combined with multivariate statistical analysis. Molecules 22(12):2122
88. Oh J et al (2013) Extracellular signal-regulated kinase is a direct target of the anti-inflammatory compound amentoflavone derived from Torreya nucifera. Mediators of inflammation 2013:761506

89. Shingnaisui K, Dey T, Manna P, Kalita J (2018) Therapeutic potentials of Houttuynia cordata Thunb. against inflammation and oxidative stress: a review. J Ethnopharmacol 220:35–43
90. Chiow K, Phoon M, Putti T, Tan BK, Chow VT (2016) Evaluation of antiviral activities of Houttuynia cordata Thunb. extract, quercetin, quercetrin and cinanserin on murine coronavirus and dengue virus infection. Asian Pac J Trop Med 9(1):1–7
91. Endo Y, Osada Y, Kimura F, Fujimoto K (2006) Effects of Japanese torreya (Torreya nucifera) seed oil on lipid metabolism in rats. Nutrition 22(5):553–558
92. Xiao M, Cao N, Fan J, Shen Y, Xu Q (2011) Studies on flavonoids from the leaves of Lindera aggregata. J Chinese Med Mat 34(1):62–64
93. Jung S-H et al (2017) Inhibition of collagen-induced platelet aggregation by the secobutanolide secolincomolide A from Lindera obtusiloba Blume. Front Pharmacol 8:560
94. Zhao M et al (2018) Quality evaluation of Rhei Radix et Rhizoma decoction. China J Chinese Materia Medica 43(5):861–867
95. Yang Y, Islam MS, Wang J, Li Y, Chen X (2020) Traditional Chinese medicine in the treatment of patients infected with 2019-new coronavirus (SARS-CoV-2): a review and perspective. Int J Biol Sci 16(10):1708
96. Shen C-Y, Jiang J-G, Zhu W, Ou-Yang Q (2017) Anti-inflammatory effect of essential oil from Citrus aurantium L. var. amara Engl. J Agric Food Chem 65(39):8586–8594
97. Ziaei S, Heidari M, Amin G, Kochmeshki A, Heidari M (2009) Inhibitory effects of germinal angiotensin converting enzyme by medicinal plants used in Iranian traditional medicine as antihypertensive. J Kerman Univer Med Sci 16(2):134–143
98. Kim M, Nguyen D-V, Heo Y, Park KH, Paik H-D, Kim YB (2020) Antiviral activity of fritillaria thunbergii extract against human influenza virus H1N1 (PR8) in vitro, in ovo and in vivo. J Microbiol Biotechnol 30(2):172–177
99. Liang L, Xu J, Liang Z-T, Dong X-P, Chen H-B, Zhao Z-Z (2018) Tissue-specific analysis of secondary metabolites creates a reliable morphological criterion for quality grading of polygoni multiflori radix. Molecules 23(5):1115
100. Ho T-Y, Wu S-L, Chen J-C, Li C-C, Hsiang C-Y (2007) Emodin blocks the SARS coronavirus spike protein and angiotensin-converting enzyme 2 interaction. Antivir Res 74(2):92–101
101. Stohs SJ (2017) Safety, efficacy, and mechanistic studies regarding Citrus aurantium (bitter orange) extract and p-synephrine. Phytother Res 31(10):1463–1474
102. Pimenta FCF et al (2016) Anxiolytic effect of Citrus aurantium L. on patients with chronic myeloid leukemia. Phytother Res 30(4):613–617
103. Marhoume FZ et al (2019) Anti-aggregant effect of butanolic extract of Rubia tinctorum L on platelets in vitro and ex vivo. J Ethnopharmacol 241:111971
104. Xiong Y, Yang Y, Xiong W, Yao Y, Wu H, Zhang M (2019) Network pharmacology-based research on the active component and mechanism of the antihepatoma effect of Rubia cordifolia L. J Cell Biochem 120(8):12461–12472
105. Lajkó E, Bányai P, Zámbó Z, Kursinszki L, Szőke É, Kőhidai L (2015) Targeted tumor therapy by Rubia tinctorum L.: analytical characterization of hydroxyanthraquinones and investigation of their selective cytotoxic, adhesion and migration modulator effects on melanoma cell lines (A2058 and HT168-M1). Cancer Cell Int 15(1):119
106. Console L et al (2019) Human mitochondrial carnitine acylcarnitine carrier: molecular target of dietary bioactive polyphenols from sweet cherry (Prunus avium L.). Chem Biol Interact 307:179–185
107. Sharifi N, Souri E, Ziai SA, Amin G, Amanlou M (2013) Discovery of new angiotensin converting enzyme (ACE) inhibitors from medicinal plants to treat hypertension using an in vitro assay. DARU J Pharmaceut Sci 21(1):74
108. Renda G, Arzu Ö, Barut B, Korkmaz B, Yayli N (2018) In vitro protection by Crataegus microphylla extracts against oxidative damage and enzyme inhibition effects. Turkish J Pharmaceut Sci 15(1):77

109. Hosseinimehr SJ, Mahmoudzadeh A, Azadbakht M, Akhlaghpoor S (2009) Radioprotective effects of Hawthorn against genotoxicity induced by gamma irradiation in human blood lymphocytes. Radiat Environ Biophys 48(1):95
110. Sabahi Z, Farmani F, Soltani F, Moein M (2018) DNA protection, antioxidant and xanthine oxidase inhibition activities of polyphenol-enriched fraction of Berberis integerrima Bunge fruits. Iranian J Basic Med Sci 21(4):411
111. Tung NH, Kwon H-J, Kim J-H, Ra JC, Kim JA, Kim YH (2010) An anti-influenza component of the bark of Alnus japonica. Arch Pharm Res 33(3):363–367
112. Won TH et al (2015) Bioactive metabolites from the fruits of Psoralea corylifolia. J Nat Prod 78(4):666–673
113. Schneiderová K, Šmejkal K (2015) Phytochemical profile of Paulownia tomentosa (Thunb) Steud. Phytochem Rev 14(5):799–833
114. Tian C, Zhang Z, Wang H, Guo Y, Zhao J, Liu M (2019) Extraction technology, component analysis, and in vitro antioxidant and antibacterial activities of total flavonoids and fatty acids from Tribulus terrestris L. fruits. Biomed Chromatogr 33(4):e4474
115. Chopra B, Dhingra AK, Dhar KL (2013) Psoralea corylifolia L.(Buguchi)—folklore to modern evidence. Fitoterapia 90:44–56
116. Kim DW et al (2014) Phenolic phytochemical displaying SARS-CoV papain-like protease inhibition from the seeds of Psoralea corylifolia. J Enzyme Inhibit Med Chem 29(1):59–63
117. Ştefănescu R, Tero-Vescan A, Negroiu A, Aurică E, Vari C-E (2020) A comprehensive review of the phytochemical, pharmacological, and toxicological properties of Tribulus terrestris L. Biomolecules 10(5):752
118. Song YH et al (2014) Papain-like protease (PLpro) inhibitory effects of cinnamic amides from Tribulus terrestris fruits. Biol Pharm Bull 37(6):1021–1028
119. Hawkins J, Baker C, Cherry L, Dunne E (2019) Black elderberry (Sambucus nigra) supplementation effectively treats upper respiratory symptoms: a meta-analysis of randomized, controlled clinical trials. Complement Ther Med 42:361–365
120. Bhuiyan FR, Howlader S, Raihan T, Hasan M (2020) Plants metabolites: possibility of natural therapeutics against the COVID-19 pandemic. Front Med 7:444
121. Shin WJ, Lee KH, Park MH, Seong BL (2010) Broad-spectrum antiviral effect of Agrimonia pilosa extract on influenza viruses. Microbiol Immunol 54(1):11–19
122. Gansukh E, Muthu M, Paul D, Ethiraj G, Chun S, Gopal J (2017) Nature nominee quercetin's anti-influenza combat strategy—demonstrations and remonstrations. Rev Med Virol 27(3):e1930
123. Khan RI, Abbas M, Goraya K, Zafar-ul-Hye M, Danish S (2020) Plant derived antiviral products for potential treatment of COVID-19: a review. Phyton 89(3):438
124. Zhang P et al (2018) Astragalus polysaccharides inhibit avian infectious bronchitis virus infection by regulating viral replication. Microb Pathog 114:124–128
125. Kumar VS, Navaratnam V (2013) Neem (Azadirachta indica): prehistory to contemporary medicinal uses to humankind. Asian Pac J Trop Biomed 3(7):505–514
126. Hussein G, Miyashiro H, Nakamura N, Hattori M, Kakiuchi N, Shimotohno K (2000) Inhibitory effects of Sudanese medicinal plant extracts on hepatitis C virus (HCV) protease. Phytother Res 14(7):510–516
127. Murugan NA, Pandian CJ, Jeyakanthan J (2020) Computational investigation on Andrographis paniculata phytochemicals to evaluate their potency against SARS-CoV-2 in comparison to known antiviral compounds in drug trials. J Biomol Struct Dyn:1–12. https://doi.org/10.1080/07391102.2020.1777901
128. Park J-Y et al (2016) Chalcones isolated from Angelica keiskei inhibit cysteine proteases of SARS-CoV. J Enzyme Inhibit Med Chem 31(1):23–30
129. Warowicka A, Nawrot R, Goździcka-Józefiak A (2020) Antiviral activity of berberine. Arch Virol 165(9):1935–1945
130. Rastogi S, Pandey MM, Rawat AKS (2015) Medicinal plants of the genus Betula—traditional uses and a phytochemical–pharmacological review. J Ethnopharmacol 159:62–83
131. Kala CP (2012) Leaf juice of Carica papaya L, A remedy of dengue fever. Med Aromat Plan Theory 1:109

132. Zhang H, Penninger JM, Li Y, Zhong N, Slutsky AS (2020) Angiotensin-converting enzyme 2 (ACE2) as a SARS-CoV-2 receptor: molecular mechanisms and potential therapeutic target. Intensive Care Med 46(4):586–590
133. Usachev EV, Pyankov OV, Usacheva OV, Agranovski IE (2013) Antiviral activity of tea tree and eucalyptus oil aerosol and vapour. J Aerosol Sci 59:22–30
134. Khan MY, Kumar V (2019) Mechanism & inhibition kinetics of bioassay-guided fractions of Indian medicinal plants and foods as ACE inhibitors. J Tradit Complement Med 9(1):73–84
135. Wen C-C et al (2011) Traditional Chinese medicine herbal extracts of Cibotium barometz, Gentiana scabra, Dioscorea batatas, Cassia tora, and Taxillus chinensis inhibit SARS-CoV replication. J Tradit Complement Med 1(1):41–50
136. Kojoma M et al (2010) In vitro proliferation and triterpenoid characteristics of licorice (Glycyrrhiza uralensis Fischer, Leguminosae) stolons. Plant biotechnology 27(1):59–66
137. Pal D (2011) Sunflower (Helianthus annuus L.) Seeds in health and nutrition. In: Nuts and seeds in health and disease prevention. Elsevier, London, pp 1097–1105
138. Li W, Zhou P, Zhang Y, He L (2011) Houttuynia cordata, a novel and selective COX-2 inhibitor with anti-inflammatory activity. J Ethnopharmacol 133(2):922–927
139. Shahrajabian MH, Sun W, Shen H, Cheng Q (2020) Chinese herbal medicine for SARS and SARS-CoV-2 treatment and prevention, encouraging using herbal medicine for COVID-19 outbreak. Acta Agric Scand B Soil Plant Sci 70:437–443
140. Chen C-J et al (2008) Toona sinensis Roem tender leaf extract inhibits SARS coronavirus replication. J Ethnopharmacol 120(1):108–111
141. Orhan IE, Deniz FSS (2020) Natural products as Potential Leads against coronaviruses: could they be encouraging structural models against SARS-CoV-2? Natural Products and Bioprospecting 10(4):171–186
142. Loizzo MR et al (2008) Phytochemical analysis and in vitro antiviral activities of the essential oils of seven Lebanon species. Chem Biodivers 5(3):461–470
143. Kasprzak D, Jodlowska-Jedrych B, Borowska K, Wojtowicz A (2018) Lepidium meyenii (Maca)–multidirectional health effects–review. Curr Issues Pharmacy Med Sci 31(3):107–112
144. Maideen NMP (2020) Prophetic medicine-Nigella Sativa (Black cumin seeds)–potential herb for COVID-19? J pharmacopuncture 23(2):62
145. Gonçalves J et al (2005) In vitro anti-rotavirus activity of some medicinal plants used in Brazil against diarrhea. J Ethnopharmacol 99(3):403–407
146. Li Y-H, Lai C-Y, Su M-C, Cheng J-C, Chang Y-S (2019) Antiviral activity of Portulaca oleracea L. against influenza A viruses. J Ethnopharmacol 241:112013
147. Luo W et al (2009) Anti-SARS coronavirus 3C-like protease effects of Rheum palmatum L. extracts. Bioscience Trends 4:3
148. Chen C et al (2014) Sambucus nigra extracts inhibit infectious bronchitis virus at an early point during replication. BMC Vet Res 10(1):1–12
149. Rothan H, Zulqarnain M, Ammar Y, Tan E, Rahman N, Yusof R (2014) Screening of antiviral activities in medicinal plants extracts against dengue virus using dengue NS2B-NS3 protease assay. Trop Biomed 31(2):286–296
150. Choi H-J, Song J-H, Kwon D-H, Baek S-H, Ahn Y-J (2008) Antiviral activity of Zanthoxylum species against influenza virus. Korean J Med Crop Sci 16(4):273–278
151. Donma MM, Donma O (2020) The effects of Allium Sativum on immunity within the scope of COVID-19 infection. Med Hypotheses 144:109934
152. Tapsell LC, et al (2006) Health benefits of herbs and spices: the past, the present, the future. Med J Aus 185(4):4–24
153. Schafer G, Kaschula CH (2014) The immunomodulation and anti-inflammatory effects of garlic organosulfur compounds in cancer chemoprevention. Anticancer Agents Med Chem 14(2):233–240
154. Pittler MH, Ernst E (2007) Clinical effectiveness of garlic (Allium sativum). Mol Nutr Food Res 51(11):1382–1385
155. Anywar G, Kakudidi E, Byamukama R, Mukonzo J, Schubert A, Oryem-Origa H (2020) Medicinal plants used by traditional medicine practitioners to boost the immune system in people living with HIV/AIDS in Uganda. Eur J Integr Med 35:101011

156. Sahoo B, Banik B (2018) Medicinal plants: source for immunosuppressive agents. Immunol Curr Res 2(106):2
157. Lissiman E, Bhasale AL, Cohen M (2014) Garlic for the common cold. Cochrane Database Syst Rev (11)
158. Srivastava A, Chaurasia J, Khan R, Dhand C, Verma S (2020) Role of medicinal plants of traditional use in recuperating devastating COVID-19 situation. Med Aromat Plants (Los Angeles) 9(359):2167-0412.20
159. Rather MA, Dar BA, Sofi SN, Bhat BA, Qurishi MA (2016) Foeniculum vulgare: a comprehensive review of its traditional use, phytochemistry, pharmacology, and safety. Arab J Chem 9:S1574–S1583
160. He W, Huang B (2011) A review of chemistry and bioactivities of a medicinal spice: Foeniculum vulgare. J Med Plants Res 5(16):3595–3600
161. Mohamad RH et al (2011) Antioxidant and anticarcinogenic effects of methanolic extract and volatile oil of fennel seeds (Foeniculum vulgare). J Med Food 14(9):986–1001
162. Dorra N, El-Berrawy M, Sallam S, Mahmoud R (2019) Evaluation of antiviral and antioxidant activity of selected herbal extracts. J High Ins Pub Health 49(1):36–40
163. Islam M et al (2017) Nigellalogy: a review on Nigella sativa. MOJ Bioequiv Availab 3(6):00056
164. Tariq M (2008) Nigella sativa seeds: folklore treatment in modern day medicine. Saudi J Gastroenterol 14(3):105
165. Shamim Molla M, Azad AK, Al Hasib MAA, Hossain MM, Ahammed MS, Rana S, & Islam, MT (2019) A review on antiviral effects of Nigella sativa L. Pharmacology Online, Newsletter, 2:47–53
166. Dissanayake KGC, Liyanage WA, Waliwita C, Liyanage RP (2020) A review on medicinal uses of Zingiber officinale (Ginger). Int J Health Sci Res 10(6):142–148
167. Habeballa RS, Ahmedani EI, Awad NS, Abdein MA (2020) In vitro antiviral activity of Illicium verum and Zingiber officinale ethanolic extracts. Med Sci 24(105):3469–3480
168. Kaushik S, Jangra G, Kundu V, Yadav JP, Kaushik S (2020) Anti-viral activity of Zingiber officinale (Ginger) ingredients against the Chikungunya virus. VirusDis 31:3
169. Arora R et al (2011) Potential of complementary and alternative medicine in preventive management of novel H1N1 flu (Swine flu) pandemic: thwarting potential disasters in the bud. Evid Based Complement Alternat Med 2011:586506
170. Wang L, Yang R, Yuan B, Liu Y, Liu C (2015) The antiviral and antimicrobial activities of licorice, a widely-used Chinese herb. Acta Pharm Sin B 5(4):310–315
171. Asl MN, Hosseinzadeh H (2008) Review of pharmacological effects of Glycyrrhiza sp. and its bioactive compounds. Phytother Res 22(6):709–724
172. Seki H et al (2008) Licorice β-amyrin 11-oxidase, a cytochrome P450 with a key role in the biosynthesis of the triterpene sweetener glycyrrhizin. Proc Natl Acad Sci 105(37):14204–14209
173. Fiore C et al (2008) Antiviral effects of Glycyrrhiza species. Phytother Res 22(2):141–148
174. Ming LJ, Yin ACY (2013) Therapeutic effects of glycyrrhizic acid. Nat Product Commun 8(3):1934578X1300800335
175. Al-Snafi AE (2018) Glycyrrhiza glabra: a phytochemical and pharmacological review. IOSR J Pharmacy 8(6):1–17
176. Lemon Balm C. Lemon balm (Melissa officinalis l.) an herbal medicinal plant with broad therapeutic uses and cultivation practices: a review. 2015.
177. Astani A, Heidary Navid M, Schnitzler P (2014) Attachment and penetration of Acyclovir-resistant Herpes Simplex virus are inhibited by Melissa officinalis Extract. Phytother Res 28(10):1547–1552
178. Ulbricht C et al (2005) Lemon balm (Melissa officinalis L.): an evidence-based systematic review by the Natural Standard Research Collaboration. J Herb Pharmacother 5(4):71–114
179. Li Y, Liu Y, Ma A, Bao Y, Wang M, Sun Z (2017) In vitro antiviral, anti-inflammatory, and antioxidant activities of the ethanol extract of Mentha piperita L. Food Sci Biotechnol 26(6):1675–1683

180. Nolkemper S, Reichling J, Stintzing FC, Carle R, Schnitzler P (2006) Antiviral effect of aqueous extracts from species of the Lamiaceae family against Herpes simplex virus type 1 and type 2 in vitro. Planta Med 72(15):1378–1382
181. Battistini R et al (2019) antiviral activity of essential oils against hepatitis A virus in soft fruits. Food Environ Virol 11(1):90–95
182. Shin H-B et al (2013) Antiviral activity of carnosic acid against respiratory syncytial virus. Virol J 10(1):1–11
183. Hudson J, Vimalanathan S (2011) Echinacea—a source of potent antivirals for respiratory virus infections. Pharmaceuticals 4(7):1019–1031
184. Chiang LC, Ng LT, Cheng PW, Chiang W, Lin CC (2005) Antiviral activities of extracts and selected pure constituents of Ocimum basilicum. Clin Exp Pharmacol Physiol 32(10):811–816
185. Hamidpour M, Hamidpour R, Hamidpour S, Shahlari M (2014) Chemistry, pharmacology, and medicinal property of sage (Salvia) to prevent and cure illnesses such as obesity, diabetes, depression, dementia, lupus, autism, heart disease, and cancer. J Tradit Complement Med 4(2):82–88
186. Geuenich S et al (2008) Aqueous extracts from peppermint, sage and lemon balm leaves display potent anti-HIV-1 activity by increasing the virion density. Retrovirology 5(1):27
187. Santoyo S, Jaime L, García-Risco MR, Ruiz-Rodríguez A, Reglero G (2014) Antiviral properties of supercritical CO2 extracts from oregano and sage. Int J Food Prop 17(5):1150–1161
188. Šmidling D, MITIć-ćULAfIć D, VUkOVIć-GAčIć B, Simić D, Knežević-Vukčević J (2008) Evaluation of antiviral activity of fractionated extracts of sage Salvia officinalis L.(Lamiaceae). Arch Biol Sci 60(3):421–429
189. Blank DE, de Oliveira Hübner S, Alves GH, Cardoso CAL, Freitag RA, Cleff MB (2019) Chemical composition and antiviral effect of extracts of Origanum vulgare. Adv Biosci Biotechnol 10(07):188
190. Duke JA (2002) Handbook of medicinal herbs. CRC press, Boca Raton
191. Wang X et al (2020) Senna alexandrina extract supplementation reverses hepatic oxidative, inflammatory, and apoptotic effects of cadmium chloride administration in rats. Environ Sci Pollut Res 27(6):5981–5992
192. Leelavathi V, Udayasri P. Qualitative and quantitative analytical studies for the screening of phytochemicals from the leaf extracts of Senna alexandrina Mill. International Journal of Pharmaceutical and Clinical Research, 10(8):210–215
193. Fan Y et al (2020) Food as medicine: a possible preventive measure against coronavirus disease (COVID-19). Phytother Res 34:3124
194. Hosseinzadeh H, Asl MN, Parvardeh S (2005) The effects of carbenoxolone, a semisynthetic derivative of glycyrrhizinic acid, on peripheral and central ischemia-reperfusion injuries in the skeletal muscle and hippocampus of rats. Phytomedicine 12(9):632–637
195. Tattelman E (2005) Health effects of garlic. Am Fam Physician 72(1):103–106
196. Somasundaram S, Edmund NA, Moore DT, Small GW, Shi YY, Orlowski RZ (2002) Dietary curcumin inhibits chemotherapy-induced apoptosis in models of human breast cancer. Cancer Res 62(13):3868–3875
197. Bedrood Z, Rameshrad M, Hosseinzadeh H (2018) Toxicological effects of Camellia sinensis (green tea): a review. Phytother Res 32(7):1163–1180
198. Chandrasekara A, Shahidi F (2018) Herbal beverages: bioactive compounds and their role in disease risk reduction-a review. J Tradit Complement Med 8(4):451–458
199. Ekor M (2014) The growing use of herbal medicines: issues relating to adverse reactions and challenges in monitoring safety. Front Pharmacol 4:177
200. Kostoff RN, Briggs MB, Porter AL, Aschner M, Spandidos DA, Tsatsakis A (2020) COVID-19: post-lockdown guidelines. Int J Mol Med 46(2):463–466
201. Vilanova-Sanchez A et al (2018) Are Senna based laxatives safe when used as long term treatment for constipation in children? J Pediatr Surg 53(4):722–727

Chapter 3
Traditional Chinese Medicines as Possible Remedy Against SARS-CoV-2

Saqib Mahmood, Tariq Mahmood, Naeem Iqbal, Samina Sabir, Sadia Javed, and Muhammad Zia-Ul-Haq

Introduction

Contagious outbreak of COVID-19 has raised intense attention of microbiologist, pathologist, and ethnobiologist. In January, it has been declared as Public Health Emergency of International Concern by the World Health Organization (2020) [1]. Ethnobiologist deals with life using living tools as therapeutic source, developing dynamic relation among different components of ecosystem. In general, the use of traditional medicines being safer and economical and with possibility of least side effect is an acceptable practice. In current scenario, the absence of any vaccine against COVID-19 has intensified the role of traditional remedies.

From China to date, seven issues of guidelines about trial clinical application against COVID-19 have been published. In the seventh edition, which is the most recent, the absence of the discovery of precise treatments of COVID-19 has been mentioned officially by the National Health Commission [2]. Chinese government has allowed to use TCM at clinical level against COVID-19 in the third version of *COVID-19 Treatment Guidelines* (published on January 23, 2020) [3]. Here is an overview of TCM with reported potential against a number of viruses in general, followed by literature pointing out possible role of TCM against other members of SARS-CoV and COVID-19.

S. Mahmood · N. Iqbal · S. Sabir
Department of Botany, Government College University Faisalabad, Faisalabad, Punjab, Pakistan

T. Mahmood
ABWA Hospital & Research Centre, Faisalabad, Pakistan

S. Javed (✉)
Department of Biochemistry, Government College University, Faisalabad, Pakistan

M. Zia-Ul-Haq
Office of Research, Innovation and Commercialization, Lahore College for Women University, Lahore, Pakistan

© The Author(s), under exclusive license to Springer Nature Switzerland AG 2021
M. Zia-Ul-Haq et al. (eds.), *Alternative Medicine Interventions for COVID-19*,
https://doi.org/10.1007/978-3-030-67989-7_3

Traditional Chinese Medical Practices

TCM are part of the tradition of China for more than 3000 years. According to the National Center for Complementary and Integrative Health [3], here are different practices under this head:

- Chinese use herbs as remedy directly in the form of some herbs/extracts/decoctions.
- Acupuncturing technique.
- Cupping treatment for bloodletting.
- Skin scrapping known as gua sha.
- Pushing, rubbing, and squeezing patient body known as massage or tui na.
- Setting of bone known as bonesetter (die-da).
- Exercise (qigong).
- Dietary therapy.

In general, usage of Chinese medicines is further categorized into three basic categories including detoxifications, eliminating dampness, and clearance of heat as mentioned in Fig. 3.1. In this chapter, we have emphasized detoxification category exclusively from abovementioned Chinese medication traditions. Henceforward, we have discussed plant metabolites responsible for antiviral activities, followed by possible correlation of existing data with the potential of these metabolites/herbs/decoction to cure COVID-19.

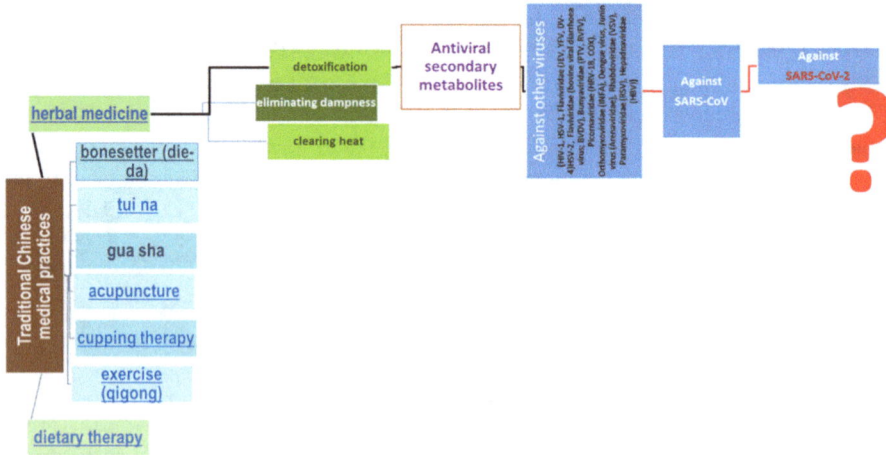

Fig. 3.1 General types of traditional Chinese practices (ignoring subcategories of all except herbal medicine)

Laws for TCM Formulation

TCM preventive laws have been launched by the National Administration of Traditional Chinese Medicine of China, with the recommendations to follow certain rules during formulation of traditional medicines according to the age group. For ideal and safer formulation of TCM, knowledge of phytoconstituents present in herbs is required along with their bioactivity within host. In addition, appropriate understandings of viral structure, its pathogenesis, and immunity of host may guide to explore therapies in the form of drugs. Here we are discussing active plant metabolites with their antiviral activities reviewed from literature, along with therapeutic justification for the inhibition of viruses. Data has been collected from literature about the mode of antiviral actions induced in infected patients in vivo/vitro analysis. This information may lead toward their utilization against other viruses including COVID-19.

Antiviral Potential of Plant Metabolites

Plants being sessile need active endogenous defensive machinery to fight against biotic and abiotic adversaries. To combat with them, nature has blessed plants with endogenous defensive machinery. It is mediated by thousands of plant secondary metabolites (PSMs) including phenolics, alkaloids, polyketides, flavonoid, etc. [4]. Viruses are one the serious threats to life at globe. Previously a number of secondary metabolites have been reported for partial or complete inhibition of different viruses. It includes HIV-1, HSV-1, HSV-2, *Flaviviridae*, *Bunyaviridae*, *Picornaviridae*, *Orthomyxoviridae*, *Dengue virus*, *Junin virus*, *Rhabdoviridae*, *Paramyxoviridae*, and *Hepadnaviridae*. A variety of PSMs to date have been reported to combat these viruses such as different alkaloids, coumarins, and polyphenols: flavonoids, isoflavonoids, lignans, tannins, triterpenes, essential oils, flavans and derivatives, diphyllin, and justicidin (Fig. 3.2).

TCM has historical evidences for success stories of thousands of years defeating infections. TCM proved itself beneficial in 2003 when severe acute respiratory syndrome has become epidemic [6]. Previously serious infections of SARS in 2009 and pandemic H1N1 influenza were succesfully coped with the use of TCM. TCM also helped to fight pestilence in China earlier with the cure of a number of patients [7].

Structure of SARS-CoV-2

Brief introduction of generalized and then specifically SARS-CoV viruses may add up in understanding literature regarding antiviral activities and associated mechanisms.

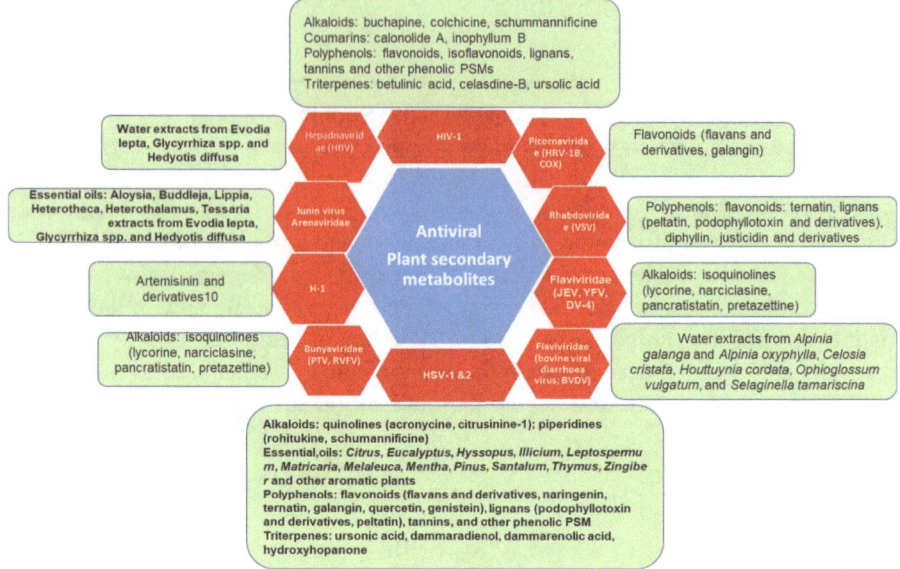

Fig. 3.2 Plant secondary metabolites (PSMs) with antiviral potential [5]

In comparison to prokaryotes or eukaryotes, viral structure is very simple comprising of nucleic acid (RNA/DNA) and nucleocapsid (nucleic acid-binding proteins). The capsid is bordered by a membranous system that is derived from host endoplasmic reticulum (additional residence of a number of viral proteins) helping its attachment with cellular receptors of host [8] (Fig. 3.3).

SARS-CoV-2 is a member of *Coronaviridae* [9, 10]. It has a genetic material in the form of single strand (RNA+ sense); it could perform as mRNA itself. Therefore, no need for transcription here during translation. Its mRNA is used to form nucleocapsid. Envelope of virus with its spike proteins is derived from host cell's endoplasmic reticulum. This envelope is then used for binding of virus with receptors. After entry into host cell with the help of polymerase enzyme, viral mRNA makes its copies, which are further translated into proteins (embedded into the ER membrane).

Relevance of SARS-CoV with SARS-CoV-2

Entry of SARS-CoV vs. SARS-CoV-2 into Host Cell

Entry of SARS-CoV and SARS-CoV-2 in host cell is mediated by receptors ACE2 and serine protease TMPRSS2. Therefore, it is suggested that clinically approved protease inhibitor of these receptors for SARS-CoV may be tried for SARS-CoV-2

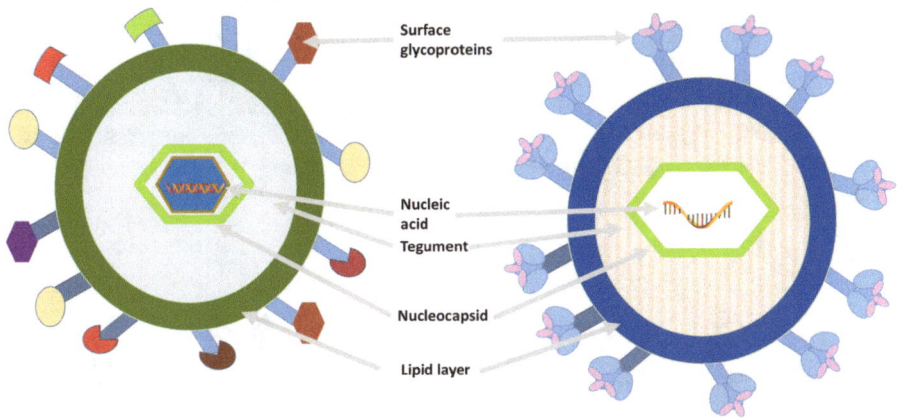

Fig. 3.3 Generalized virus cell (left) versus COVID-19 cell (right) [8]

also at clinical level. Hoffmann et al. [11] provided evidences for the use of clinically approved inhibitors of TMPRSS2 for blockage of virus entry in the host cell as cure of corona. This study has unveiled similarities between COVID-19 and other member of SARS, i.e., CoV-2. There opened a path toward identification of potential targets and invention of antiviral intervention. Therefore, the discovered antibodies with inhibition of SARS-CoV spikes can be possible solutions against SARS-CoV-2.

With respect to the immune system and pathophysiology, SARS-CoV-2-infected patients have been observed with a number of similarities with the patients of other SARS-CoV [12].

TCM and SARS-CoV

Therefore, due to the abovementioned similarity of behavior between other SARS-CoV and SARS-CoV-2 (Fig. 3.4), previous studies on the use of TCM against SARS-CoV may lead in search of antiviral TCM against SARS-CoV-2 too.

Series of research on pandemic SARS-CoV-1 has explored secondary metabolites as potent antiviral agents. Keeping in view the close relation of SARS-CoV-2 and SARS-CoV-1, above mentioned findings may help to find remedies against SARS-CoV-2 too for future application (Fig. 3.5).

Above data is in active use to find vaccines against these viruses. Based upon similarity of COVID-19 entry in host cell, it is supposed that all these active ingredient may be tried against COVID-19 too.

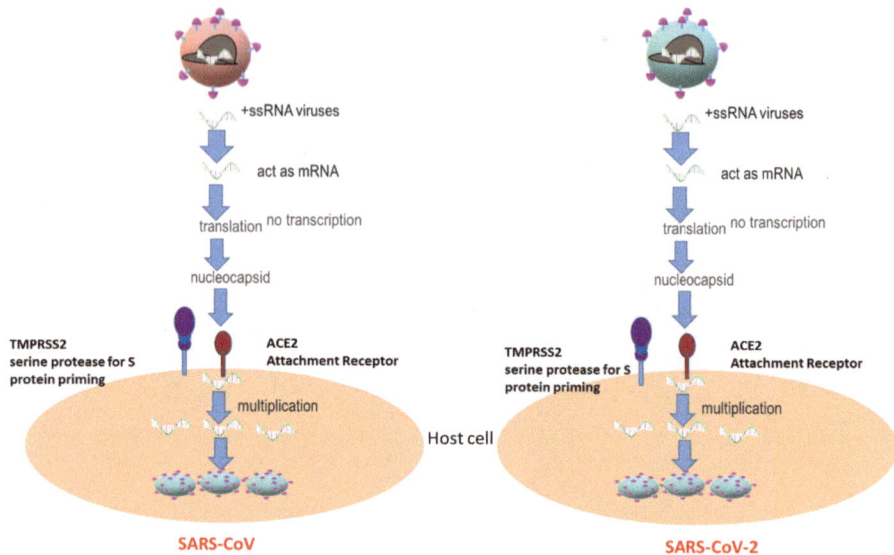

Fig. 3.4 Similarity between other SARS-CoV and SARS-CoV-2 from single-stranded mRNA translation to the receptor (ACE2) attachment and serine protease (TMPRSS2) mediation at host cell [10]

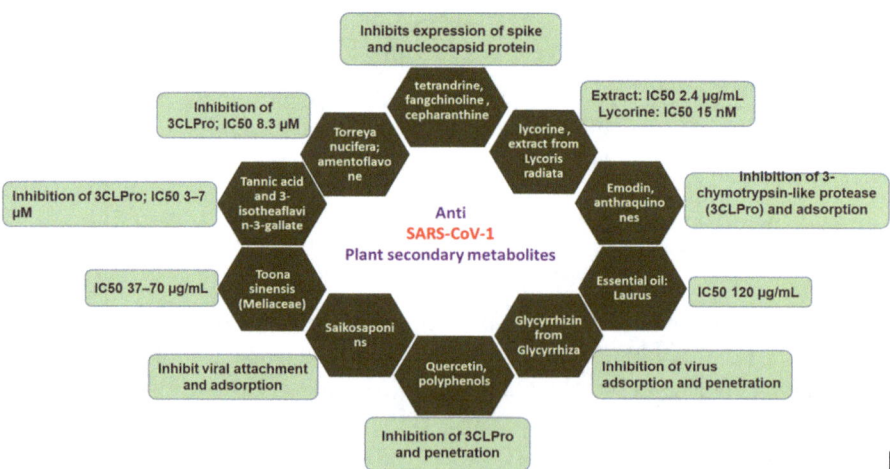

Fig. 3.5 Examples of plant extracts (rich in secondary metabolites) with their reported activities against SARS [5, 12–19]

TCM and SARS-CoV-2 (COVID-19)

In view of previous success stories of TCM, Chinese had started using them to fight against COVID-19. Oral ingestion of preventive herbal formulation and fumigation of house with herbal medicines are general practices in China.

Traditional Chinese medicine (TCM) has been employed against a number of viruses as therapeutic strategy [9]. Currently a number of studies are on the way at clinical level using complex TCM drugs to treat patients of COVID-19 in China, among which some are already nearly to be cured. Plant secondary metabolites need to be isolated to explore their target-oriented effect.

Most Practiced Herbs in Formulations Against COVID-19

Against COVID-19, TCM are in active practice whether independent or in combination with other plants or with some western medicines (Figs. 3.6, 3.7, and 3.8). That ranges from Glycyrrhizae Radix et Rhizoma to Armeniacae Semen Amarum, Ephedrae Herba, Gypsum Fibrosum, Scutellariae Radix, Atractylodis Rhizoma, Poria Sclerotium, Citri Reticulatae Pericarpium, Pinelliae Praeparatum cum Zingiberis, Forsythiae Fructus, Magnoliae Officinalis, Agastachis Herba, Platycodonis Radix, Agastachis Herba, Amomi Tsao-ko Fructus, Lepidii seu Descurainiae Semen, Coicis Semen, Agastachis Herba, Corni Fructus, Rhei Radix et Rhizoma, Ginseng Radix, and Trichosanthis Fructus.

Utilization Frequency of Herbs in the Formulation of (COVID-19) Decoctions

It has been noted that 12 herbs are more frequently used with average frequency of about 30 in most practiced decoctions of China, among them Armeniacae Semen Amarum and Ephedrae Herba got superiority upon others at all stages of patients (mild/moderate/severe stages). The Gypsum Fibrosum and Lepidii seu Descurainiae Semen followed them which are preferred in moderate and severe stages. Other most practiced herbs include Agastachis Herba, Forsythiae Fructus, Atractylodis Rhizoma, and Scutellariae Radix. Glycyrrhizae Radix et Rhizoma are with greater application at all four stages and for overall use (Fig. 3.9):

Fig. 3.6 Herbs most frequently used in TCM formulations

3 Traditional Chinese Medicines as Possible Remedy Against SARS-CoV-2

Fig. 3.7 Herbs most frequently used in TCM formulations

Fig. 3.8 Herbs most frequently used in TCM formulations

Approved/Proposed Decoctions with Successful Clinical Trials

A number of formulations in China are in practice against COVID-19. Some are restricted to only domestic level; some are used in clinics. Few are well approved by the Government of China, whereas some are unapproved due to under review studies. To date, the following decoctions have been reported in literature (Fig. 3.10):

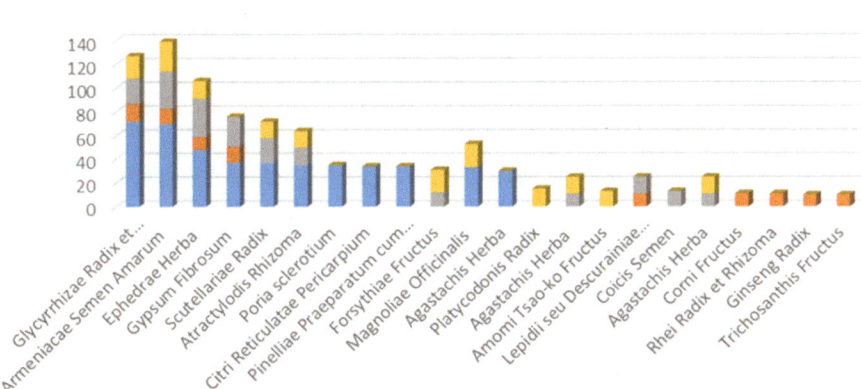

Fig. 3.9 Graphical presentation of the review about frequency of most practiced herbs in decoctions for treatment of COVID-19 patients with varying stages of severity (mild, moderate stage, severe stage, and all stages) [20]

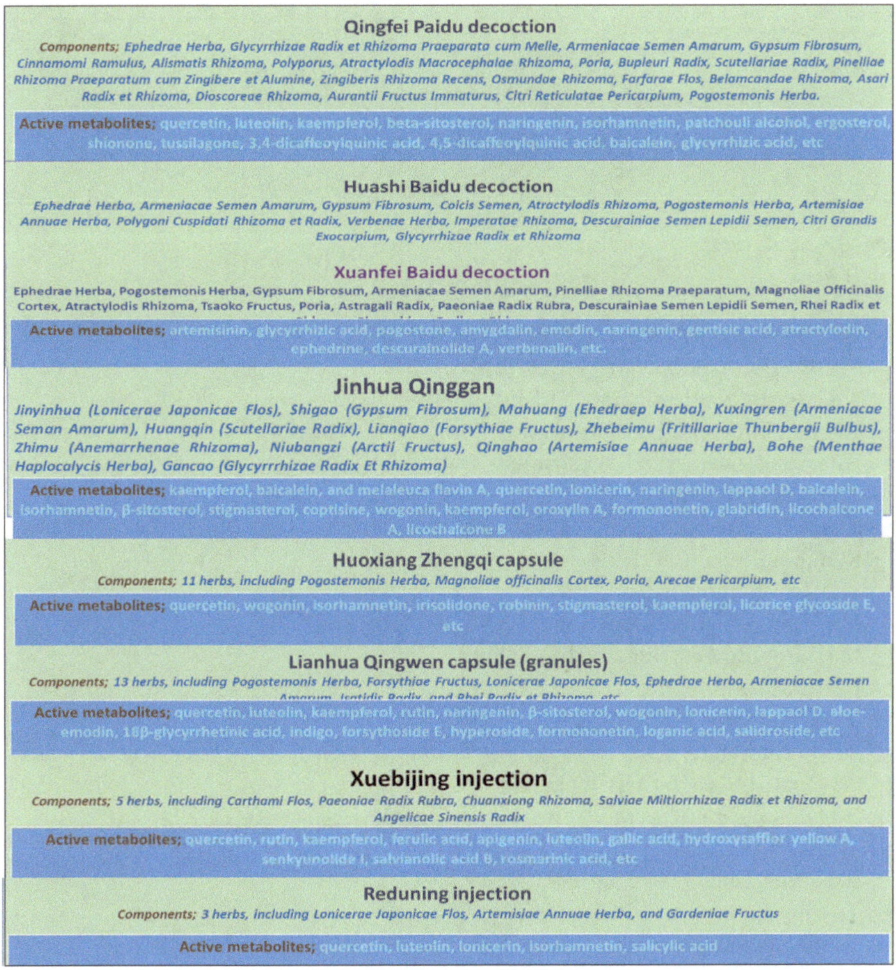

Qingfei Paidu decoction
Components: Ephedrae Herba, Glycyrrhizae Radix et Rhizoma Praeparata cum Melle, Armeniacae Semen Amarum, Gypsum Fibrosum, Cinnamomi Ramulus, Alismatis Rhizoma, Polyporus, Atractylodis Macrocephalae Rhizoma, Poria, Bupleuri Radix, Scutellariae Radix, Pinelliae Rhizoma Praeparatum cum Zingibere et Alumine, Zingiberis Rhizoma Recens, Osmundae Rhizoma, Farfarae Flos, Belamcandae Rhizoma, Asari Radix et Rhizoma, Dioscoreae Rhizoma, Aurantii Fructus Immaturus, Citri Reticulatae Pericarpium, Pogostemonis Herba.
Active metabolites: quercetin, luteolin, kaempferol, beta-sitosterol, naringenin, isorhamnetin, patchouli alcohol, ergosterol, shionone, tussilagone, 3,4-dicaffeoylquinic acid, 4,5-dicaffeoylquinic acid, baicalein, glycyrrhizic acid, etc

Huashi Baidu decoction
Components: Ephedrae Herba, Armeniacae Semen Amarum, Gypsum Fibrosum, Coicis Semen, Atractylodis Rhizoma, Pogostemonis Herba, Artemisiae Annuae Herba, Polygoni Cuspidati Rhizoma et Radix, Verbenae Herba, Imperatae Rhizoma, Descurainiae Semen Lepidii Semen, Citri Grandis Exocarpium, Glycyrrhizae Radix et Rhizoma

Xuanfei Baidu decoction
Components: Ephedrae Herba, Pogostemonis Herba, Gypsum Fibrosum, Armeniacae Semen Amarum, Pinelliae Rhizoma Praeparatum, Magnoliae Officinalis Cortex, Atractylodis Rhizoma, Tsaoko Fructus, Poria, Astragali Radix, Paeoniae Radix Rubra, Descurainiae Semen Lepidii Semen, Rhei Radix et
Active metabolites: artemisinin, glycyrrhizic acid, pogostone, amygdalin, emodin, naringenin, gentisic acid, atractylodin, ephedrine, descurainolide A, verbenalin, etc.

Jinhua Qinggan
Jinyinhua (Lonicerae Japonicae Flos), Shigao (Gypsum Fibrosum), Mahuang (Ehedraep Herba), Kuxingren (Armeniacae Seman Amarum), Huangqin (Scutellariae Radix), Lianqiao (Forsythiae Fructus), Zhebeimu (Fritillariae Thunbergii Bulbus), Zhimu (Anemarrhenae Rhizoma), Niubangzi (Arctii Fructus), Qinghao (Artemisiae Annuae Herba), Bohe (Menthae Haplocalycis Herba), Gancao (Glycyrrhizae Radix Et Rhizoma)
Active metabolites: kaempferol, baicalein, and melaleuca flavin A, quercetin, lonicerin, naringenin, lappaol D, baicalein, isorhamnetin, β-sitosterol, stigmasterol, coptisine, wogonin, kaempferol, oroxylin A, formononetin, glabridin, licochalcone A, licochalcone B

Huoxiang Zhengqi capsule
Components: 11 herbs, including Pogostemonis Herba, Magnoliae officinalis Cortex, Poria, Arecae Pericarpium, etc
Active metabolites: quercetin, wogonin, isorhamnetin, irisolidone, robinin, stigmasterol, kaempferol, licorice glycoside E, etc

Lianhua Qingwen capsule (granules)
Components: 13 herbs, including Pogostemonis Herba, Forsythiae Fructus, Lonicerae Japonicae Flos, Ephedrae Herba, Armeniacae Semen Amarum, Isatidis Radix, and Rhei Radix et Rhizoma, etc
Active metabolites: quercetin, luteolin, kaempferol, rutin, naringenin, β-sitosterol, wogonin, lonicerin, lappaol D, aloe-emodin, 18β-glycyrrhetinic acid, indigo, forsythoside E, hyperoside, formononetin, loganic acid, salidroside, etc

Xuebijing injection
Components: 5 herbs, including Carthami Flos, Paeoniae Radix Rubra, Chuanxiong Rhizoma, Salviae Miltiorrhizae Radix et Rhizoma, and Angelicae Sinensis Radix
Active metabolites: quercetin, rutin, kaempferol, ferulic acid, apigenin, luteolin, gallic acid, hydroxysafflor yellow A, senkyunolide I, salvianolic acid B, rosmarinic acid, etc

Reduning injection
Components: 3 herbs, including Lonicerae Japonicae Flos, Artemisiae Annuae Herba, and Gardeniae Fructus
Active metabolites: quercetin, luteolin, lonicerin, isorhamnetin, salicylic acid

Fig. 3.10 Most practiced TCM (decoctions) with their active metabolites and components

Yupingfeng San
Astragalus 20g +Fangfeng 15g +Atractylodes 15g

Yin Qiao San decoction
*Lonicerae Flos, Forsythiae Fructus, Platycodonis Radix, Menthae Herba, Lophatheri Herba, Schizonepetae Spica, Glycine Semen Preparatum, Arctii Semen, Phragmitis Rhizoma

Xiang Su San decoction
Cyperi Rhizoma, Perillae Folium, Citri Reticulatae Pericarpium, Glycyrrhizae Radix et Rhizoma, Bupleuri Radix, Cinnamomi Ramulus, Saposhnikoviae Radix, Osterici seu Notopterygii Radix et Rhizoma

Buhuan Jin Zhengqi San decoction
*Citri Reticulatae Pericarpium, Atractylodis Rhizoma, Magnoliae Cortex, Glycyrrhizae Radix et Rhizoma, Amomi Tsao-ko Fructus, Pinelliae Rhizoma, Agastachis Herba Severe

Xuanbai Chengqi Tang Ganlu Xiaodu Dan decoction
* Gypsum Fibrosum, Rhei Radix et Rhizoma, Armeniacae Semen Amarum, Trichosanthis Fructus, Talcum, Scutellariae Radix, Artemisiae Scopariae Herba, Acori Tatarinowii Rhizoma, Fritillariae Cirrhosae Bulbus, Akebiae Caulis, Agastachis Herba, Forsythiae Fructus, Amomi Fructus Rotundus, Menthae Herba, Belamcandae Rhizoma

San Ren Tang decoction
Armeniacae Semen Amarum 10 g, Amomi Fructus Rotundus 5 g, Coicis Semen 15 g, Pinelliae Rhizoma Praeparatum 10 g, Magnoliae Cortex×15 g, Talcum 10 g, Stachyuri Medulla Helwingiae Medulla 5 g, Agastachis Herba 10 g, Poria Sclerotium 15 g, Arecae Pericarpium 15 g, Scutellariae Radix 10 g, Artemisiae Annuae Herba 21 g

Liu Junzi Tang + Yu Ping Feng San decoction
*Ginseng Radix, Atractylodis Macrocephalae Rhizoma, Poria Sclerotium, Glycyrrhizae Radix et Rhizoma, Citri Reticulatae Pericarpium, Pinelliae Rhizoma, Saposhnikoviae Radix, Astragali Radix

San Ren Tang + Sheng Jiang San + Xin Jia Xiang Ru Yin
Armeniacae Semen Amarum 12 g, Talcum 15 g, Tetrapanacis Medulla 6 g, Amomi Fructus Rotundus 5 g, Lophatheri Herba 10 g, Magnoliae Officinalis Cortex 6 g, Coicis Semen 20 g, Pinelliae Rhizoma Praeparatum 6 g, Batryticatus Bombyx 6 g, Curcumae Longae Rhizoma 9 g, Cicadae Periostracum 6 g, Atractylodis Rhizoma 6 g, Scutellariae Radix 10 g, Artemisiae Annuae Herba 10 g, Bupleuri Radix 15 g, Moslae Herba 12 g, Forsythiae Fructus 15 g

San Ren Tang + Sheng Jiang San + Xin Jia Xiang Ru Yin
Armeniacae Semen Amarum 12 g, Talcum 15 g, Tetrapanacis Medulla 6 g, Amomi Fructus Rotundus 5 g, Lophatheri Herba 10 g, Magnoliae Officinalis Cortex 6 g, Coicis Semen 20 g, Pinelliae Rhizoma Praeparatum 6 g, Batryticatus Bombyx 6 g, Curcumae Longae Rhizoma 9 g, Cicadae Periostracum 6 g, Atractylodis Rhizoma 6 g, Scutellariae Radix 10 g, Artemisiae Annuae Herba 10 g, Bupleuri Radix 15 g, Moslae Herba 12 g, Forsythiae Fructus 15 g

Yupingfeng San
Astragalus 20g +Fangfeng 15g +Atractylodes 15g

Fig. .10 (continued)

- Qingfei Paidu decoction (QFPD) (approved)
- Huashi Baidu decoction (HBD) (approved)
- Xuanfei Baidu decoction (XFBD) (approved)
- Jinhua Qinggan (JQ) (approved)
- Huoxiang Zhengqi (HZ) capsule
- Lianhua Qingwen (LQ)
- Xuebijing injection (XbI) (approved)
- Reduning injection (RdI)
- Shufeng Jiedu capsule (SFJC)
- Ma Xing Shi Gan Decoction (MXSGD) (approved)
- Ma Xing Shi Gan Tang**(MXSGT)
- Keli/Jiaonang (approved)
- Yin Qiao San (YQS)
- Xiang Su San (XSS)
- Bu Huan Jin Zheng Qi San (BJZS)
- Xuanbai Chengqi Tang Ganlu Xiaodu Dan decoction (XCTGXD)
- San Ren Tang (SRT)
- Liu Junzi Tang + Yu Ping Feng San
- Yu Ping Feng San (YPFS)
- Toujie Quwen granules (TQG)

All the abovementioned TCM have been discussed below along with their formulation, medical unit and outcomes, active ingredients, and potential metabolic targets. Among them, some have been approved for clinical trials, whereas some are under study.

**Name of the herbal formula MXSGT not reported yet or properly registered. Named by authors suggested using the dictionary of TCM formulas.

Literature reveals possible role of these decoctions in targeting certain signaling cascade or some gene expression proving antiviral activity. This activity is further associated with plant secondary metabolites. Keeping in view the limit of these studies, it needs comprehensive series of well-designed in vitro and in vivo experimental work. It may range from omics level lab work to clinical trial of pharmaceutical to pharmacological level.

Qingfei Paidu Decoction (QFPD) (Approved)

QFPD was practiced at clinical level in 66 institutes of medical science. One thousand, two hundred, and sixty three patients were under trial for effectiveness assessment of this formula, among which 1214 cases became healthy and were discharged (96.1%). Among the severely affected, 57 patients were supplemented with this TCM along with western medicine, where 73.7% was curing rate (with 42 patients). Comparison of their chest radiograph expressed that the consumption of QFPD for two 6-day course (2) was with reduction of lung lesions with 93.0% cure rate (53 patients) (Figs. 3.11, 3.12, 3.13, 3.14, 3.15, 3.16, 3.17, 3.18, 3.19, 3.20, 3.21, 3.22, 3.23, and 3.24).

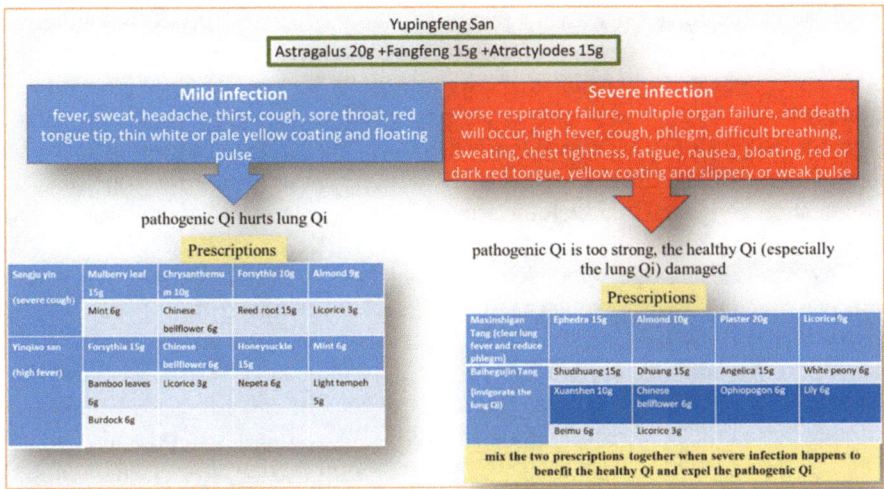

Fig. 3.11 Potential and prescriptions of YPFS decoction against COVID-19 as reported by Xu et al. [53]

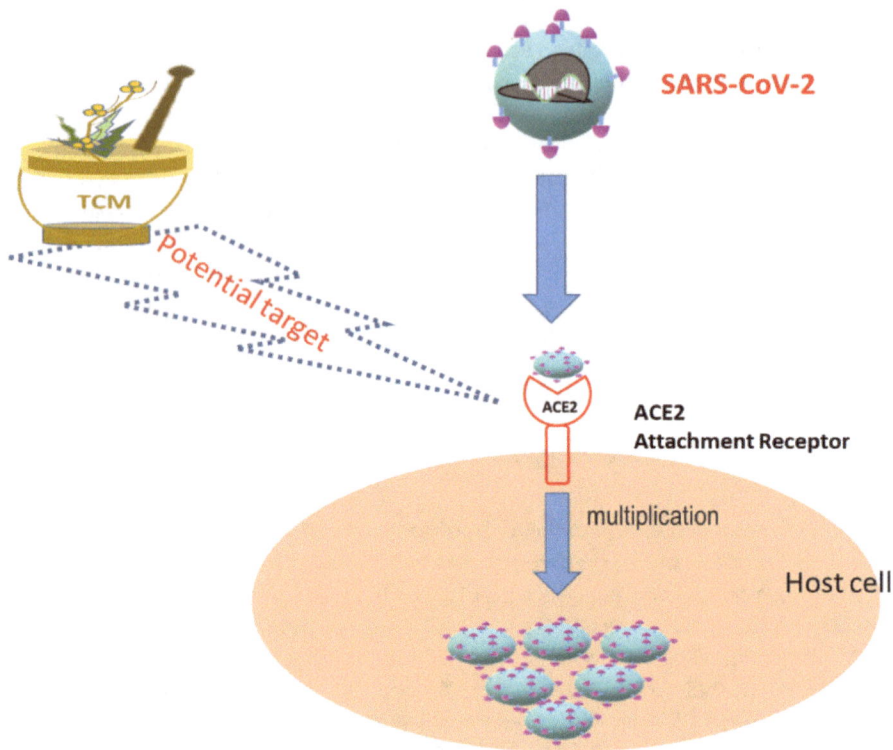

Fig. 3.12 Conceptual diagram presenting possible mechanism of TCM against COVID-19 directly targeting ACE2 (receptor of SARS-CoV-2 on human cell)

QFPD in combination with western medicine has been practiced in China too, where 63 patients with positive results for COVID-19 virus were under study. Different symptoms of patients showed marked improvement of recovery in response to medication, including circulation of white blood cells, counting of lymphocytes, and glutamic-oxaloacetic transaminase levels. All these parameters were compared in patients treated with QFPD alone, western medicine alone, and the combination of both. Combination proved its anti-inflammatory potential against COVID-19 [23].

This decoction is rich with secondary metabolites like flavonoids, glycosides, carboxylic acids, and saponins (45, 15, 10, and 5%, respectively). It has become most adopted decoction individually or in combination with some other drugs. It has been followed by Yang et al. [24], Liu et al. [25], Xin et al. [23], Chen [26], Luo [27], and Gao [28]. It is now an approved decoction according to the National Administration of TCM (accessed on May 28, 2020). It has been screened out and highly recommended to treat SARS-CoV-2 infection (reviewed by Tong et al. [29] and Huang et al. [7]).

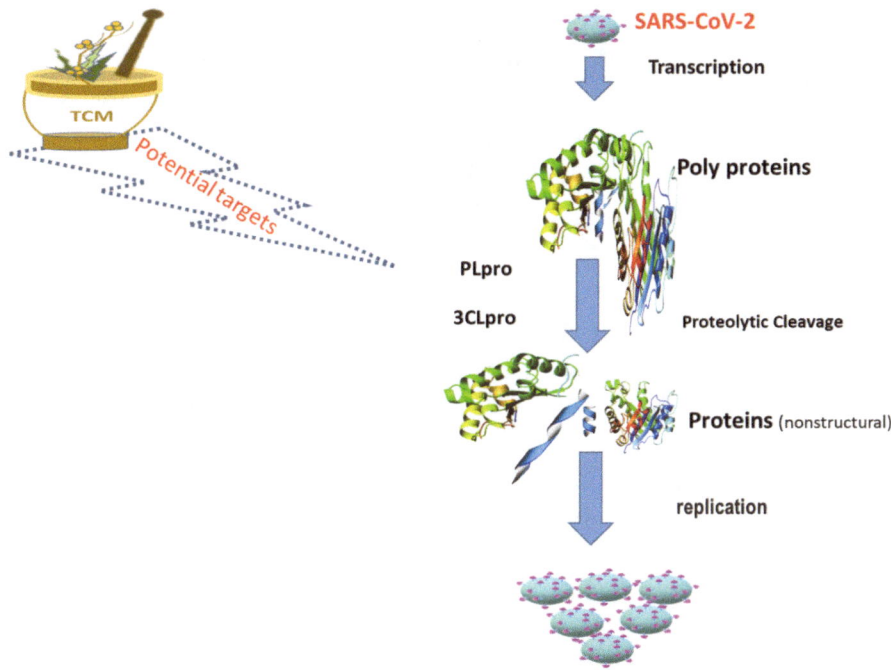

Fig. 3.13 Conceptual diagram presenting possible mechanism of TCM against COVID-19 via direct targeting proteases (enzyme that controls replication of COVID-19 in human cell)

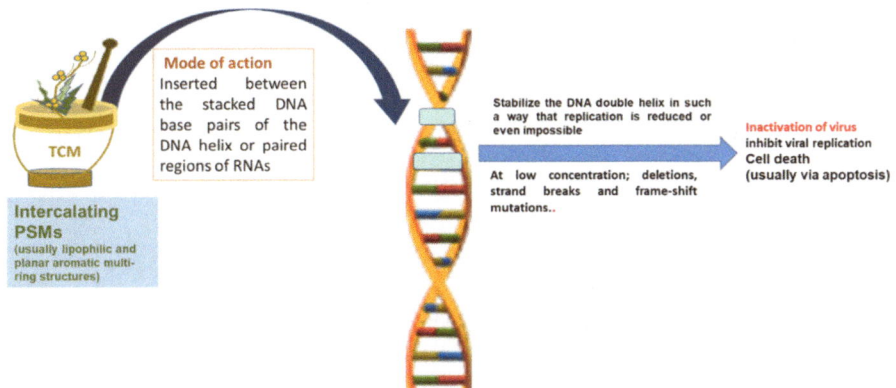

Fig. 3.14 Conceptual diagram presenting possible mechanism of TCM directly targeting DNA via intercalation

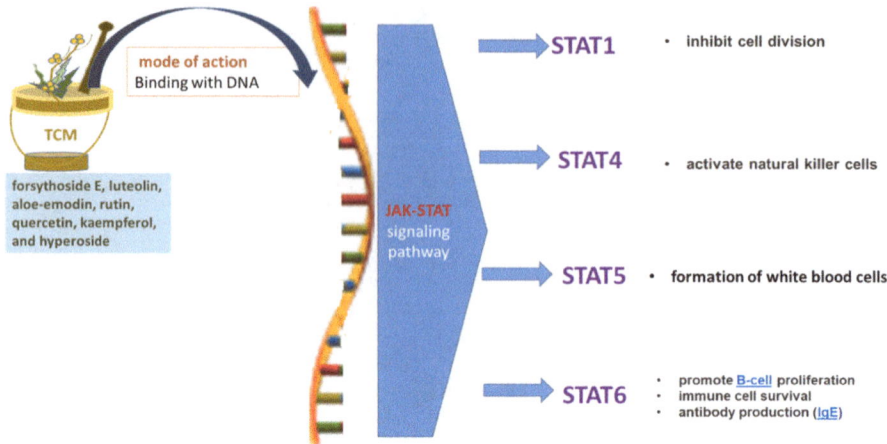

Fig. 3.15 Conceptual diagram presenting possible mechanism of TCM against COVID-19 via affecting JAK-STAT signaling pathway

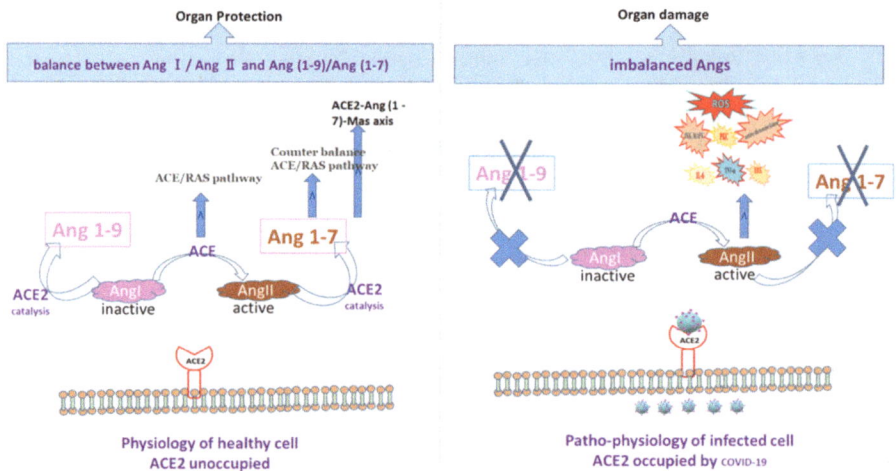

Fig. 3.16 Conceptual comparison of healthy and COVID-19-infected cells for ACE2-regulated renin-angiotensin-aldosterone system (RAAS)

Huashi Baidu Decoction (HBD) (Approved)

HBD has been practiced in a number of Chinese institutes including Wuhan Jinyintan Hospital for 75 severe patient of COVID-19, where there was improvement of symptoms particularly pulmonary inflammation along with negative conversion of nucleic acid decline. One hundred and twenty-four patient of COVID-19

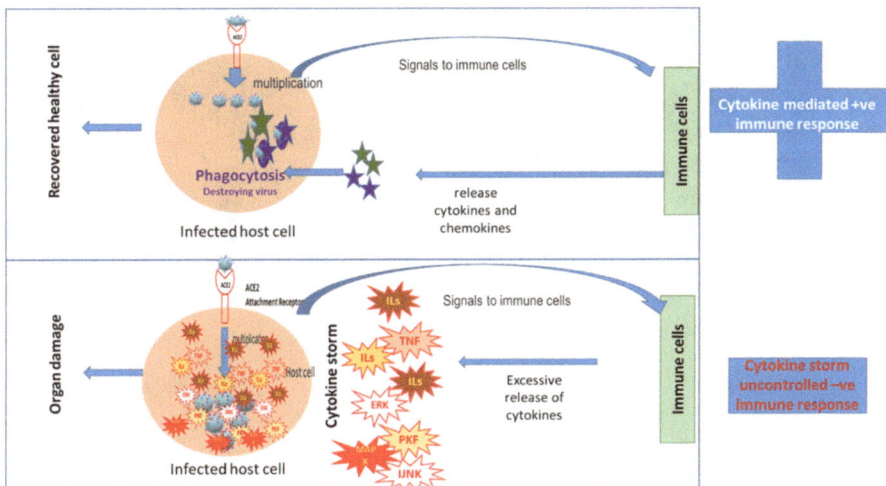

Fig. 3.17 Comparison of positive (above) and negative (below) cytokine-mediated immune responses in COVID-19-infected human cells

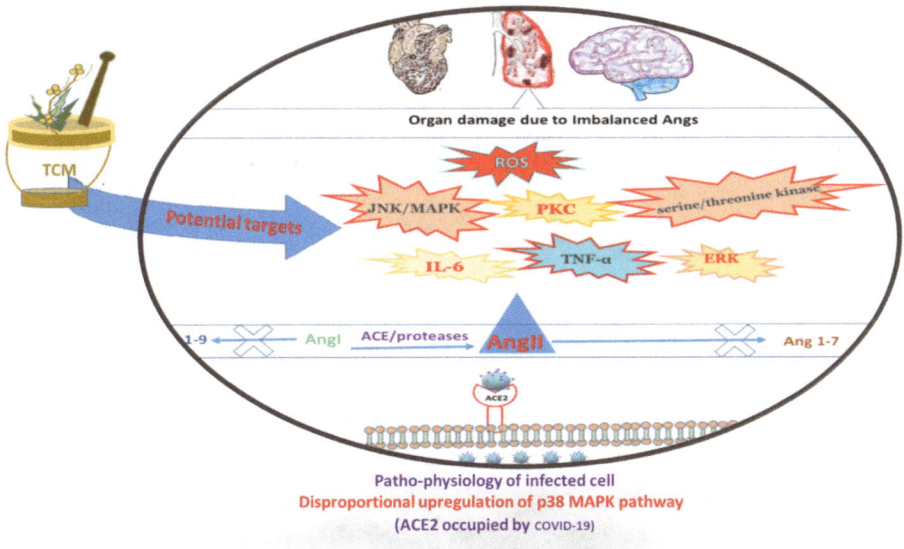

Fig. 3.18 Pathophysiology of COVID-19-infected cells, ACE2 occupied by virus, disproportional RAS pathway, conversion of Ang II into pro-inflammatory cytokine storm (IL, TNF, ERK, PKF, MAPK/JNK, etc.), disproportional p38MAPK pathway, consequent into organ damage

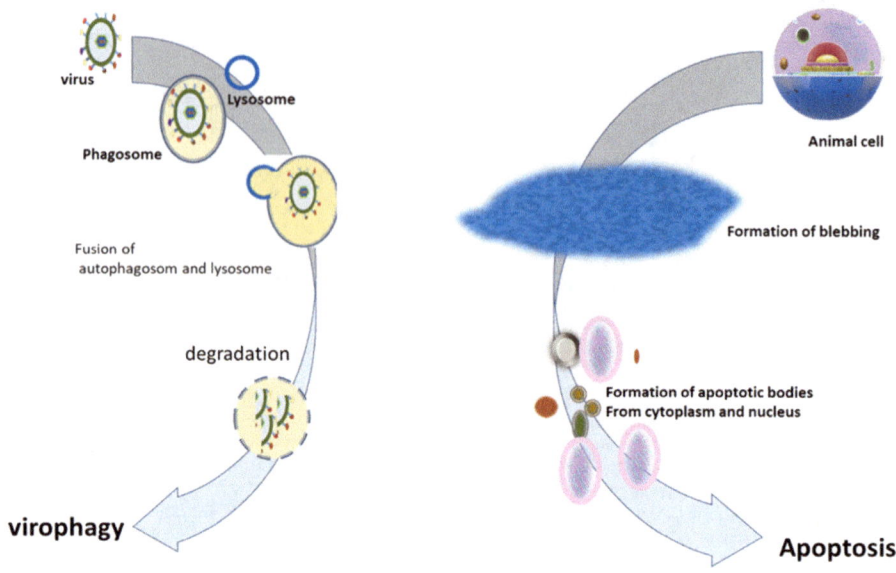

Fig. 3.19 Virophagy and apoptosis, a conceptual comparison

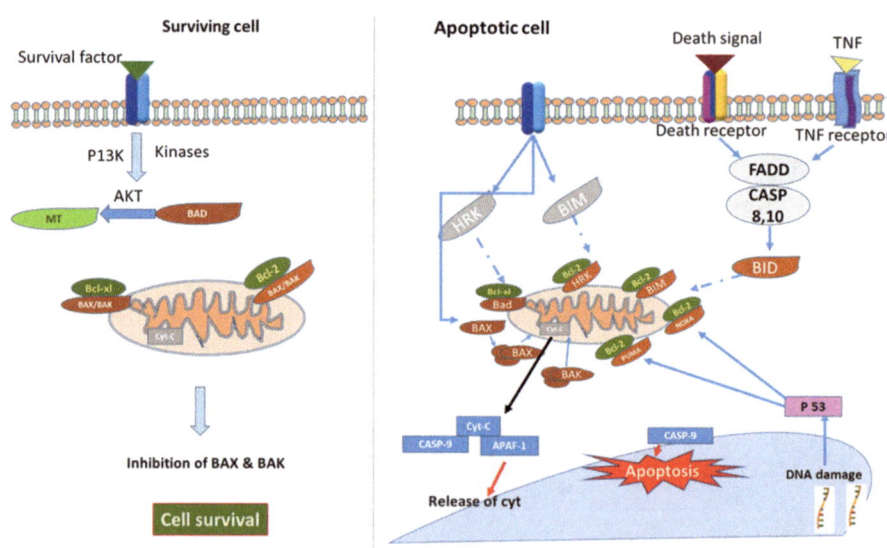

Fig. 3.20 Hypothetical comparison of an apoptotic (right) and surviving (left) cell under COVID-19 infection

were treated in Street Health Center at Jiangjun Road having moderate-level infection. Here, also negative conversion of nucleic acid decreased. Similar results were obtained in another institute, that is, Dongxihu Makeshift Cabin Hospital, where 894 COVID-19 patients with mild and moderate symptoms were used to treat (Figs. 3.12, 3.13, 3.14, 3.15, 3.16, 3.17, 3.18, 3.19, 3.20, 3.21, 3.22, 3.23, and 3.24).

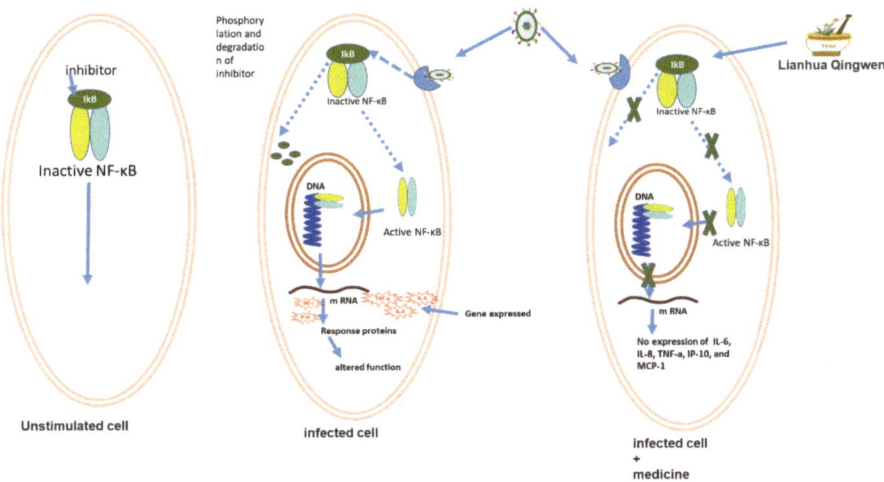

Fig. 3.21 Conceptual mechanism possibly induced by TCM in virus-infected cell. Pointing out therapeutic justification of TCM tested by in vivo and in vitro studies

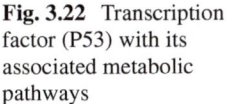

Fig. 3.22 Transcription factor (P53) with its associated metabolic pathways

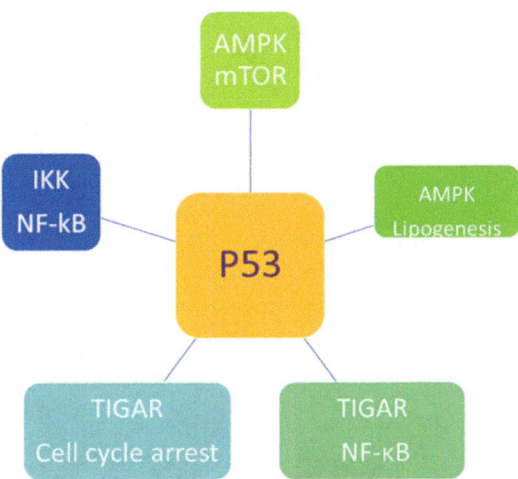

Clinical trials have proven its efficacy against COVID-19. It is considered to be multi-targeting that reduces timing of nucleic acid turning to negative. It is additional to clinical and chemical symptoms observed in abovementioned cases. Therefore, it got approval of its uses in clinics by the National Administration of Traditional Chinese Medicine (2020).

Pharmacological activities of HBD against COVID-19 (Figs. 3.12, 3.13, 3.14, 3.15, 3.16, 3.17, 3.18, 3.19, 3.20, 3.21, 3.22, 3.23, and 3.24) have been correlated with its active metabolites like licorice phenol, baicalein, etc. [30]. It has been screened out and highly recommended to treat SARS-CoV-2 infection [7, 29]

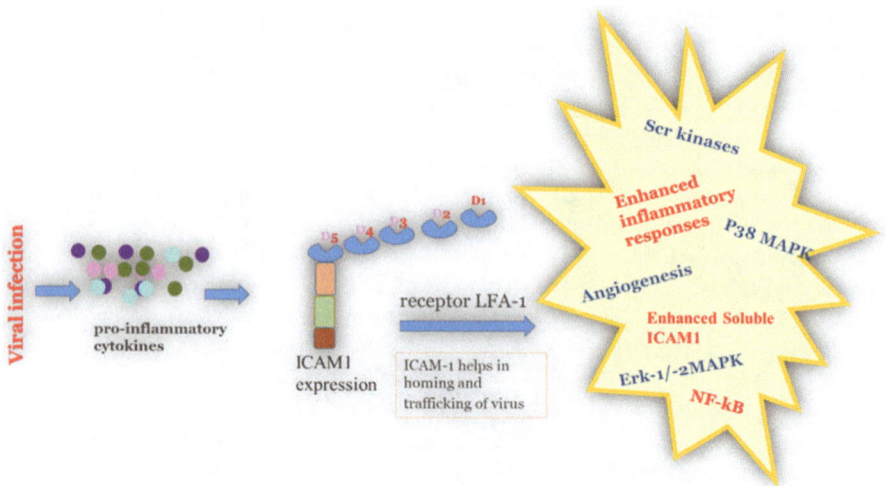

Fig. 3.23 Hypothetical model cytokine-mediated inflammation in terms of ICAM1 expression

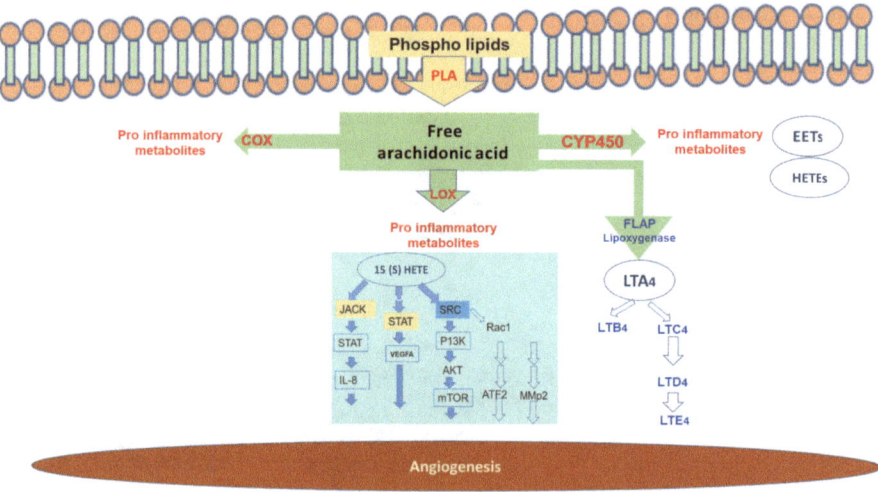

Fig. 3.24 Conceptual diagram of arachidonic acid metabolism based upon literature [21, 22]

Xiaochaihu Decoction (XCHD)

Sun et al. [31] have reported anti-SARS-CoV-2 potential of XCHD. They have correlated its efficacy with its secondary metabolites, particularly baicalein, formononetin, and quercetin. Reported targets include IL-6, NOS2, ESR1, and TNF signaling pathway.

Xuanfei Baidu Decoction (XFBD) (Approved)

XFBD was practiced at clinical level in different hospitals including Wuhan Hospital of TCM, Hubei Provincial Hospital, etc. It was applied on 70 patients of COVID-19. Here about 22% was the curing rate. Overall in case of mild and moderate cases, there was relief of clinical symptoms, whereas in severe cases it was with almost negligible improvement.

Its pharmacological targets (Figs. 3.12, 3.13, 3.14, 3.15, 3.16, 3.17, 3.18, 3.19, 3.20, 3.21, 3.22, 3.23, and 3.24) have been reported [32] in multiple articles; still direct mechanism of action has not been well explored. Yet direct relation with mechanism associated with COVID-19 needs to be unveiled. It has been screened out and highly recommended to treat SARS-CoV-2 infection (reviewed by Tong et al. [29] and Huang et al. [7]).

Jinhua Qinggan (JQ) (Approved)

JQ is one of the registered TCM formulae against other viruses including H1N1 influenza. Against COVID-19, it proved its antiviral activity in terms of detoxification and relieve of temperature in COVID-19 patients [33–37]. It lowered chances of COVID-19 entry in host cell by affecting its receptor ACE2 [38]. This activity was correlated with its PSM mentioned above as active metabolites. Huang et al. [7] confirmed its efficiency for AEC2 and 3CL protein targeting. Therefore, it could inhibit inflammation via CASP3 signaling transduction pathway (Figs. 3.12, 3.13, 3.14, 3.15, 3.16, 3.17, 3.18, 3.19, 3.20, 3.21, 3.22, 3.23, and 3.24). In addition, it may be linked with its PSM mentioned above.

Literature regarding bioactivity of its PSM against COVID-19 proved positive activation of signaling pathways related to the inactivation of COVID-19. Major targets reported for COVID-19 inhibition have been mentioned under direct and indirect mechanism of action (Fig). It has been now recommended to treat SARS-CoV-2 infection [7, 29].

Huoxiang Zhengqi (HZ) Capsule

About activity of HZ capsule, there are a number of research articles supporting the idea of its utilization for COVID-19 inhibition. Targets reported with its application (Figs. 3.12, 3.13, 3.14, 3.15, 3.16, 3.17, 3.18, 3.19, 3.20, 3.21, 3.22, 3.23, and 3.24) proved its potential to effect multiple pathways associated with COVID-19 infection [39–41]

Lianhua Qingwen (LQ)

A protein complex NF-κB found in inactive form in the normal cell due to inhibitory proteins if stimulated by COVID-19 become activated, removing inhibitory proteins. LQ capsule/granule could block activation of NF-kB in COVID-19 patients reducing expression of relevant genes IL-6, IL-8, TNF-a, IP-10, and MCP-1. It results in reduction of inflammatory cytokines too (Figs. 3.12, 3.13, 3.14, 3.15, 3.16, 3.17, 3.18, 3.19, 3.20, 3.21, 3.22, 3.23, and 3.24). In other studies, a number of other signaling pathways were positively affected with its application [39, 42–46]. It has been screened out and highly recommended to treat SARS-CoV-2 infection (reviewed by Tong et al. [29] and Huang et al. [7]).

Xuebijing Injection (XbI) (Approved)

Multiple reports regarding molecular basis of COVID-19 inhibition are available, highlighting pharmacological role of XbI. XbI-based organ protection has been correlated with different signaling pathways (Figs. 3.12, 3.13, 3.14, 3.15, 3.16, 3.17, 3.18, 3.19, 3.20, 3.21, 3.22, 3.23, and 3.24) [38, 39, 47–50]. It has been screened out and highly recommended to treat SARS-CoV-2 infection [7, 29] due to its potential to target PI3K-Akt signal and NF-κB signaling pathway in addition to a number of other metabolic targets (Fig).

Reduning Injection (RdI)

RdI proved itself to be a potent medicine against COVID-19. Its metabolic targets have been pointed out in figures (Figs. 3.12, 3.13, 3.14, 3.15, 3.16, 3.17, 3.18, 3.19, 3.20, 3.21, 3.22, 3.23, and 3.24). It may affect arachidonic acid metabolism (targeting COX-2), apoptosis (targeting CASP3), and cytokine storm (targeting IL4, MAPK1, IL1B, MAPK14, IL-17) as well as JAK-STAT signaling pathway (targeting EGFR) (Figs. 3.12, 3.13, 3.14, 3.15, 3.16, 3.17, 3.18, 3.19, 3.20, 3.21, 3.22, 3.23, and 3.24). It can also target FOS and CXCL10, yet majority of pathways responsible for COVID-19 inhibition are not very clear [38, 39, 51].

Shufeng Jiedu Capsule (SFJC)

SFJC has been published for its targeting potential to affect arachidonic acid metabolism (targeting COX-2), apoptosis (targeting CASP3, BCL2L), and NF-κB and cytokine storm (targeting IL6, IL1B, CCL2, IL2, IL4, IL1B, MAPK8,1,14). It also affects JAK-STAT signaling pathway (targeting EGFR) (Figs. 3.12, 3.13, 3.14, 3.15,

3.16, 3.17, 3.18, 3.19, 3.20, 3.21, 3.22, 3.23, and 3.24) [52–54]. Studies about SFJC and COVID-19-related signaling are still incomplete and need more confirmations.

Ma Xing Shi Gan Decoction (MXSGD) (Approved)

MXSGD positively affected pulmonary inflammation and reduced nucleic acid deactivation. Different metabolic targets made it effective for related pathways including arachidonic acid metabolism (targeting COX-2), apoptosis (targeting CASP3, Akt), NF-κB and cytokine storm (targeting interleukins, MAPK, TNF, p53, RAS pathway) (Fig.), and JAK-STAT signaling pathway (targeting EGFR).

In addition, Wang et al. [43] reported other least explored targets too. That include CDK, dipeptidyl peptidase 4 (DPP4), estrogen receptor 1 (ESR1), prothrombin (F2), PTGS2, TUBGCP, VCAM1, and VEGFA. Heat-shock protein 90 alpha of families A (HSP90AA1) and B (HSP90AB1) were targeted by this decoction too. Moreover, RAC-alpha serine/threonine-protein kinase (Akt1), androgen receptor (AR), and carbonic anhydrase II (CA2) may also be affected by its application (Figs. 3.12, 3.13, 3.14, 3.15, 3.16, 3.17, 3.18, 3.19, 3.20, 3.21, 3.22, 3.23, and 3.24).

Yin Qiao San (YQS)

YQS may reduce TLR7, MyD88, IRAK4, and NF-κB expressions. Hence, it could regulate TLR7/NF-κB signal pathway [55], boosting the immune system against viral pathogens.

Yupingfeng San (YPFS)

Science of TCM trusts on Qi as the main factor deciding balance of human body functioning. That may be healthy or pathogenic. The healthy Qi are responsible in maintenance of normal body functions, while pathogenic ones are involved in harms related to healthy body. YPFS (patent medicine) protect lung Qi with avoidance of pathogenic Qi. It is formulated by using three herbs: astragalus, Fangfeng, and *Atractylodes*. It supports the lung Qi and lower phlegm. Its use is case-specific decided with the severity of infection as described in Fig. 3.11.

Some of the decoctions are in practice but need official approvals. A number of clinical trials are under practice using the decoctions mentioned in Fig. 3.10 and expect some beneficial links with COVID-19 inhibitory mechanisms. These decoctions include Keli/Jiaonang, Ma Xing Shi Gan Tang, Toujie Quwen granules, etc. Among these, Keli/Jiaonang has got approval for application.

Therapeutic Logic of TCM as COVID-19 Inhibitor

Mechanism of Direct Inhibition

Some of decoctions like JQ granules, LQ capsules, XbI, QFPD, HBD, and Xuanfei Baidu have been approved after a series of therapeutic studies against different viruses. These are also recommended against COVID-19 [7, 29].

Plant secondary metabolites against a number of other diseases have been reported for certain mechanics that could help in viral inhibition. These including DNA intercalation, viral protein binding, etc. based upon previous pharmacological efficiency PSM have been studied in detail with reference to their efficiency against COVID-19. Available data has been organized to make it simple into two major categories, i.e., with direct or indirect mechanism of action. Furthermore, to make all mechanisms understandable, small description of the relevant mechanism as target of TCM has been discussed.

To get the upper hand on COVID-19, there is an urge for better understanding of its mechanism of action and the possible targets of TCM. Starting from COVID-19 infection to cell death, there is involvement of a number of signaling pathways. Below is the data regarding different TCMs and their possible signaling pathways involved in organ protection along with simplified description of the pathway to make it understandable for nontechnical reader.

Possible mechanism of TCM against COVID-19 may be divided into two major categories: direct and indirect actions. According to the literature [13, 53, 56], both direct and indirect targets are being studied as inhibitory mechanism against COVID-19.

TCM Targeting ACE2

COVID-19 enters in human cell with the facilitation of its receptor on human cell. Angiotensin-converting enzyme 2 (ACE2) is one of the COVID-19 acceptors. A number of TCM fight against COVID-19 directly via affecting its receptors on host cells [13, 53, 56]. Hence, ACE2 is considered as potential target (Fig.) restricting reception of virus within host cell.

Simayi [38] and Shen [52] reported the efficiency of kaempferol, quercetin, luteolin, baicalein, oroxylin A, licochalcone B, and glyasperin C to bind with ACE2. These are active ingredients of different decoctions like JQ granules. Targeting ACE2 by JQ has been confirmed by Huang et al. [7] and Ling [44] too for COVID-19 patients. It was all correlated with ACE2-based mechanism and active ingredients like 18β-glycyrrhetinic acid, indigo, β-sitosterol, and naringenin. QFPD could also target ACE2 and inactivate COVID-19 possibly [53, 57, 58].

Quinine is an alkaloid. It has become evident that quinine, chloroquine, and hydroxychloroquine have affinity for ACE2 receptor. They bind with amino acid residue Lys353 and therefore could block entry of COVID-19 in host cell via mechanism mentioned in Fig. 3.6. Currently the decoctions with these ingredients are in practice to fight SARS-CoV-2 viral infections as possible antidote [59].

TCM Targeting Protease

3C-like protease (3CL pro) is an enzyme that controls replication of COVID-19 and plays a major role in its life cycle. Polyproteins in general need proteases for proteolytic breakdown to produce smaller peptide chains during processing. For this proteolytic activity, COVID-19 has 3C-like protease (3CL pro) as main protease like other (+)ssRNA viruses. This protease is currently one of the good targets in drug discovery against COVID-19, as it has key importance in translation of virus.

QFPD works via affecting protease (3CL pro) as mentioned in Fig. 3.7 [51, 57, 58]. HZ also reported for inhibition of 3CL pro of COVID-19 [40, 41].

HBD binds with Mpro (main protease of COVID-19) and, hence, controls COVID-19 with inhibited replication and blocked binding sites [30]. Lianhua Qingwen capsule has been observed to use this mechanism of action against COVID-19 [42]. It has been ended with decline in virions after decoction application. Huang et al. [7] reported similar efficiency of JQ regarding COVID-19 entry mediated by 3CL pro.

Luteolin and chloroquine have affinity to bind with main protease of COVID-19. It may result in the rise of late endosomal and lysosomal pH. That consequent into impaired removal of virus from endosome/lysosome because virus needs low pH environment for its binding. So virus will become unable to release its genetic material in cell for replication [60].

Coronavirus 3CL hydrolase (Mpro) is a proteolytic enzyme in the presence of which COVID-19 matures. There are some evidences about correlation of kaempferol, quercetin, baicalein, luteolin, and rhubarb wogonin with this enzymatic activity-related inhibition of COVID-19 of Tanreqing injection formulation [49]

TCM Directly Targeting DNA Via Intercalation

Another way of drugs to inhibit viruses is DNA intercalation. During this process, PSM may insert itself among the base pairs of viral DNA. It stabilizes the double helix of DNA and restricts its replication. Some time it results in strand breakdown or mutation. Being a single-stranded DNA, how intercalating PSM could work it needs keen series of studies.

Quinine is an alkaloid having planar ring system that has the potential of intercalation with DNA. Against a number of viruses, it has proved its antiviral potential including malarial virus [61, 62]. It may intercalate COVID-19 DNA to inhibit its activity. Therefore, it could be part of TCM.

Emetine is another alkaloid with known potential of DNA intercalation formerly to inhibit HIV with reduced reverse transcriptase [63]. Therefore, it may be used against COVID-19 too for inhibition of its replication [64].

TCM Directly Targeting DNA via JAK-STAT Signaling Pathway

Wang [43–45] correlated the inhibition of COVID-19 with PSM of TCM particularly forsythoside E, luteolin, aloe emodin, rutin, quercetin, kaempferol, and hyperoside. The pathway suggested by them for this inactivation of virus was JAK-STAT signaling pathway. It may be involved in cytokine reception of immune cells. This pathway activates a number of STATs, which bind DNA and result in inflammation, natural killer cell activation, promotion of white blood cells, B-cell proliferation, and antibody (IgE) production [65]. Currently via modulating this pathway, induction of immunity in COVID-19 patients is considered beneficial. For that purpose, a number of TCM has been reported to affect this pathway targeting different genes, enzymes, etc.

TCM Targeting sEH

LQ capsule/granule inactivate COVID-19 [42–46] targeting sEH, hence interfering JAK-STAT pathway. In addition, XFBD [32] and RdI [39, 51] may also target JAK1 and STAT1

Targeting VEGF

LQ capsule/granule can also interfere JAK-STAT pathway targeting EGFR and, therefore, pointed out as potential source of medication against COVID-19 in some report [42–46]. In some other reports, XbI [47–49] reported similar anti-inflammatory role of XI in COVID-19 patients. LQ capsule/granule was also reported for such potential [39, 42–46]. In detailed studies of VEGF signaling pathways, Yang [24, 66] had also confirmed the efficiency of QFPD as potential candidate to treat COVID-19.

Targeting DPP4

TCM may be effective for JAK-STAT pathway targeting DPP4. According to current reports, SFJ capsule [52–54] and LQ [42–44, 46] could target DPP4. That was correlated with quercetin and luteolin later on [39]. Wang et al. [46] had explored the efficiency MXSG decoction against similar targets.

Targeting JUN

Literature preserves evidences for QFPD [39, 43, 56, 66, 67] and LQ capsule/granule [39, 42–46] to target JUN, hence being effective for JAK-STAT pathway in patients of COVID-19

Targeting EGFR

Some TCM target metabolism of patients suffering COVID-19 via targeting EGFR of JAK-STAT pathway. Such modulation has been observed by QFPD [39, 43, 56, 57, 66, 67], JQ granules [35], SFJ capsule [52–54], Tanreqing injection [49], and LQ [39, 42–46]

Targeting IL1B

ILIB is another reported target that could be part of JAK-STAT pathway modulation by drugs. Some TCM showed their potential to target it too. These include QFPD [39, 43], SFJ capsule [52–54], and LQ capsule/granule [39, 42–46].

Mechanism of Indirect Mechanisms

Anti-inflammation and Immune Regulation to Avoid Cytokine Storm

One of the deleterious effects of COVID-19 is related to a series of inflammation. It is further linked with over-activation of the immune system called cytokine storm. The understanding of this mechanism may lead to better understanding of potential PSM related to downregulation of this mechanism.

Host cells after viral infections release signals for recruitment of immune system's activity. As a result immune cells started the secretion of a number of cytokines (regulatory proteins involved in human immunity responses) and chemokines (a type of cytokine). Immune cells after activation become able to destroy virus cells. It is made possible by phagocytosis and release of inflammatory mediators. Excessive activation of this immune system may lead to the cytokine secretions above than normally required that is called cytokine storm. It is acute hyperinflammatory response. Cytokine storm due to excessive release in COVID-19 patients' lungs induce series of pharmacological issues including alteration in vascular permeability, blockage of airways, edema, hypoxia, target organ damaging, etc. That can even lead to death [68, 69]. TCM regulate positively different steps of above-mentioned sequence of events. It includes variation of cytokines, macrophages, monocytes, neutrophils and active T cells [70].

TCM Protect Organ Damage in COVID-19 Patients

TCM made possible the protection of organ from COVID-19 in patients via a number of pathways.

Regulation of Renin-Angiotensin-Aldosterone System (RAAS) and Potential TCM Targets

ACE2 counterbalances the deleterious effect of the ACE/RAS pathway. Renin-angiotensin-aldosterone system (RAAS) maintains balance between Ang I/Ang II and Ang (1–9)/Ang (1–7) to protect organs.

The critical role of RAS has been shown in the pathogenesis of metabolic inflammatory diseases [71]. ACE can regulate the renin-angiotensin-aldosterone system (RAAS). It may cause deterioration of metabolic system. ACE2 in addition to the reception of COVID-19 have the potential of counterbalancing these deleterious effects of the ACE/RAS pathway. ACE2 catalyze the angiotensin (polypeptide hormone) splitting, where it splits angiotensin (Ang) I to produce Ang 1–9. Whereas, from Ang II the Ang 1–7 are produced. That in turn regulates ACE2-angiotensin (1–7)-Mas axis for protection against lung injuries [72]. In COVID-19 patients, the virus attaches itself with ACE2; therefore, the balance between Ang I/Ang II and Ang (1–9)/Ang (1–7) becomes disturbed [73] that results in uncontrolled ROS production leading to disturbed permeability of cell membranes with ultimate organ damaging [74]. A number of organs including alveolar epithelial cells, lung tissue, heart, kidneys, and liver may experience damages. Additionaly, imbalance of T-helper-1 (Th1) and Th2 cells is triggered leading to cytokine storm and MODS ultimately [75].

QFPD support lungs and a number of other organs [39, 43, 56–58, 66, 67, 76, 77]. Mao et al. [35] worked on JQ granules and noted its potential to regulate RAS pathway. HBD has been also reported for similar potential in other studies [7, 30]. HBD in other studies was confirmed for this capability [7, 30].

Cytokine-Mediated Positive and Negative Immunity

RAS pathway was affected by QFPD supporting against organ damage [57, 58, 77]. Mao et al. [35] observed similar therapeutic response of JQ granules in COVID-19 patients.

TCM Targeting MAPK (Cytokine)

MAPK is a kinase enzyme that is activated by p38 mitogen. It has crucial part in decision of death or survival of infected cells. In short, P38-MAPK is a phenomenon with promotion of life cycle of COVID-19. Therefore, therapeutic potential of any drug or TCM can be judged based upon its ability to inhibit p38 MAPK. Recent literature about in practice TCMs revealed the positive potential of a number of formulae. Amongst these TCMs, QFPD fights COVID-19 targeting MAPK1, MAPK3, MAPK8, and MAPK14 [39, 43, 56, 57, 66, 67]. SFJ capsule could target 1, 8, and 14 MAPK in patients of COVID-19 [76]. In some other reports, MAPK 1, 3, and 8 were suggested as targets of HBD [30]; 1, 14, and 8 of JQ granule [7]; and 1 and 8 of LQ [39, 42–46].

Evidences about MAPK targeting by XB [47–49, 51] RdI [39, 51], LQ [45], and Tanreqing injection [49] are also available in literature.

Shen [52] and Cao [54] suggested positive correlation of MAPK activation and bioactive PSM used during formulation of decoctions. These known metabolites include licochalcon, β-sitosterol, acacetin, quercetin,, kaempferol, isorhamnetin, wogonin, 5,7,4′-trihydroxy-8-methoxyflavone, and luteolin.

TCM Targeting Platinum Drug Resistance (Cytokine)

JQ granules (Mao [35], Xu [53]) and SFJ capsule (Xu [53], Shen [52], and Cao [54]) could target cytokine platinum drug resistance.

TCM Targeting Tyrosine Kinase Resistance (Cytokine)

JQ granules (Mao [35], Xu [53]) and SFJ capsule [52–54] could target cytokine tyrosine kinase resistance.

TCM Targeting Interleukins (ILs) (Cytokine)

Interleukin are natural proteins that are involved in immune responses of human. It is one of the angiotensin (Ang II)-induced stressful products that could lead to the organ damage if not controlled. Therefore, it is one of the potential targets of drugs and TCM. IL-6 is a known pro-inflammatory cytokine. QFPD target IL-6 and protect organ damage in COVID-19 patients [43, 56, 57, 66].

There is a variety of ILs targeted by different decoctions successfully. IL-6 and IL-2 were targeted by Shenmai injection [78], IL-17 and IL-6 by HBD [30], and IL-17 and IL-10 by XFBD [32]. Similarly, SFJ capsule could attack on IL6 and IL1B [39, 52–54], and Tanreqing injection could target TNF and IL-6 [49].

In patients of COVID-19, chemotactic and inflammatory response in terms of CCL20, CXCL1, IL-1B, IL-6, CXCLs, and TNF expression can be observed [79]. LQ capsules reduce pro-inflammatory cytokines including TNF-α, IL-6, CCL-2/MCP-1, and CXCL-10/IP-10 [42]. Efficiency of LQ has been related to its quercetin, luteolin, and kaempferol [44, 57].

TCM Targeting TNF (Cytokines)

TNF is a factor that is expressed after activation of p38-MAPK. QFPD protects lungs [67] and a number of other organs due to its potential of targeting TNF [39, 43, 56–58, 66, 67, 77]. Ren [39] explored TNF-mediated inhibition of COVID-19 with HZ capsules, JQ granules, and LQ capsules, XbI, RdI, and Tanreqing injection. It has been correlated with its PSM including quercetin, luteolin, and kaempferol.

Reports are available for similar anti-inflammation by XbI [42, 47–49], Tanreqing injection [49], Shenmai injection [77], XFBD [32], and LQ capsule/granule [39, 42–46, 51].

TCM Targeting Lymphocyte-Mediated Adaptive Immune System

Human system is naturally blessed with adaptive immune system to combat virus-like problems. B and T cells are known component of this adaptive system. The greater the number of these cells, the better will be the immunity of patient against COVID-19. T and B cells are actually T and B lymphocytes (immune cells). The efficiency of these cells is correlated with their migration rate. Their penetration is further facilitated by leukocyte function associated with antigen 1 (LFA-1/β2 integrin/ITGB2). A number of TCM have been reported with the production of these two adaptive immune cells in COVID-19 patients. QFPD could enhance B-cell [40–42, 52] and T-cell receptor [39, 43, 57, 66, 67]. XFBD [32] could target ITGB2 in COVID-19 patients.

Regulation of Apoptosis and Potential TCM Targets

Cross talk between apoptosis and autophagy in an infected cell decides the pattern of disease perception, whether to perceive or to resist the viral infection. Xia et al. [79] has reported the role of TCM on this cross talk. It is being applied on patients with complaints of cough, nasal congestion, fatigue, etc. It is supposed to work against COVID-19.

At the time of exaggeration of immune inflammation (harmful for host cell), autophagy is needed to block virus-induced immune responses. Fu et al. [11] have noted similarity in pathophysiology and immune system of SARS-CoV-2 with other SARS-CoV. Therefore, the medicines that can affect upon this cross talk may support the COVID-19 patients with better immunity. Autophagy has a central role in immunity of a host against viruses. Currently, a number of researchers had started to manipulate autophagy with the aim to control COVID-19 [80]. It is a catabolic reaction during which unwanted components are eliminated by enclosing them in double membrane-walled organelles known as autophagosomes. Fusion of these autophagosomes with lysosomes resulted in degeneration of the content. Degraded products may be released in cytoplasm. Similar mechanism when involved with the elimination of some infectious element like a virus is called xenophagy or being more precise to virus may be called as virophagy. It has been successfully employed to fight against a number of viruses. To date we are not sure whether it is going to work against COVID-19 or not [38–40, 52].

Apoptosis is a type of programmed cell death where in response to some problem host cell may become bleb, nucleus become condensed, and all cytoplasmic and nuclear component become parted into apoptotic bodies with independent boundaries. It results into abnormal functioning of cell. Any possible application that can reduce apoptosis and enhance autophagy may be a promising tool against the infection like viruses. SFJ capsule/granule have been proposed to have this potential so far being used to fight COVID-19.

SFJ capsule/granule is suggested to lower viral inflammation with reduction of apoptosis and elevating autophagy [39, 43, 56, 57, 66, 67]. Here are some possible targets associated with apoptosis/virophagy.

TCM Targeting BCL Proteins

Bcl-2 and other member proteins of this family play a critical regulatory role in programmed cell death with affecting cell mitochondrial permeability. Those include anti-apoptotic and promotives of apoptosis: Bcl-2, Bcl-Xl, Bax, Bak, etc. Bcl-2, Bcl-Xl, BCL-w, BFL-1, and BCL-B are anti-apoptotic. BID, BAD, BIK, BIM, Bax, Bak, etc. are involved in the promotion of apoptosis. BCL2 genes encode B-cell lymphoma 2 (Bcl-2) that participates in apoptotic regulation in human cells, as apoptosis plays part in immune system regulation when functional drugs or TCM that could target BCL2 show their positive therapeutic activity against any viral disease. Against COVID-19, JQ granules [38, 52] and LQ capsule/granule [39, 42–46] target BCL2 and induce viral inactivation.

TCM Targeting CASP

QFPD [39, 43, 56, 57, 66, 67], SFJ capsule [52–54], XFBD [32], XbI [38, 39, 47, 51, 56, 57], RdI [39, 51], and JQ granules [35, 38, 52, 49] can combat COVID-19 via targeting CASP3. Wang [43], Ling [44], and Wang [46] have correlated quercetin, luteolin, and kaempferol of LQ with CASP3 targeted antiviral activity of LQ capsule/granule (Li et al. [42]). HZ targets E2F1 signaling pathway [40, 41]. It has role in some function involved in cell proliferation and p53-dependent/independent apoptosis [81].

7TCM Targeting Akt or Protein Kinase B (PKB)

This signaling pathway is associated with apoptosis inhibition. It becomes activated due to a kinase enzyme known as P13K (in turn induced by some factor). Therefore, Akt can participate directly/indirectly in promotion of cell survival. A pro-apoptotic protein (BAD) becomes phosphorylated dissociating it from Bcl-2/Bcl-X complex. That in turn results in loosening of the pro-apoptotic activities. Akt may also affect NF-κB activating with regulation of IκB kinase (IKK) leading to cell survival.

To date reported decoctions with potential to modulate apoptosis in COVID-19 patients targeting Akt signaling pathway include QFPD [39, 43, 57, 66, 67], XbI, HZ [44, 45, 48, 50–57], XbI, and JQ granules (Mao) [35].

PIK3C2G is a gene that encodes for an enzyme that belongs to PI3-kinase family. HZ targets PIK3CG signaling pathway [40, 41]. Hence, LQ capsule/granule [38, 39, 42–46] and XBI [39, 47, 49] could inactivate COVID-19.

Regulation of NF-κB Pathway and Potential TCM Targets

A protein complex NF-κB has important role in response to animal cells against viruses and some other stressors. It regulates immune response against the infections. Therefore, via activation this complex can defend human body against COVID-19. It is present in inactive form in the cell bound with some inhibitory proteins. Any stimulatory source that dissociates the complex by removing inhibitory protein or degenerating it may result in activation of NF-κB. It is considered most responsive toward harmful stimuli. Therefore in unstimulated cell, inactive form of NF-κB is responsible for absence of any NF-κB-induced negative responses. In case of stimulated cell, virus becomes attached with its specific receptor, and a cascade of signal may start. Phosphorylation of inhibitor results into its degeneration of ikB leaving active NF-κB. It would enter into the nucleus and may become part of genetic machinery responsible for altered response of stimulated cells mediated by response proteins (synthesized by DNA having attachment of NF-κB). LQ capsule/granule possibly blocks this activation of NF-kB in virus-infected cells and, therefore, may reduce expression of those genes that are responsible for disease symptoms including IL-6, IL-8, TNF-a, IP-10, and MCP-1.

A number of TCM have been reported to target this pathway against COVID-19. These TCM include HBD [30], SFJ capsule [42, 53, 54], and QFPD [24, 39, 43, 47–50, 56, 57, 66, 67, YQS 55]. Some of these TCM directly affect this protein complex. Some TCM indirectly target pro-inflammatory cytokines that include IL, TNF, ERK, PKF, MAPK/JNK, etc.

TCM Targeting TP53

An enzyme TIGAR is involved in human cell DNA repair and the breakdown of its own organelles. Hence, it could participate in cell protection. P53 is a transcription factor that could attach itself with more than one binding sites at TIGAR gene of human. It is generally known as tumor suppressor but has potential to coordinate with other stressors too. These genes are also responsible for mediation of cell division, growth, and mechanism of survival along with programmed cell death [82], therefore activating its expression. This factor (p53) has been found recently involved in downregulation of COVID-19 replication. HBD [83] and XFBD [32] have been reported to target TP53 and inactivate COVID-19 successfully (MXSG) [32].

- p53 may increase expression of TIGAR and restrict the activity of NF-κB.
- IKK phosphorylates and inactivates IkB (the inhibitor of NF-kB).
- AMPK drives catabolic responses such as the inhibition of lipogenesis and mTOR.
- TIGAR-mediated cell cycle arrest in stressed cells.

TCM Targeting ICAM1

Innate and adaptive immune system in COVID-19 decides the level of severity in patients [84]. Intercellular adhesion molecule 1 (ICAM1) protein in human body is expressed on immune cell. It is involved in innate and adaptive responses against antigens as an antibody glycoprotein also known as immunoglobulin. It participates in human defense activities via recognition and attachment of antigens (like viruses). Some viruses may exploit it for their entry in the respiratory system. Hence it may facilitate viral destruction. Therefore, its gene (ICAM1gene) if manipulated for its regulation may be potential target for inactivation of COVID-19 via some drug or TCM.

Wang et al. [32] had observed the potential of XFBD to target ICAM1 gene in COVID-19 patients. Melo et al. [78] confirmed the role of this gene in chemotactic and inflammatory responses of COVID-19 patients. The level of ICAM-1 expression is also positively correlated with pro-inflammatory cytokines including TNFα, IL-1, and IFNγ. Therefore, TCM with ability to target these cytokine regulations may indirectly effect upon this immunoglobulin protein.

Lymphocyte function-associated antigen-1 (LFA-1) is responsible for regulation of ICAM-1 gene expression. It also has crosstalk with other signals like STAT-1 and STAT-3 (Fig), NF-κB (Fig), RelA (Fig), and Ras-MAPK pathway (Fig).

TCM Targeting RELA

Literature reveals that XI [39, 47–49, 51] and LQ capsule/granule inactivate COVID-19 [39, 42–46] targeting RELA and play a part in regulation of NF-κB pathway.

Regulation of Arachidonic Acid Metabolism and Potential TCM Targets

Arachidonic acid is an important component of cell membrane derived by the action of PLAs on membrane phospholipids. Three enzymes including cyclooxygenase (COX), lipoxygenase (LOX), and cytochrome P450 (CYP450) catabolize its degradation.

From host membrane, arachidonic acid is released under stress by phospholipase A_2 (PLA_2) and phospholipase C (PLC). It is converted into free arachidonic acids that produce pro-inflammatory bioactive intermediates. Three metabolic pathways started by three enzymes named with their key enzymes:

1. Cyclooxygenase (COX) pathway produces prostaglandins (PGs) and thromboxanes (TXs).
2. Lipoxygenase (LOX) pathway produces leukotrienes (LTs) and lipoxins (LXs).
3. Cytochrome P450 (CYP450) pathway generating epoxyeicosatrienoic acids (EETs) or hydroxyeicosatetraenoic acids (HETEs) collectively all these products are called as eicosanoids with bioactivity for mediating inflammation and other pathogenic activity in patients [22]

Arachidonic acid is found to be associated with COVID-19 resistance immunity in patients. It is produced endogenously in COVID-19 patients and could perform as antiviral compound. Host immune response can be improved with negative control of all three pathways (COX, KOX, and CYP450) that degrade it into inflammatory intermediaries. COVID-19 may also be controlled via targeting different enzymes associated with their pathways.

Literature reveals that inhibition of microsomal prostaglandin E synthase-1 (mPGES-1) may control COVID-19. It is also correlated with COX-2 inhibitors [85]. Ren [39] worked on HZ capsules, JQ granules, LQ capsules, QFPD, XI, RdI, and Tanreqing injection and noted their potential to target arachidonic acid metabolic pathway in COVID-19 patients. It was further supported by a number of authentic studies [39, 43–46] about arachidonic acid metabolism of COVID-19 patients. Huang confirmed activity of JQ for inhibition of COVID-19 via arachidonic acid metabolism modulation.

TCM Targeting cPLA2α of Arachidonic Acid Metabolism

Müller et al. [86] had proposed the view that cytosolic phospholipase A2-α ($cPLA2\alpha$) can be a potential target against COVID-19. This enzyme is involved in early steps of arachidonic acid metabolism. In patients of COVID-19, viral replication was found to be associated with $cPLA2\alpha$. LQ capsule/granule have been reported to target this enzyme. Therefore, it has been supposed to inactivate this virus [39, 42–48, 57]. It was later confirmed with application of other TCM too [49].

TCM Targeting Cyclooxygenase-2 of Arachidonic Acid Metabolism

Cyclooxygenase-2 is an enzyme that is involved in pain and series of inflammatory responses in patients (with breakdown of arachidonic acid). In 1987 first time Lipfert et al. [87] suggested it as a target for drug activity. They suggested that the selectivity of COX-2 could be reduced via some drug to lower pain and inflammation. From the day onward, it is being used as therapeutic target. Currently in search of therapeutic TCM against COVID-19, LQ [39, 42–46] and QFPD [39, 43, 56, 57, 66, 67] have proved its potential to target COX-2. RdI may also target it [39, 51].

TCM Targeting LOX of Arachidonic Acid Metabolism

Limited reports are available about the efficiency of TCM against these enzymes. Successful targeting of 12-LOX has been noted by XI treatment in COVID-19 patients [39, 47–49]. It pointed out possible control of arachidonic acid breakdown into such a group of metabolites that in turn mediate series of inflammation.

TCM Targeting Cytochrome P450

Systemic inflammatory responses and immunity play major role in establishment of certain acute/chronic diseases. It may also effect drug or TCM pharmacokinetics; the efficiency of such TCM vary with their interaction with metabolizing enzymes like cytochrome P450.

These are associated with disease-drug interactions of a number of diseases [88]. Therefore, it has been suggested as good target against COVID-19 [89]. Lee et al. [90] worked to find out correlation between plant secondary metabolites and cytochrome p450 while working against HIV. They found *Lonicera japonica*, *Panax notoginseng*, *Prunella vulgaris*, *Scutellaria baicalensis*, *Ricinus communis*, *Sophora flavescens*, and Trichosanthes kirilowii effective for this pathway. Hence, it is not out of logics to try the decoctions having these components for COVID-19 inactivation too.

TCM Targeting Leukotrienes (LTs)

Leukotrienes are lipid messengers produced from 5-lipoxygenase (5-LO) pathway. This pathway play major role in adaptive immune processes of human. Epoxide LTA4 is key intermediate of inflammation. LTA4 hydrolase hydrolyzes it into LTB4. With the action of LTC4 synthase, it may form LTA4 conjugate with LTC4, which may be converted to LTD4 and LTE4. XUB when applied to COVID-19 patients affected leukotrienes [39, 47–49]. Afterward HZ showed its potential to target these messengers [39–41].

TCM Targeting CALM

There is not well established data about this target and potential TCM. Yet SFJ capsule [52–54], LQ [43, 44, 46], and XI [47–49] have some evidences in this regard.

In some other studies [39], HZ capsules, JQ granules, LQ capsules, QFPD, XbI, RdI, and Tanreqing injection have been noted antiviral against COVID-19 through cytokine regulation following the arachidonic acid metabolic pathway. Core target proteins of COVID-19 were associated with responses including HSP90AA1 and ELA. Literature about HSP90AB1 targeting MXSG and about HSP90AA1 targeting HZ [39–41] is in support to the use of these decoctions as a remedy against COVID-19.

TCM Targeting HIF-1 Signaling Pathways

In another study, Sun [51] worked on RdI and noted lowering of inflammation against COVID-19. The target for that mechanism was C-type lectin receptor and HIF-1 signaling pathways. XI was [47–49] reported for anti-inflammation related to HIF-1.

TCM Targeting Endocrine Resistance

JQ granules [35, 53] and SFJ capsule [52–54] target endocrine resistance.

TCM Targeting FOS

SFJ capsule [52–54] and JQ granules [35, 53] target FOS.

TCM Targeting PTGS2

Possible mechanism of antiviral activity against COVID-19 in relation to PTGS2 is not very clear to date. yet limited data is available pointing out possible correlation of PTGS2 and anti viral activities of some TCMs including QFPD [39, 43, 56, 57, 66, 67], JQ granules [38, 52]. SFJ capsule [52–54], XbI [47–49] and HZ capsule [39–41].

Regulation of Polyamine Metabolism and Potential TCM Targets

Viruses are dependent upon host metabolic machinery for their replication and infections. This dependence may be used as a tool against them to indirect inhibition. Polyamine pathway is one of those pathways in human that can inhibit virus if properly regulated by some drug, vaccine, or TCM. Therefore, a number of researchers are working on inhibitors of this pathway for improved antiviral medication.

TCM Targeting NOS2 and AOC1

Infection of COVID-19 proceeds in patient's body under major contribution of NOS2 and AOC genes. They express for enzymes inducible nitric oxide synthase (iNOS) and diamine oxidase (DAO), respectively. Both are part of polyamine metabolism. Both are involved in viral replication. Via downregulation of these enzymes, viral replication can be controlled. Additionally, negative correlation of NOS2 with COVID-19 accepter has been reported earlier. Hence, with the inhibition or lowered expression of NOS2 and AOC1 host may become susceptibile against this virus. keeping in view such susceptibility and infection therapeutic targets of both genes (NOS2 and AOC1) are under study. Any remedy that could overexpress these genes could proove its antiviral potential. SFJ capsule [52–54], QFPD [39, 43, 47, 48, 56, 57, 67], XbI [47–49]. HZ [39–41], and RdI [39, 51] showed positive correlation with these pathways.

TCM Targeting CCL2

One of the chemotactic and inflammatory responses is possibly CCLs (CCL20, CXCL1, CXCL2) [78]. The se of SFJ capsule [51–54] and RdI [39, 51] has been used to target CCLs. Their mechanism is being supported by a number of PSM.

Meta-analysis

It is a formal design of epidemiology for systemic and quantitative assessment of preexisting findings of other researchers. It helps in derivation of some results from literature and clinical trials. Here we have narrated some meta-analysis studies about COVID-19.

Currently the western medicines and TCM are working hand in hand to combat global fight against COVID-19. Existing clinical evidences and lab studies have been pooled below in the form of available and proposed meta-analysis.

Meta-analysis-based reviews of the following author has been summarized below in the form of figures (Fig. 3.25):

Miao et al. [30] used 716 citations after identification from electronic database. Here the duplication was removed, so there remained only 697 articles. By reading titles and abstracts, 42 articles were downloaded. Due to certain issues of design, 77 were also excluded later on.

They concluded that the TCM in combination with conventional treatment proved themselves good choice for patients with COVID-19 pneumonia. Even then credibility needs further clinical verifications (Figs. 3.26, 3.27, and 3.28).

Current evidence shows that CHM, as an adjunct treatment with standard care, helps to improve treatment outcomes in COVID-19 cases [58].

Here are some meta-analysis proposed to study (Fig. 3.29).

Chi et al. [92] has proposed a protocol for meta-analysis comparing the application of western medicine with the combination of western medicine and TCM. It may provide the existing suggestion about TCM application in clinics all over the world. It will guide in treating COVID-19 patients with better care and efficiency of traditional Chinese medicine. The review is not in press yet (Fig. 3.30).

Zhang et al. [93] have proposed a protocol for meta-analysis with the objective to search reference base to be utilized in clinics to optimize some treatment using TCM for patients suffering with COVID-19 (Fig. 3.31).

Liang et al. [44] has proposed meta-analysis (registration number: CRD42020187422) with the aim of assessment of the effect and security of TCM with respect to the symptoms in COVID-19 patients in the upper respiratory tract: TCM therapy. Control interventions would be western medicine therapy. Seven databases were searched to collect studies about the clinical characteristics of COVID-19 from January (Fig. 3.32).

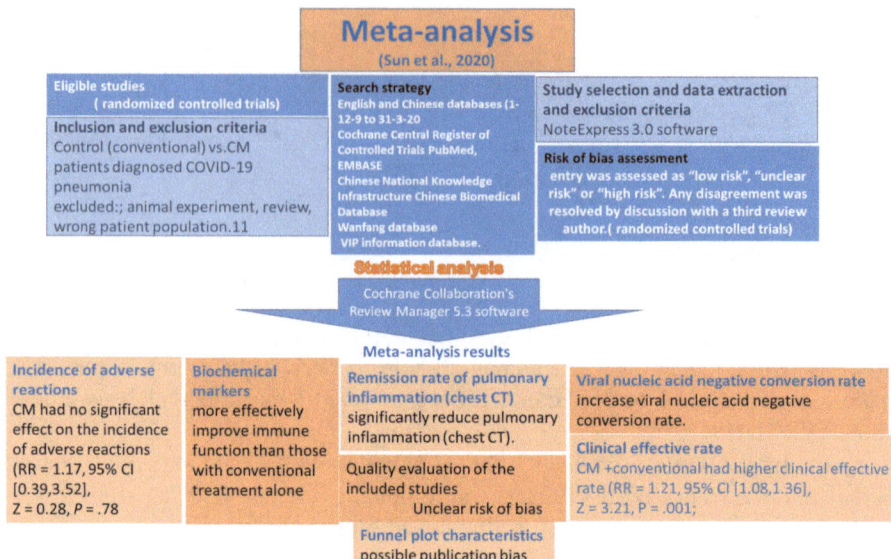

Fig. 3.25 Meta-analysis scheme followed by Sun et al. [91] to evaluate the role of Chinese medicine in COVID-19 pneumonia

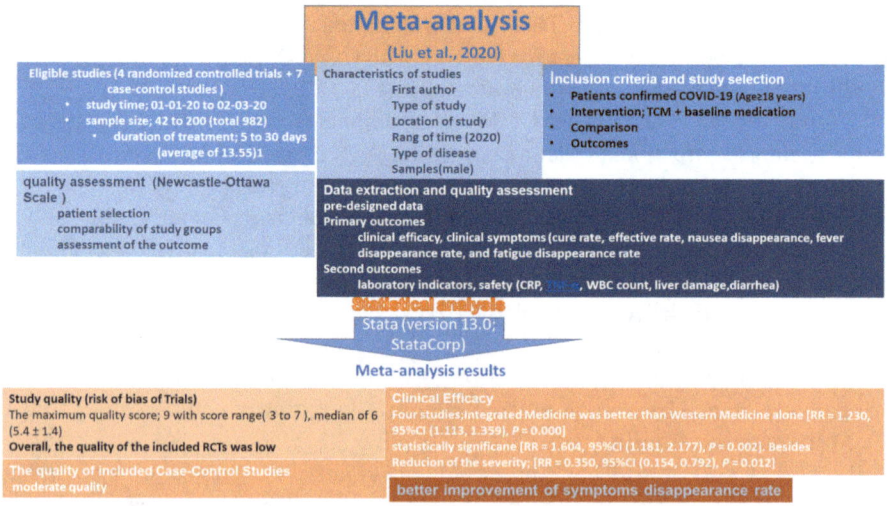

Fig. 3.26 Meta-analysis scheme followed by Liu et al. [25] to evaluate the role of TCM in COVID-19 pneumonia

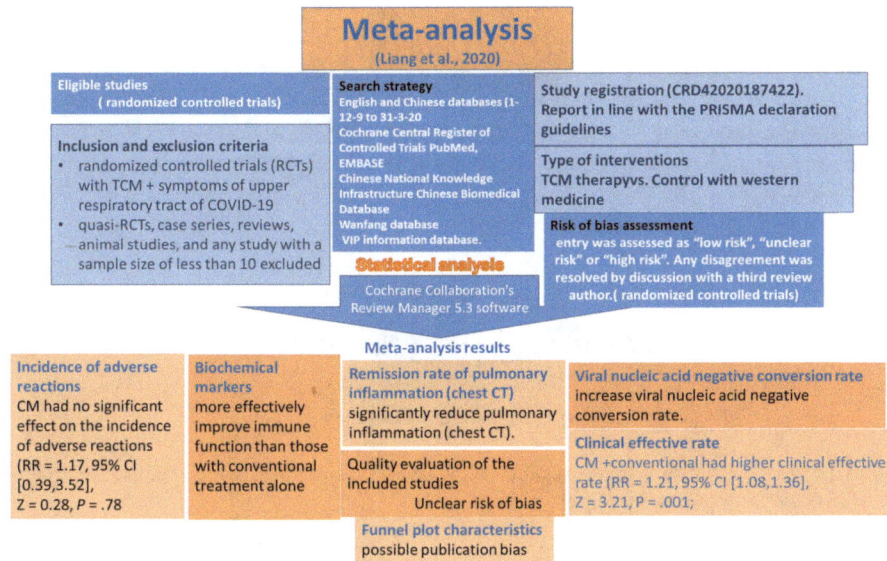

Fig. 3.27 Meta-analysis scheme followed by Liang et al. [44] to evaluate the role of TCM in COVID-19 pneumonia

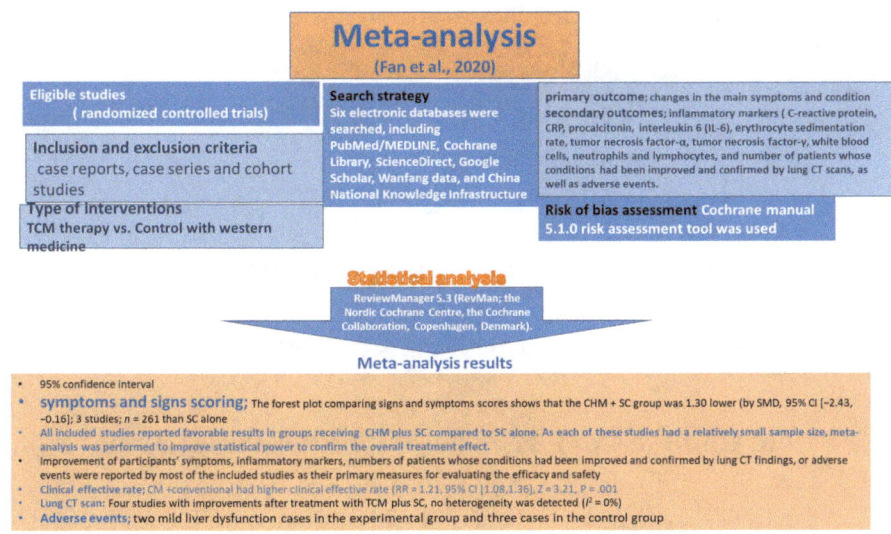

Fig. 3.28 Meta-analysis scheme followed by Fan et al. [58] to evaluate the role of TCM in COVID-19 pneumonia

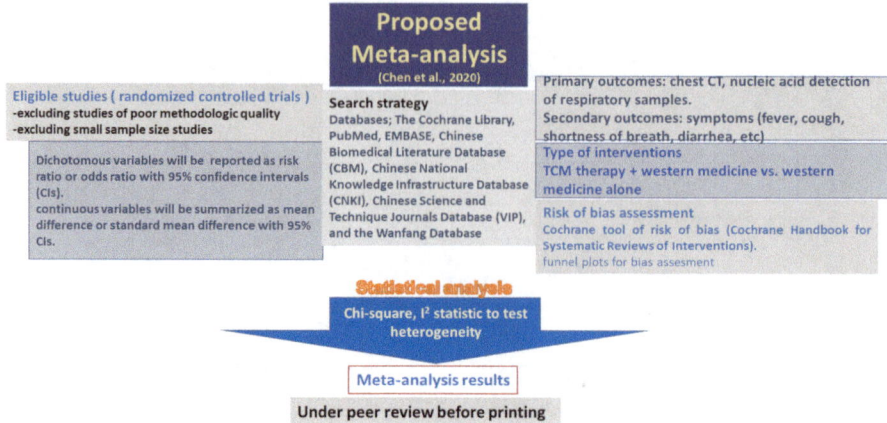

Fig. 3.29 Proposed meta-analysis scheme followed by Chen et al. [26] to evaluate the role of TCM in COVID-19 pneumonia

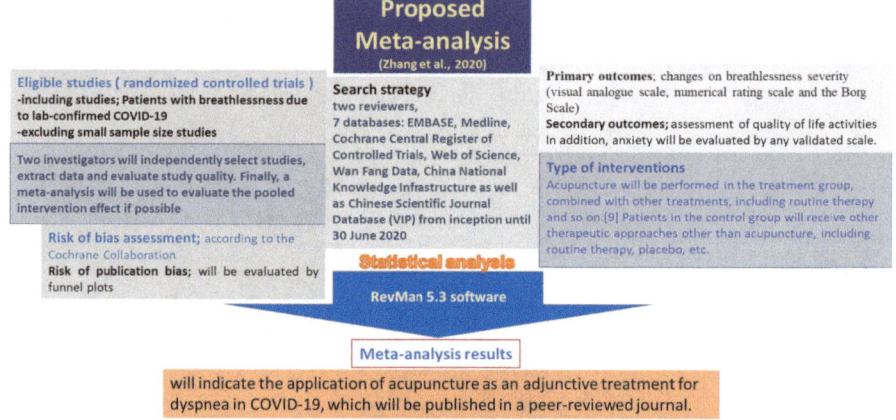

Fig. 3.30 Proposed meta-analysis scheme by Zhang et al. [93] to evaluate the role of TCM in COVID-19 pneumonia

TCM from Chinese Combat Zone to World Combat Zone

As the pandemic has crossed the border of China, the use of TCM has also got familiarity worldwide. With minor variations, TCM are in practice globally.

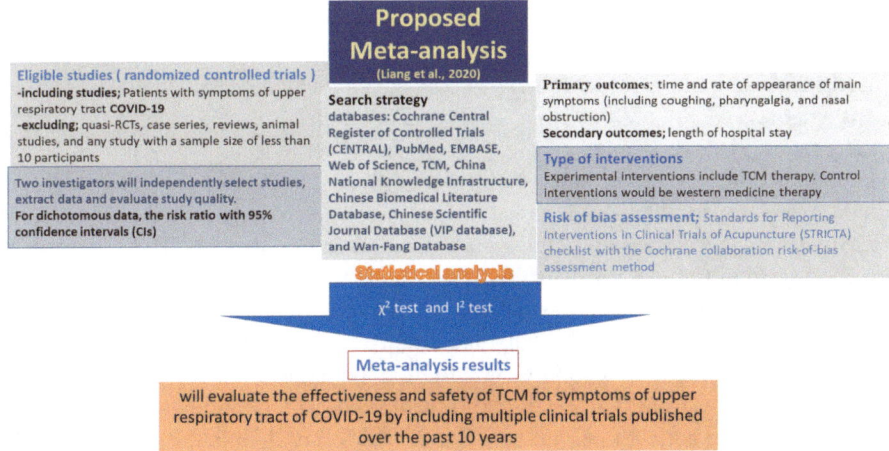

Fig. 3.31 Proposed meta-analysis scheme followed by Liang et al. [44] to evaluate the role of TCM in COVID-19 pneumonia

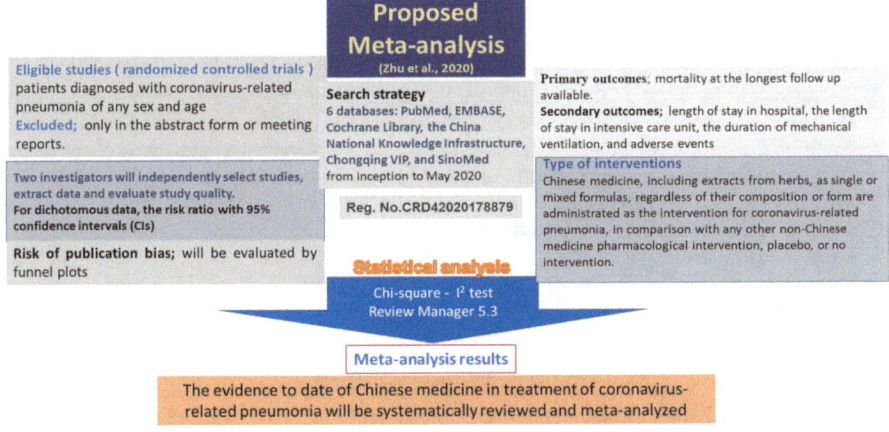

Fig. 3.32 Proposed meta-analysis scheme followed by Zhu et al. [94] to evaluate the role of TCM in COVID-19 pneumonia

Shortcomings of Antiviral Medicine on COVID-19

To date collected data about TCM against COVID-19 has range of shortcomings. It is related with some unveiled metabolic pathways. Detailed pharmacological studies are the need of time.

Conclusions

Pharmaceutical and pharmacological studies have revealed the mode of TCM treatment against SARS-CoV-2, with a variety of compounds involved as well as their mechanisms. An overlapping of the pattern was observed indicating that different compounds could target the same target with different pathways or different targets via similar pathway. That leads toward a series of controversy regarding the use of TCM with respect to certain pathway specifically. In the nutshell, TCM are one of the good sources to fight against COVID-19 due to their therapeutic contributions in human metabolism. Being accessible and economical, they could be easily followed all over the world including low-income countries. Still possible side effects and limit of certain metabolic level information may generate controversy about its use with limited knowledge. Till the time of invention of proper authentic vaccines and till the time of its access to common people, TCM may maintain its worth to fight COVID-19.

References

1. Organization, W. H. (2020) Coronavirus disease 2019 (COVID-19): situation report, 72
2. Commission, N. H. (2020) National Administration of Traditional Chinese Medicine. Diagnosis and treatment program for COVID-19 (trial version 7)
3. Traditional Chinese Medicine, National Center for Complementary and Integrative Health, Traditional Chinese Medicine: an introduction archived 26 June 2015 at the Wayback Machine
4. Mahmood S, Parveen A, Hussain I, Javed S, Iqbal M (2014) Possible involvement of secondary metabolites in the thermotolerance of maize seedlings. Int J Agric Biol 16(6):1075–1082
5. Wink M (2020) Potential of DNA intercalating alkaloids and other plant secondary metabolites against SARS-CoV-2 causing COVID-19. Diversity 12(5):175
6. Xiao X-H, Wang J-B, He C-H (2003) On the rational exertion for the prescriptions and drugs of TCM in preventing and treating SARS. Zhongguo Zhong yao za zhi/Zhongguo zhongyao zazhi/China J Chin Mater Med 28(7):664–668
7. Huang Y-F, Bai C, He F, Xie Y, Zhou H (2020) Review on the potential action mechanisms of Chinese medicines in treating Coronavirus Disease 2019 (COVID-19). Pharmacol Res 158:104939
8. Gelderblom HR (1996) Structure and classification of viruses. In: Medical microbiology, 4th edn. University of Texas Medical Branch at Galveston, Galveston
9. Yang Y, Islam MS, Wang J, Li Y, Chen X (2020) Traditional Chinese medicine in the treatment of patients infected with 2019-new coronavirus (SARS-CoV-2): a review and perspective. Int J Biol Sci 16(10):1708
10. Hoffmann M, Kleine-Weber H, Schroeder S, Krüger N, Herrler T, Erichsen S et al (2020) SARS-CoV-2 cell entry depends on ACE2 and TMPRSS2 and is blocked by a clinically proven protease inhibitor. Cell 181:271–280.e8
11. Fu Y, Cheng Y, Wu Y (2020) Understanding SARS-CoV-2-mediated inflammatory responses: from mechanisms to potential therapeutic tools. Virol Sin 35(3):266–271
12. Yang L, Cui H, Liu X, Wen W, Wang H (2020) Feasibility of Xiaochaihu decoction on fever induced by coronavirus disease 2019 (COVID-19) based on network pharmacology. Chin Trad Herbal Drugs 7:1761–1775
13. Chen C-N, Lin CP, Huang K-K, Chen W-C, Hsieh H-P, Liang P-H et al (2005) Inhibition of SARS-CoV 3C-like protease activity by theaflavin-3, 3′-digallate (TF3). Evid Based Complement Alternat Med 2:209–215

14. Wen C-C, Shyur L-F, Jan J-T, Liang P-H, Kuo C-J, Arulselvan P et al (2011) Traditional Chinese medicine herbal extracts of Cibotium barometz, Gentiana scabra, Dioscorea batatas, Cassia tora, and Taxillus chinensis inhibit SARS-CoV replication. J Tradit Complement Med 1(1):41–50
15. Ryu YB, Jeong HJ, Kim JH, Kim YM, Park J-Y, Kim D et al (2010) Biflavonoids from Torreya nucifera displaying SARS-CoV 3CLpro inhibition. Bioorg Med Chem 18(22):7940–7947
16. Cheng PW, Ng LT, Chiang LC, Lin CC (2006) Antiviral effects of saikosaponins on human coronavirus 229E in vitro. Clin Exp Pharmacol Physiol 33(7):612–616
17. Loizzo MR, Saab AM, Tundis R, Statti GA, Menichini F, Lampronti I et al (2008) Phytochemical analysis and in vitro antiviral activities of the essential oils of seven Lebanon species. Chem Biodivers 5(3):461–470
18. Chen C-J, Michaelis M, Hsu H-K, Tsai C-C, Yang KD, Wu Y-C et al (2008) Toona sinensis Roem tender leaf extract inhibits SARS coronavirus replication. J Ethnopharmacol 120(1):108–111
19. Kim H-Y, Shin H-S, Park H, Kim Y-C, Yun YG, Park S et al (2008) In vitro inhibition of coronavirus replications by the traditionally used medicinal herbal extracts, Cimicifuga rhizoma, Meliae cortex, Coptidis rhizoma, and Phellodendron cortex. J Clin Virol 41(2):122–128
20. Ang L, Lee HW, Choi JY, Zhang J, Lee MS. Herbal medicine and pattern identification for treating COVID-19: a rapid review of guidelines. Integrative Medicine Research. 2020 Jun 1;9(2):100407
21. Hedi H, Norbert G (2004) 5-Lipoxygenase pathway, dendritic cells, and adaptive immunity. Biomed Res Int 2004(2):99–105
22. Singh NK, Rao GN (2019) Emerging role of 12/15-Lipoxygenase (ALOX15) in human pathologies. Prog Lipid Res 73:28–45
23. Xin S, Cheng X, Zhu B, Liao X, Yang F, Song L et al (2020) Clinical retrospective study on the efficacy of Qingfei Paidu decoction combined with Western medicine for COVID-19 treatment. Biomed Pharmacother 129:110500
24. Yang Y, Islam MS, Wang J, Li Y et al (2020) Traditional Chinese medicine in the treatment of patients infected with 2019-new coronavirus (SARS-CoV-2): a review and perspective. Int J Biol Sci 16(10):1708
25. Liu N, Li S, Fan K et al (2020) The prevention and treatment of COVID-19 with Qingfei Paidu decoction in shanxi China. TMR Mod Herb Med 3:1–5
26. Chen H, Xie Z, Zhu Y et al (2020) Chinese medicine for COVID-19: a protocol for systematic review and meta-analysis. Medicine 99(25):e20660
27. Luo E, Zhang D, Luo H, Liu B, Zhao K, Zhao Y et al (2020) Treatment efficacy analysis of traditional Chinese medicine for novel coronavirus pneumonia (COVID-19): an empirical study from Wuhan, Hubei Province. China Chin Med 15:1–13
28. Gao K, Song Y-P, Chen H et al (2020) Therapeutic efficacy of Qingfei Paidu decoction combined with antiviral drugs in the treatment of corona virus disease 2019: a protocol for systematic review and meta-analysis. Medicine 99(22):e20489
29. Tong ZD, Tang A, Li KF, Li P, Wang HL, Yi JP et al (2020) Potential presymptomatic transmission of SARS-CoV-2, Zhejiang province, China, 2020. Emerg Infect Dis 26(5):1052
30. Sun KB, Zhang XY, Liu J et al (2020) Network pharmacological analysis and mechanism prediction of Xiaochaihu Decoction in treatment of COVID-19 with syndrome of pathogenic heat lingering in lung and obstructive cardinalate. Chin Tradit Herb Drug 51(7)
31. Miao S, Jialei T, Shaoju X (2020). The molecular mechanism of treating COVID-19 with Huashi Baidu Formula based on network pharmacology. J Chinese Med Mat:1–7
32. Wang Y, Li X, Zhang JH, Xue R, Qian JY, Zhang XH et al (2020) Mechanism of Xuanfei Baidu Tang in treatment of COVID-19 based on network pharmacology. Zhongguo Zhong yao za zhi/Zhongguo zhongyao zazhi/China J Chin Mater Med 45(10):2249–2256
33. Runfeng L, Yunlong H, Jicheng H, Weiqi P, Qinhai M, Yongxia S et al (2020) Lianhuaqingwen exerts anti-viral and anti-inflammatory activity against novel coronavirus (SARS-CoV-2). Pharmacol Res 156:104761

34. Duan C, Xia W, Zheng C (2020) Clinical observation on the treatment of novel coronavirus pneumonia with Jinhua qinggan granules. J Tradit Chin Med:1–5
35. Mao Y, Su Y, Xue P et al (2020) Discussion on the mechanism of Jinhua Qinggan Granule in the treatment of novel coronavirus pneumonia. J Chin Med Mater 11:2843–2849
36. Gong PY, Guo YJ, Li XP, et al (2020) Exploring active compounds of Jinhua Qinggan Granules for prevention of COVID-19 based on network pharmacology and molecular docking. Chin Tradit Herb Drug 51(7)
37. Jimilikhan SMY, Mohammad MN, Ainival W et al (2020) Study on the active components of Jinhua Qinggan Granule in adjuvant treatment of novel coronavirus (COVID-19) based on network pharmacology and molecular docking. J Chin Med Mater (05):1275–1283. (Chinese)
38. Simayi J, Noormaimaiti M, Wumaier A, Yusufu M, Noor M, Mahemuti N et al (2020) Study on the active components in the adjuvant treatment of novel coronavirus pneumonia (COVID-19) with Jinhua Qinggan granules based on network pharmacology and molecular docking. J Chin Med Mater. http://kns.cnki.net/kcms/detail/44.1286.R.20200323.1926.002.html. [Google Scholar]
39. Ren Y, Yao MC, Huo XQ, Gu Y, Zhu WX, Qiao YJ et al (2020) Study on treatment of "cytokine storm" by anti-2019-nCoV prescriptions based on arachidonic acid metabolic pathway. Zhongguo Zhong yao za zhi/Zhongguo zhongyao zazhi/China J Chin Mater Med 45(6):1225–1231
40. Du H, Wang P, Ma Q, Li N, Ding J, Sun T et al (2020) Preliminary study on the effective components and mechanisms of Huoxiang Zhengqi decoction inhibiting the replication of new coronavirus. Modernization of traditional Chinese medicine and materia materia-world science and technology
41. Deng YJ, Liu BW, He ZX, Liu T, Zheng RL, Di Yang A et al (2020) Study on active compounds from Huoxiang Zhengqi oral liquid for prevention of coronavirus disease 2019 (COVID-19) based on network pharmacology and molecular docking. Chin Tradit Herb Drug 51(5)
42. Li Q, Wang H, Li X, Zheng Y, Wei Y, Zhang P et al (2020) The role played by traditional Chinese medicine in preventing and treating COVID-19 in China. Front Med 14(5):681–688
43. Wang Y, Ma J, Wang S, Zeng Y, Zhou C, Ru Y et al (2020) Utilizing integrating network pharmacological approaches to investigate the potential mechanism of Ma Xing Shi Gan Decoction in treating COVID-19. Eur Rev Med Pharmacol Sci 24(6):3360–3384
44. Liang F, Dong L, Zhou L et al (2020) Traditional Chinese medicine for symptoms of upper respiratory tract of COVID-19: a protocol for systematic review and meta-analysis. Medicine 99(30):e21320
45. Ye CH, Gao MN, Lin WQ, Yu KQ, Li P, Chen GH (2020) Theoretical study of the anti-NCP molecular mechanism of traditional Chinese medicine Lianhua-Qingwen Formula (LQF). Polar 2(21.52):10–68
46. Wang D, Hu B, Hu C, Zhu F, Liu X, Zhang J et al (2020) Clinical characteristics of 138 hospitalized patients with 2019 novel coronavirus–infected pneumonia in Wuhan, China. JAMA 323(11):1061–1069
47. Shi Y, Wei J, Liu M, Jin X, Zhou H, Zhu W et al (2020) Study on the overall regulation of Xuebijing injection in treating corona virus disease 2019. Shanghai J Tradit Chin Med 54:1–7
48. He T, Duan C, Li X et al (2020) Potential mechanism of Xuebijing injection in treatment of coronavirus pneumonia based on network pharmacology and molecular docking. Chin J Mod Appl Pharm 37(4):398–405
49. Kong Y, Lin LL, Chen Y et al (2020) Modernization of traditional Chinese medicine and materia medica-world science and technology. Exploring the mechanism of Xuebijing injection in treating COVID19 based on network pharmacology. http://kns.cnki.net/kcms/detail/11.5699.r.20200411.2157.008.html. [Google Scholar]
50. Commission, G. O. o. t. N. H. (2020) Office of the State Administration of Traditional Chinese Medicine. Notice on printing and distributing the diagnosis and treatment program of new coronavirus pneumonia (trial seventh edition). Medical letter of the state health office(184)

51. Sun X, Zhang Y, Liu Y et al (2020) Study on mechanism of Reduning Injection in treating novel coronavirus pneumonia based on network pharmacology. J Chin Med Mater:1–9
52. Shen F, Fu Z, Wu Y, Kuang G, Li L, Zhu K et al (2020) The potential targets and mechanisms of Shufeng Jiedu Capsule for novel coronavirus pneumonia (COVID-19) based on network pharmacology and molecular docking. Guiding J Tradit Chin Med Pharm 26(5):8–15
53. Xu X, Chen P, Wang J, Feng J, Zhou H, Li X et al (2020) Evolution of the novel coronavirus from the ongoing Wuhan outbreak and modeling of its spike protein for risk of human transmission. Sci China Life Sci 63(3):457–460
54. Cao C, Cui Y, Chu Y-X, Shi Y-Y, Wu X-H, Wang X-Y et al (2020) Investigation on mechanism and active components of Shufeng Jiedu Capsule in treatment of COVID-19 based on network pharmacology and molecular docking. Chin Trad Herbal Drugs 51:2283–2296
55. Fu Y-J, Yan Y-Q, Qin H-Q, Wu S, Shi S-S, Zheng X et al (2018) Effects of different principles of traditional Chinese medicine treatment on TLR7/NF-κB signaling pathway in influenza virus infected mice. Chin Med 13(1):42
56. Zhang H, Penninger JM, Li Y, Zhong N, Slutsky AS (2020) Angiotensin-converting enzyme 2 (ACE2) as a SARS-CoV-2 receptor: molecular mechanisms and potential therapeutic target. Intensive Care Med 46(4):586–590
57. Wu H, Wang J, Yang Y, Li T, Cao Y, Qu Y et al (2020) Preliminary exploration of the mechanism of Qingfei Paidu decoction against novel coronavirus pneumonia based on network pharmacology and molecular docking technology. Acta Pharm Sin 55:374–383
58. Fan AY, Gu S, Alemi SF (2020) Chinese herbal medicine for COVID-19: current evidence with systematic review and meta-analysis. Integr Med 18(5):285–394
59. Lestari K, Sitorus T, Instiaty SM et al (2020) Molecular docking of quinine, chloroquine and hydroxychloroquine to angiotensin converting enzyme 2 (ACE2) receptor for discovering new potential COVID-19 antidote. J Adv Phar Educ Res|Apr–Jun 10(2):1–4
60. Fredericksen BL, Wei BL, Yao J, Luo T, Garcia JV (2002) Inhibition of endosomal/lysosomal degradation increases the infectivity of human immunodeficiency virus. J Virol 76(22):11440–11446
61. Wink M, Schmeller T, Latz-Brüning B (1998) Modes of action of allelochemical alkaloids: interaction with neuroreceptors, DNA, and other molecular targets. J Chem Ecol 24(11):1881–1937
62. Wink M (2007) Molecular modes of action of cytotoxic alkaloids: from DNA intercalation, spindle poisoning, topoisomerase inhibition to apoptosis and multiple drug resistance. Alkaloids Chem Biol 64:1–47
63. Jin XJ, Guan RN, Mao JJ et al (2020) Exploration on material basis of Qingfei Paidu Decoction with multi-target system against COVID-19 based on CADD. Chin Tradit Herb Drugs 51:1984–1995. (Chinese)
64. Choy K-T, Wong AY-L, Kaewpreedee P, Sia S-F, Chen D, Hui KPY et al (2020) Remdesivir, lopinavir, emetine, and homoharringtonine inhibit SARS-CoV-2 replication in vitro. Antiviral Res 178:104786
65. Schindler C, Levy DE, Decker T (2007) JAK-STAT signaling: from interferons to cytokines. J Biol Chem 282(28):20059–20063
66. Ruan Q, Yang K, Wang W et al (2020) Clinical predictors of mortality due to COVID-19 based on an analysis of data of 150 patients from Wuhan, China. Intensive Care Med 46(5):846–848
67. Zhao J, Tian S-S, Yang J et al (2020) Investigating mechanism of Qing-Fei-Pai-Du-Tang for treatment of COVID-19 by network pharmacology. Chin Trad Herbal Drugs 51:829–835
68. Huang C, Wang Y, Li X, Ren L, Zhao J, Hu Y et al (2020) Clinical features of patients infected with 2019 novel coronavirus in Wuhan, China. The lancet 395(10223):497–506
69. Xu Z, Shi L, Wang Y, Zhang J, Huang L, Zhang C et al (2020) Pathological findings of COVID-19 associated with acute respiratory distress syndrome. Lancet Respir Med. 2020. https://doi.org/10.1016/S2213-2600(20)

70. Chen Q, Liu J, Wang W, Liu S, Yang X, Chen M et al (2019) Sini decoction ameliorates sepsis-induced acute lung injury via regulating ACE2-Ang (1-7)-Mas axis and inhibiting the MAPK signaling pathway. Biomed Pharmacother 115:108971
71. de Kloet AD, Krause EG, Woods SC (2010) The renin angiotensin system and the metabolic syndrome. Physiol Behav 100(5):525–534
72. He LL, Gong PY, Feng Y et al (2020) Analysis on the application of Traditional Chinese Medicine in the treatment of COVID-19 by suppressing cytokine storm. Chin Tradit Herb Drug 51:1375–1385. [Google Scholar]
73. Yan B, Tian G (2017) Renin-angiotensin system and diabetic cognitive dysfunction, and effects of traditional Chinese medicine on Them. Chin J Rehab Theory Pract 23(3):270–273
74. Zhou J, Yang M, Zhang Z et al (2020) Mechanistic investigation of multiple organ dysfunction syndrome induced by 2019 novel coronavirus. Chin J Mult Organ Dis Elderly 19:226–228. [Google Scholar]
75. Chen MJ, Gao L, Tong ZQ. Network pharmacology study on screening of effective Chinese medicine for treatment of novel coronavirus pneumonia based on renin-angiotensin system. Chin Tradit Herbal Drugs. 2020. http://kns.cnki.net/kcms/detail/44.1286.R.20200326.1618.004.html. [Google Scholar]
76. Xu JH, Xue Y, Zhang W et al (2020) Study on mechanism of Shufeng Jiedu Capsule in treating COVID-19 based on network pharmacology. Chin Tradit Herb Drug 51:2015–2023. [Google Scholar]
77. Han L-W, Zhang Y-G, Li H-N, Wang H-Y, Li X-B, Wang X-J et al (2020) Network pharmacologic molecular mechanism of Shenmai Injection in treatment of COVID-19 combined with coronary heart disease. Chin Trad Herbal Drugs 9:2334–2344
78. Blanco-Melo D, Nilsson-Payant BE, Liu W-C et al (2020) Imbalanced host response to SARS-CoV-2 drives development of COVID-19. Cell 181:1036–1045.e9
79. Xia J, Rong L, Sawakami T, Inagaki Y, Song P, Hasegawa K et al (2018) Shufeng Jiedu Capsule and its active ingredients induce apoptosis, inhibit migration and invasion, and enhances doxorubicin therapeutic efficacy in hepatocellular carcinoma. Biomed Pharmacother 99:921–930
80. Zimmermann C, Krämer N, Krauter S, Strand D, Sehn E, Wolfrum U, Freiwald A, Butter F, Plachter B (2020). Autophagy interferes with human cytomegalovirus genome replication, morphogenesis, and progeny release. Autophag 2:1–7
81. Guan WJ, Ni ZY, Hu Y, Liang WH, Ou CQ, He JX et al (2020) Clinical characteristics of coronavirus disease 2019 in China. N Engl J Med 382(18):1708–1720
82. Hu W, Feng Z, Levine AJ (2012) The regulation of multiple p53 stress responses is mediated through MDM2. Genes Cancer 3(3–4):199–208
83. Tao Q, Du J, Li X, Zeng J, Tan B, Xu J et al (2020) Network pharmacology and molecular docking analysis on molecular targets and mechanisms of Huashi Baidu formula in the treatment of COVID-19. Drug Dev Ind Pharm 46(8):1345–1353
84. Wauters E, Van Mol P, Garg AD, Jansen S, Van Herck Y, Vanderbeke L et al (2020) Discriminating mild from critical COVID-19 by innate and adaptive immune single-cell profiling of bronchoalveolar lavages. BioRxiv
85. Hoxha M (2020) What about COVID-19 and arachidonic acid pathway? Eur J Clin Pharmacol 76(11):1501–1504
86. Müller C, Hardt M, Schwudke D, Neuman BW, Pleschka S, Ziebuhr J (2018) Inhibition of cytosolic phospholipase A2α impairs an early step of coronavirus replication in cell culture. J Virol 92(4)
87. Lipfert P, Seitz R, Arndt JO (1987) Studies of local anesthetic action on natural spike activity in the aortic nerve of cats. Anesthesiol J Am Soc Anesthesiol 66(2):210–213
88. Harvey RD, Morgan ET (2014) Cancer, inflammation, and therapy: effects on cytochrome P450–mediated drug metabolism and implications for novel immunotherapeutic agents. Clin Pharmacol Ther 96(4):449–457
89. El-Ghiaty MA, Shoieb SM, El-Kadi AO (2020) Cytochrome P450-mediated drug interactions in COVID-19 patients: current findings and possible mechanisms. Med Hypotheses 144:110033

90. Lee AM, Jepson C, Hoffmann E, Epstein L, Hawk LW, Lerman C et al (2007) CYP2B6 genotype alters abstinence rates in a bupropion smoking cessation trial. Biol Psychiatry 62(6):635–641
91. Sun CY, Sun YL, Li XM (2020) The role of Chinese medicine in COVID-19 pneumonia: a systematic review and meta-analysis. Am J Emerg Med 38:2153–2159
92. Chi X, Wang S, Baloch Z, Zhang H, Li X, Zhang Z et al (2019) Research progress on classical traditional Chinese medicine formula lily bulb and rehmannia decoction in the treatment of depression. Biomed Pharmacother 112:108616
93. Zhang B, Zhang K, Tang Q, Sun K, Han Z (2020) Acupuncture for breathlessness in COVID-19: a protocol for systematic review and meta-analysis. Medicine 99(27):e20701
94. Zhu Y, Jiang Z, Zhang Y, Zhang Q, Li W, Ren C, Yao R, Feng J, Ren Y, Jin L, Wang Y (2020) Assessment of Chinese medicine for coronavirus-related pneumonia: a protocol for systematic review and meta-analysis. Medicine 99(24):e20613. https://doi.org/10.1097/MD.0000000000020613

Chapter 4
Plant-Based Natural Products: Potential Anti-COVID-19 Agents

Sana Aslam, Matloob Ahmad, and Hanan A. Henidi

Introduction

Viral infections are becoming common day to day and cause many chronic human diseases. Viruses are responsible for numerous chronic diseases and hard-to-cure syndromes like HIV, hepatocellular carcinoma, HCV, type 1 diabetes, Alzheimer's disease, etc. [1–3]. The current outbreaks of globalization, increased global travel, and drug-resistant viral strains have underscored the protection of human health: the emergence of coronavirus, dengue virus, severe acute respiratory syndrome (SARS) virus, measles virus, West Nile virus outbreaks, and influenza virus [4–6].

Despite the drug and vaccine advancement, there is still a need of novel antiviral drugs or vaccine therapies that are extremely efficacious as well as economical for the control and management of viral infections. Viral infection can be avoided by either minimizing the exposure to viruses, sanitizing the skin, or boosting the immunization, and mucosal surfaces would reduce the risk of infection. However, with such a great care, there is still a need of effective treatments by virucidal or antiviral agents. Natural products especially secondary metabolites are an excellent source as therapeutic agents. They have a great antiviral potential, so are used as herbal medicines and in various pharmaceutical products from many decades. According to the

---

S. Aslam
Department of Chemistry, Government College Women University Faisalabad, Faisalabad, Pakistan

M. Ahmad (✉)
Department of Chemistry, Government College University, Faisalabad, Pakistan
e-mail: matloob.ahmad@gcuf.edu.pk

H. A. Henidi
Health Sciences Research Centre, Princess Nourah bint Abdulrahman University, Riyadh, Saudi Arabia

© The Author(s), under exclusive license to Springer Nature Switzerland AG 2021
M. Zia-Ul-Haq et al. (eds.), *Alternative Medicine Interventions for COVID-19*, https://doi.org/10.1007/978-3-030-67989-7_4

WHO report, 80% of the world's population depend upon the traditional plants (i.e., phytochemicals) to use as therapeutic agents [7].

Purified natural products and herbal medicines act either as a starting raw material or as an intermediate source for new antiviral drug development. Mechanisms of antiviral actions depicted by the natural agents have shed light on the site of binding or interaction with the viral life cycle, like viral entry, release, assembly, replication, and structure-activity relationship as well as targeting of virus host-specific interactions.

Antiviral Mechanistic Aspects of Phytochemicals

The phytochemicals were studied for their antiviral potential for more than six decades. There are a number of action mechanisms attributed to their antiviral activity, as explained in Table 4.1. The basic advantage of these plant-derived products is their nontoxicity with no or less side effects to human body as compared to the synthetic antiviral drugs. The structure-activity relationship is developed to explain the antiviral effect of these compounds.

Mechanism

Natural phytochemicals bind either directly or indirectly to the virus cell during their virucidal activities and retard the virus growth. Basically, three modes of action mechanisms were identified antiviral reagents directly:

Virucidal effect, i.e., antiviral compounds, directly inactivates the viruses.
Antiviral compounds accelerate CPE on the virus-infected cells.
Antiviral compounds inhibit the few replication steps of the virus.

For example, octyl gallate showed its antiviral potential via following all these mechanisms against HSV-1. As it is ineffective against non-enveloped poliovirus, it requires lipid envelope for its virucidal activity. According to one study, its antioxidant property helps to inactivate non-enveloped virus, which may be somehow similar to the oxidation of lipids [8].

Plant Selection for Antiviral Screening

In plant selection for antiviral screening, four basic approaches must be considered:

Random plant collection followed by mass screening
Literature-based follow-up of the existing natural products

Table 4.1 Antiviral potential of many natural products against specific strains of viruses

Sr. no.	Name of the phytochemicals/extract from plants/class	Virus	Source	Details of the study	Mechanism	References
1.	Polyphenolic compounds (PC) containing: Flavonoids, Catechins, Monne, Myricetin, Quercetin, Kaempferol, Ramnasin, Retusin			The effect of polyphenolic compounds was studied on the expression of viral proteins on the surface of virus-infected cells. The effect of viral proteins, i.e., nucleoprotein (NP), neuraminidase (NA), and hemagglutinin (HA) on the surface of infected cells was examined by ELISA (with monoclonal antibodies)	Total viral protein synthesis is inhibited by polyphenolic complex compounds, and the synthesis of hemagglutinin, neuraminidase, and nucleoprotein is also inhibited in the virus replication cycle	[9]
2.	Ent-epiafzelechin (4α→8)-epiafzelechin	Herpes simplex virus (HSV-2)	Extracted from *Cassia javanica*		Inhibits HSV-2 viral replication	[10]
3.	Excoecarianin	Herpes simplex virus (HSV)	Extracted from *Phyllanthus urinaria*		Inactivation of virus particles	[11]
4.	Chebulagic acid and punicalagin	Herpes simplex virus (HSV-1)	Derived from *Terminalia chebula* Retz.		Exhibits their antiviral potential as a GAG-competitor on the cell surface by inhibiting the viral entry (i.e., binding as well as fusion) and also by postinfection Cell-to-cell spread	[12]

(continued)

Table 4.1 (continued)

Sr. no.	Name of the phytochemicals/extract from plants/class	Virus	Source	Details of the study	Mechanism	References
5.	Meliacine	Herpes simplex virus (HSV-2)	Extracted from *Melia azedarach*	HSV-2 infection in a mouse model	Induces IFN-γ and TNF-α formation	[13]
6.	Quercetin (Q) 3-O-methylquercetin (3MQ) Luteolin (LU)	Herpes simplex virus (HSV-1)		Effect on the HSV-1 viral replication cycle	Interferes with the steps taking place between the 3 and 9 hours of replication cycle of HSV-1, i.e., transcription and translation of virus proteins	[14]
7.	Melittin	Human immunodeficiency virus (HIV-1)			Inactivates virus by disturbing the lipid envelope formation of virus	[15]
8.	Tricyclic coumarin	Human immunodeficiency virus (HIV-1)	Extracted from *Calophyllum brasiliense*		Inhibits viral replication in both chronic and acute infections by inactivation of NF-κB	[16]

Sr. no.	Name of the phytochemicals/extract from plants/class	Virus	Source	Details of the study	Mechanism	References
9.	Flavonoids complex: Amentoflavone Agathisflavone Baicalin Chrysosplenol C Coumarins Galangin (3,5,7-trihydroxyflavone) Glycosides Iridoids Morin Phenylpropanoid Rhusflavanone Robustaflavone Succedaneflavanone Theaflavin Quercetin Isoquercetin	Human immunodeficiency virus (HIV)		Effect on viral replication	HIV inhibitory activity was reported by blocking the RNA synthesis	[17–19]
10.	BCA, BA (flavonoid Baicalein)	HIV-1		Determining the action of mechanism of the antiviral effect of BA	BCA and BA (flavonoid Baicalein) both exhibited the *anti*-HIV-1 potential at the early stage of viral infection, i.e., at the time of viral entry into the host cell	[20]
10.	Extract from *Pelargonium sidoides*	Influenza virus	Isolated from *Pelargonium sidoides* roots		Inhibits viral entry and release, inhibits viral hemagglutination and neuraminidase (NA) activity	[21]

(continued)

Table 4.1 (continued)

Sr. no.	Name of the phytochemicals/extract from plants/class	Virus	Source	Details of the study	Mechanism	References
11.	Aqueous extract from dandelion	Influenza virus	Isolated from *Taraxacum officinale*		Inhibits the viral NP RNA levels and activity of polymerase	[22]
12.	Flavone (4′,5-dihydroxy 3,3′,7-trimethoxyflavone)	*Coxsackievirus Rhinovirus Picornavirus*		Show effects on viral replication	By selective inhibition RNA synthesis of virus in the cell culture as well as by inhibiting the virus replication	[23]
13.	Galangin 3-methyl ether Morin Quercetin Quercetin 7,4′-dimethyl ether Quercetin 3,7,3′4′-tetramethyl ether Quercetin 3,7,4′-trimethyl ether Quercetin 7-methyl ether Quercetin 3-methyl ether Robinin 7,4′-di-O-benzolquercetin 7-hydroxy-3,4′-dimethyl flavone 6,3′-dihydroxy-4′-methyl aurone Fisetin 4′-methyl ether	Tomato ringspot *Nepovirus* (TomRSV)		Effect on TomRSV infection in *Chenopodium quinoa*	Proposed that flavonoids delayed the early event in the virus life cycle, so there was a decreased titer and infectivity in tissue culture	[24]

Sr. no.	Name of the phytochemicals/extract from plants/class	Virus	Source	Details of the study	Mechanism	References
14.	Amentoflavone Quercetin Scutellarein	RAV-2 MMLV AMV		Study the effect on DNA synthesis	Inhibit reverse transcriptases (RT): RAV-2 RT MMLV RT AMV RT	[25]
15.	BCA, genistein	HCMV		Study of antiviral potential of genistein and baicalin against human cytomegalovirus	Proposed primary mechanism of baicalein was reported by blocking the HCMV entry into the host cell while for genistein by inhibiting the functioning of immediate-early protein 6 of HCMV	[26]

Ethnomedical approach
Chemotaxonomic approach [27]

The most preferred choices are the second and third ones, due to their cost-effective applicability. The folkloric created selection also demonstrated the five times more preferable therapeutic phytochemicals as compared to other approaches. Combining different section approaches like ethnomedical, taxonomical, and phytochemical methods collectively is also considered as the best choice. The random selection approach generally finds more novel antiviral natural compounds.

Different Classes of Phytochemicals as Antiviral Agents

Various classes of naturally occurring phytochemicals as antiviral agents are discussed in detail with reference to sources of origin, specificity, mechanistic action, structure-activity relationship (SAR), phase trials, etc.

Alkaloids

Alkaloids are a class of heterogeneous compounds having nitrogen atom linked with heterocyclic ring system. They possess basic character generally. Amino acids are usually the precursors for their biosynthesis within the plant body [28]. Handsome number of alkaloids showed potent antiviral activity. One of the studies on the 36 alkaloids isolated from *C. lanceus* or *Catharanthus roseus* as antiviral agents against polio type III and vaccinia viruses was reported. The results showed that the nine alkaloids were more potent as antiviral agents and pericalline was the most effective [29] (Fig. 4.1).

Another research group developed a structure-activity relationship (SAR) for chromone-based alkaloids, extracted from *Schumanniophyton magnificum* as anti-HSV and anti-HIV agents in Vero cells and C8166, respectively. The research group was synthesized their methyl and acyl analogs and developed their SAR. It was

Fig. 4.1 Structure of pericalline

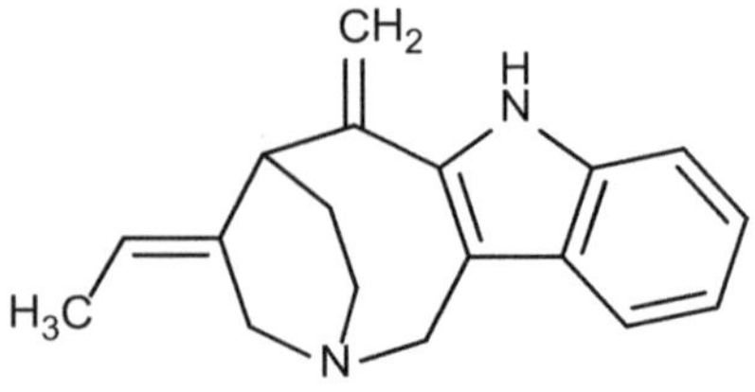

Pericalline

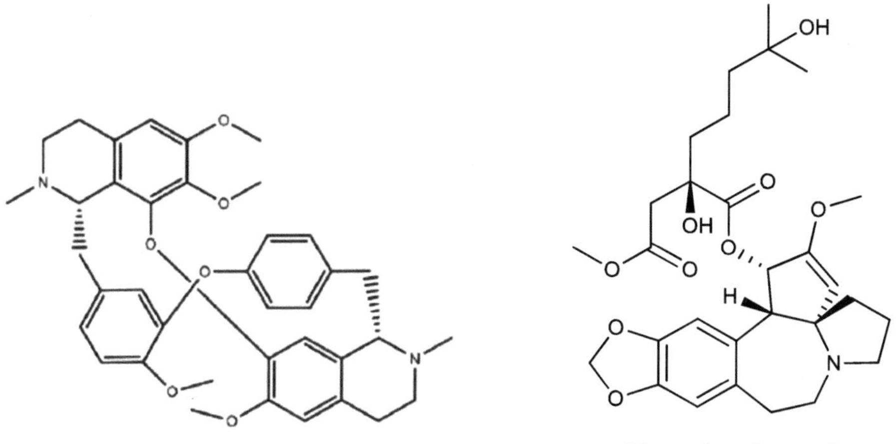

Tetrandrine

Homoharringtonine

Fig. 4.2 Structures of tetrandrine and homoharringtonine

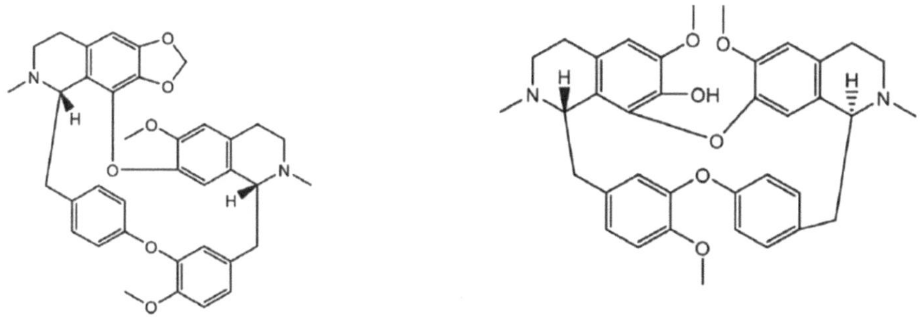

Cepharanthine

Fangchinoline

Fig. 4.3 Structures of cepharanthine and fangchinoline

concluded that the anti-HIV protentional is due to the presence of a free hydroxyl groups and piperidine ring [30] (Fig. 4.2).

Cepharanthine, an alkaloid, also showed remarkable antiviral potential by inhibiting the SARS-CoV protease enzyme at 0.5–10 μg/mL (Fig. 4.3) [31].

In another study, two alkaloids, i.e., 7-methoxycryptopleurine and tylophorine (Fig. 4.4), were isolated from *Tylophora indica* and tested for their inhibitory action for S and N protein activity, transmissible gastroenteritis virus, and enteropathogenic coronavirus replication [32]. These alkaloids showed excellent antiviral potential with IC_{50} values of <0.005 μM and 0.018 μM, respectively.

A recent research on berbamine showed its excellent antiviral activity against HCoV-NL63 with IC_{50} value 1.48 μM (Fig. 4.5).

Another research on antiviral potential of potent alkaloids, i.e., emetine, lycorine, and mycophenolate mofetil, against MERS-CoV, HCoV-NL63, HCoV-OC43,

Fig. 4.4 Structures of 7-methoxycryptopleurine and tylophorine

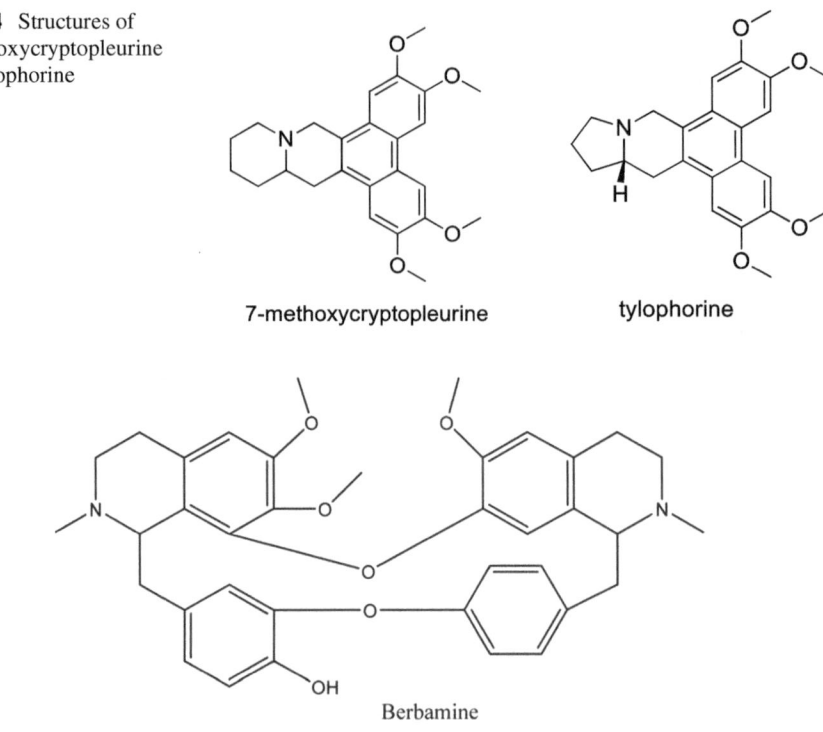

Fig. 4.5 Structure of berbamine

and MHV-A59 was reported. Emetine and lycorine exhibited their antiviral potential by inhibiting the synthesis of RNA, DNA, and protein of virus and by stopping the cell division, respectively. However, mycophenolate mofetil showed its action by suppressing the immune effect on different CoV species [33] (Fig. 4.6).

Lycorine and pretazettine were extracted from *Clivia miniata* and *Narcissus tazetta*, respectively, and reported for their anti-HSV potential via cytotoxic protein synthesis [34–36] (Fig. 4.7).

Three potent anti-HSV alkaloids, i.e., oliverine, pachystaudine, and Oxostephanine (Fig. 4.8), were isolated from *Polyalthia oliveri*, *Pachypodanthium staudi*, and *Stephania japonica*, respectively. They showed their antiviral potential by inhibiting or delayed the synthesis of protein assembly of virions [37].

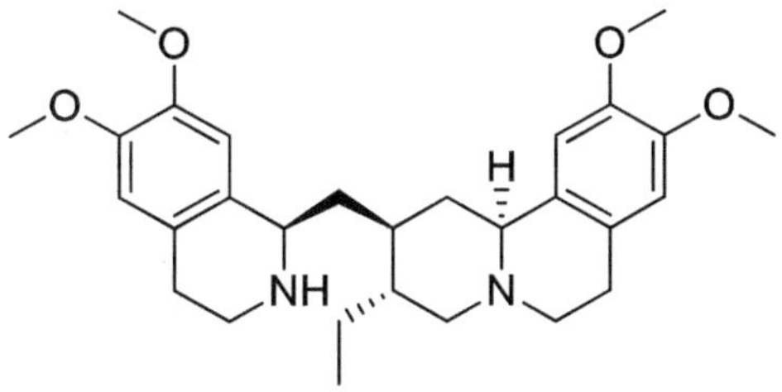

Fig. 4.6 Structures of emetine and lycorine

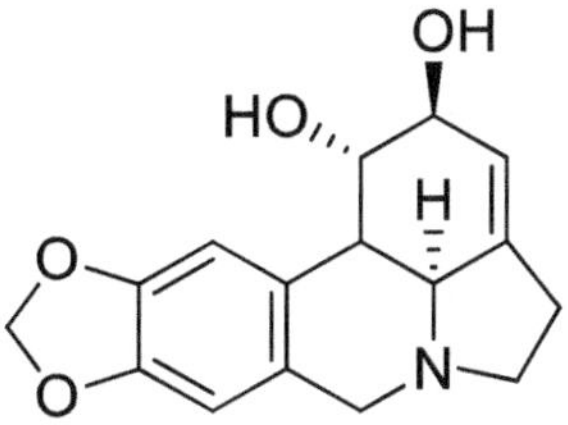

Fig. 4.7 Structure of pretazettine

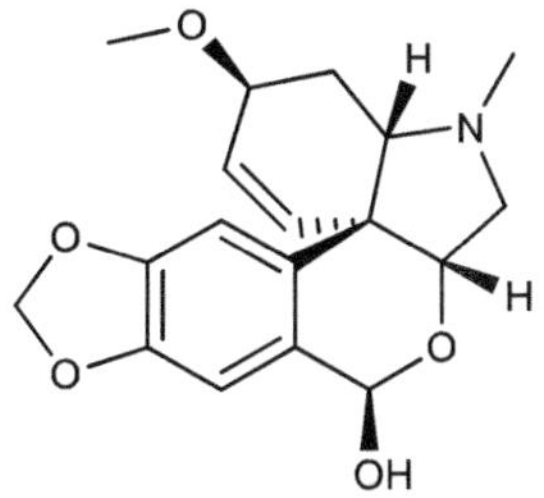

Flavonoids

Flavonoids are basically the chromone-based polyphenolic phytochemicals, consisting 15-carbon skeleton with C6-C3-C6 pattern (Fig. 4.9). However, in some flavonoids, five-membered ring replaces the six-membered heterocyclic ring C. The C2 is directly bonded to the oxygen atom to form a furan moiety called aurone.

Flavonoids are classified on the basis of substitution pattern on ring C and their mode of oxidation. The flavonoids class of phytochemicals is considered as the largest group of antiviral agents in the whole plant kingdom. Flavonoids exhibits their biochemical potential by inhibiting various enzymes like xanthine oxidase, aldose reductase, lipoxygenase, cyclooxygenase, phosphodiesterase, Ca^{+2} -ATPase, etc.

The antiviral potential of flavonols are more as compared to the flavones against HSV, and their activity order is quercetin < kaempferol < galangin [38]. Another study reported the anti-HSV-1 potential of 3,5,7-trihydroxyflavone, i.e., galangin, extracted from *Helichrysum aureonitens*. This flavone also exhibited antiviral potential against Cox B1 at 12–47/µg/ml [39]. A study showed that the natural flavonoids having molecular weight of about 2100 Daltons exhibited excellent antiviral activity against type 2 and type 1 herpes simplex virus (HSV) [40]. Structure-activity relationship for 28 flavonoids against HIV-1 and HIV-2 was developed. Results showed that the flavan-3-ol was most potent in selective inhibition of HIV due to the role of OH group on the flavone moiety [41].

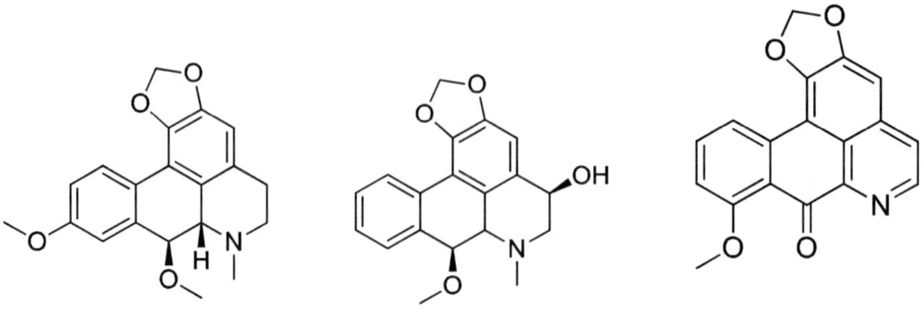

Ολιϖερινε   Pachystaudine   Oxostephanine

Fig. 4.8 Structures of oliverine, pachystaudine, and oxostephanine

Fig. 4.9 Structures of general flavonoid and aurone

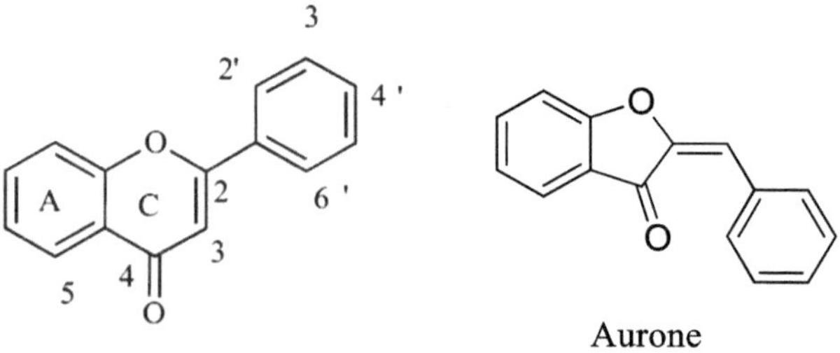

Flavonoid Numbering   Aurone

Nineteen natural compounds were extracted from *Ranunculus sceleratus* and *Ranunculus sieboldii* and were investigated for their antiviral potential against hepatitis B virus (HBV) and herpes simplex virus (HSV-1). The experimental results revealed that apigenin 7-O-betaglucopyranosyl-4′-O-alpha-rhamnopyranoside, apigenin 4'-O-alpha-rhamnopyranoside, isoscopoletin, tricin, and tricin 7-O-beta-glucopyranoside showed significant antiviral potential against HBV replication (Fig. 4.10).

In addition, protocatechuic aldehyde exhibited antiviral activity by inhibiting the HSV-1 replication [39, 42, 43] (Fig. 4.11).

Classification of Flavonoids

Chalcones

Chalcones as a major subclass of flavonoids are basically benzylideneacetophenone (1,3-diphenylpropenone) and its derivatives with basic formula ArCH=CHC(=O) Ar. Chalcones have been broadly investigated for their antiviral potential.

Chalcones also form basis for the biosynthesis of other flavonoids and isoflavonoids. Chalcones (Fig. 4.12) exhibited excellent antiviral potential [44]. They also

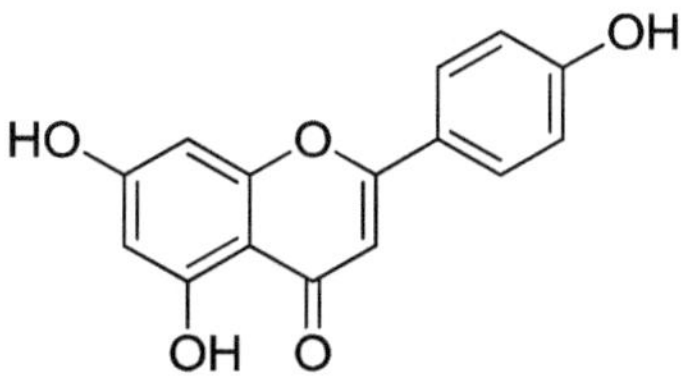

Apigenin

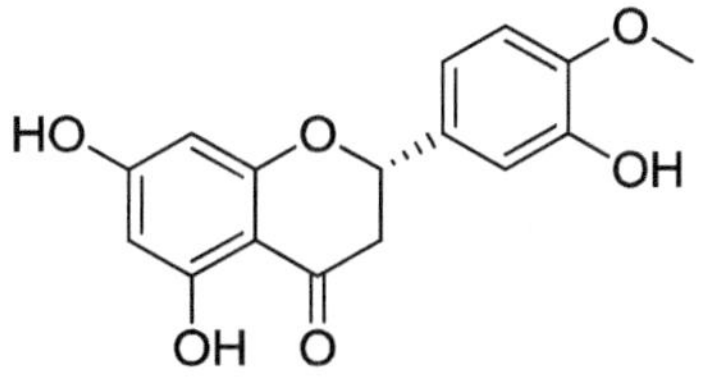

Hesperetin

Fig. 4.10 Structures of apigenin and hesperetin

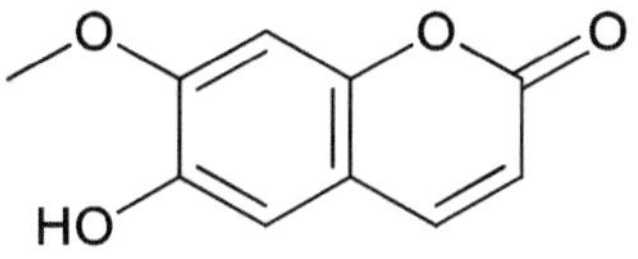

Isoscopoletin

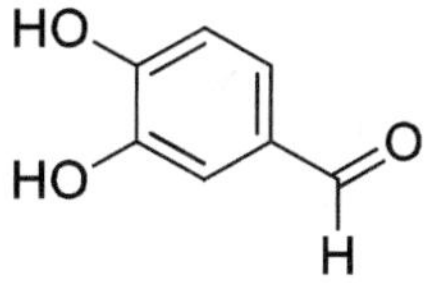

Protocatechuic aldehyde

Fig. 4.11 Structures of isoscopoletin and protocatechuic aldehyde

developed SAR for these chalcone pharmacophore models which were helpful to identify chemical signatures for the antiviral activity.

On the basis of these results, 44 chalcones exhibited inhibitory potency <100 µM, while 4 showed IC_{50} values <10 µM [45].

Different varieties of chalcones, i.e., kazinol (A, B, F, and J), broussochalcone (A and B), 3′-broussoflavan A, (3-methylbut-2-enyl)-3′,4,7-trihydroxyflavane, 4-hydroxyisolonchocarpin, and papyriflavonol A, were extracted from *Broussonetia papyrifera* and reported their antiviral activity against both SARS-CoV PL^{pro} and $3CL^{pro}$. The results reported the highest inhibition potential exhibited by papyriflavonol A against PL^{pro} with IC_{50} value 3.7 µM [46] (Fig. 4.13).

In another study, 4′-O-methylbavachalcone, psoralidin, bavachinin, corylifol, isobavachalcone, and neobavaisoflavone were extracted from *Psoralea corylifolia* and tested for their papain-like protease inhibitory action against SARS-CoV [47]. Among these isolated phytochemicals, psoralidin exhibited most potent inhibitory action against SARS-CoV, with IC_{50} value 4.2 µM (Fig. 4.14).

Dihydrochalcones

Dihydrochalcones were derived from its respective chalcone derivatives via a reduction of the C=C double bond (Figure). But as a result of reduction, its lost and chromophoric property as UV visibility is concerned as compared to its parent chalcone moiety. Dihydrochalcones extracted from *Millettia leucantha KURZ* (Leguminosae) exhibited potent anti-HSV activity [48] (Fig. 4.15).

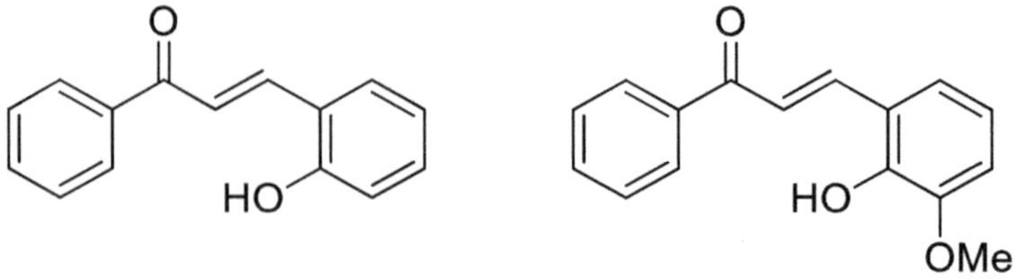

Fig. 4.12 Structure of different chalcones

Flavones

Flavones having 2-phenylchromen-4-one (2-phenyl-1-benzopyran-4-one) basic backbone constitute a significant class of flavonoid family. Flavones were mainly isolated from various plant families like Lamiaceae, Asteraceae, Apiaceae, etc. Many potent flavones were isolated from heartwood of *Artocarpus gomezianus* and studied for their anti-HSV activity. Among the identified compounds, artogomezianone (Fig. 4.16) exhibited the excellent antiherpetic properties [49].

Naringin (3′,4′-diacetoxy-5,6,7-trimethoxyflavone) having therapeutic properties especially for viral infections (e.g., HCV, HIV, respiratory virus, and *Picornavirus*). These flavones also used for the treatment of infections caused by parasites (e.g., toxoplasmosis) [50]. Methoxyflavones were studied for its SAR as anti-*Picornavirus* by means of molecular electrostatic potential (MEP) maps, and results showed that the antiviral properties are due to the negative MEP values especially in two regions, i.e., the first is in 3-methoxy (3-OMe) group, while the other is diagonally opposite to the substituent at C7 atom of the molecule [51] (Fig. 4.17).

Antiviral potent flavones and biflavones extracted from *Torreya nucifera* showed their virucidal activity against SARS-CoV 3CL[pro] [52]. Moreover, IC_{50} values of quercetin, luteolin, apigenin, and amentoflavone were 23.8, 20.2, 280.8, and 8.3 μM, respectively (Fig. 4.18).

Silybum marianum, a flavonolignan (also known as "silymarin" or "milk thistle"), was reported for its in vitro anti-HCV potential [53, 54] via exhibiting the significant effects on reducing the viral load [55–57].

Three biflavonoids, i.e., stelleranol, genkwanol C, and genkwanol B, were extracted from *Radix Wikstroemiae* and reported their effective antiviral potential against RSV [58]. Various flavone 6-C-monoglycosides were isolated from *Lophatherum gracile* leaves and showed good viricidal activity against RSV infection, determined by cytopathic effect reduction assay [59].

Flavonones

Flavonones consist of the same basic structural backbone as present in the flavones except the presence of the carbonyl moiety at the C4 carbon. Synthesis of flavanone derivatives of Abyssinone II (Fig. 4.19), a natural prenylated flavanone has been

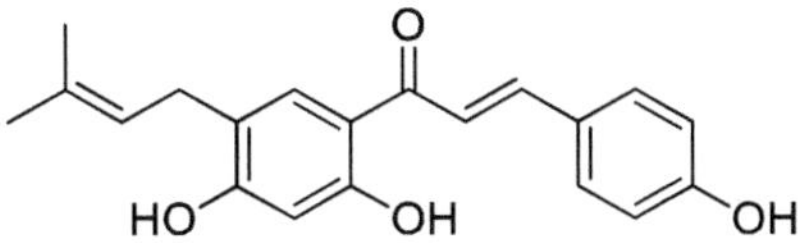

broussochalcone B

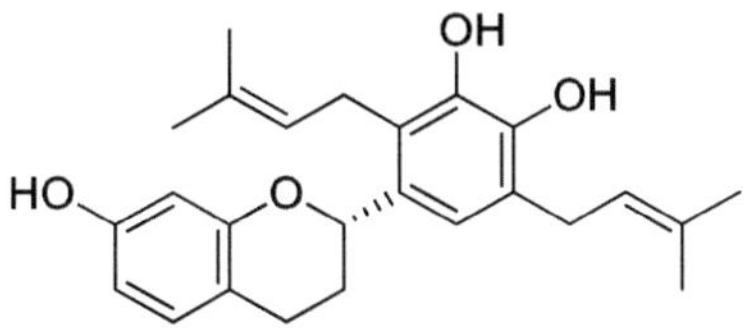

Kazinol A

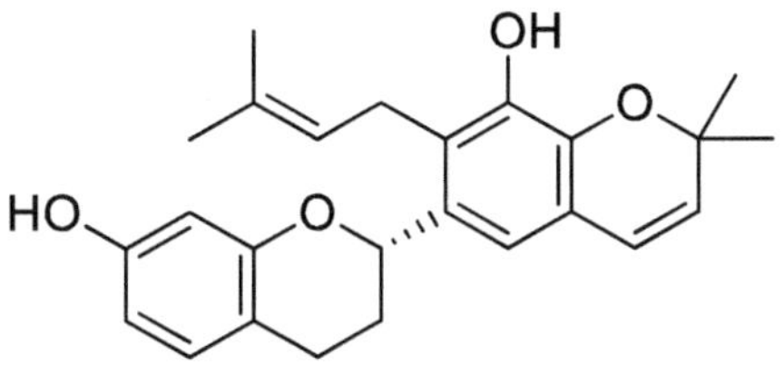

Kazinol B

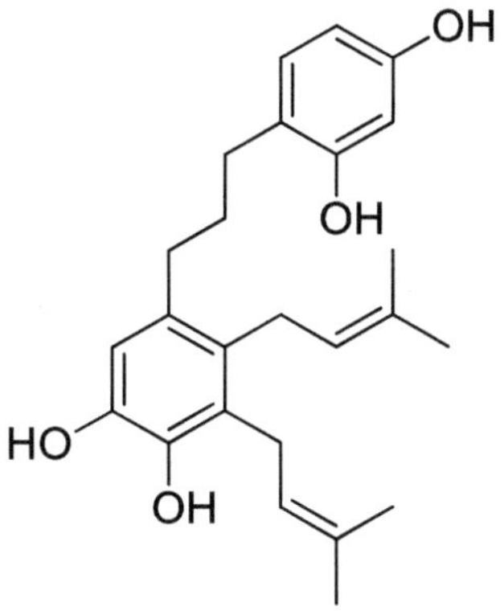

Kazinol F

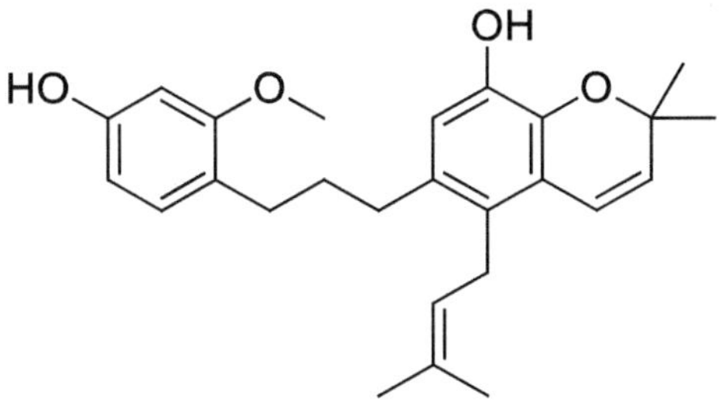

Kazinol J

Fig. 4.13 Structures of broussochalcone B and kazinol (A, B, F, and J)

reported in literature. These analogs were tested against HSV-1 in HeLa 5 cells and reported their excellent antiviral potential [60].

A number of natural plants products, i.e., 6-geranyl-4′,-5,7-trihydroxy-3′,5′-dimethoxyflavanone, different types of tomentin (A, B, C, D, and E), mimulone, diplacone, 3'-O-methyldiplacone, 4'-O-methyldiplacone, 3'-O-methyldiplacol, and

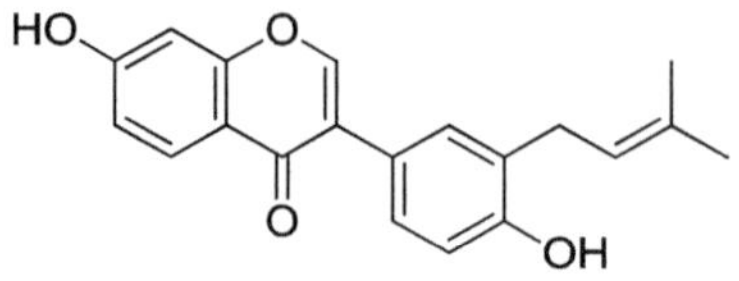

Neobavaisoflavone

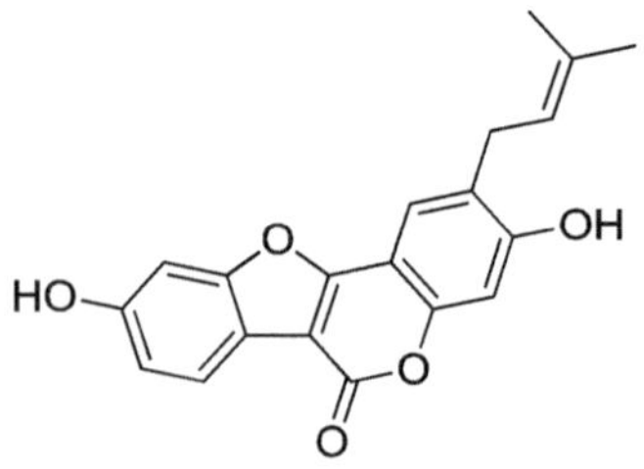

Psoralidin

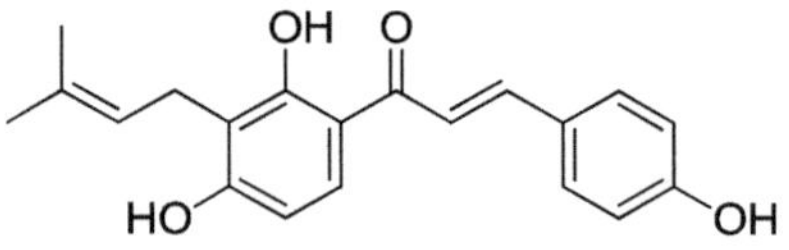

Isobavachalcone

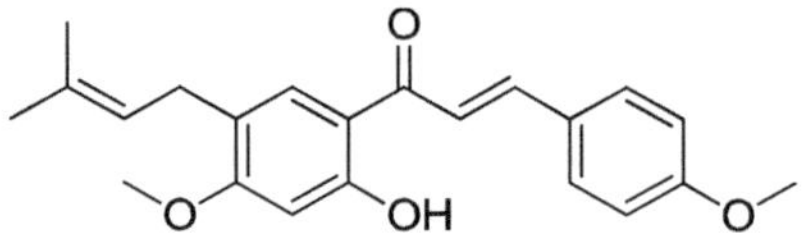

4'-O-Methylbavachalcone

Fig. 4.14 Structures of neobavaisoflavone, psoralidin, isobavachalcone, and 4'-O-methylbavachalcone

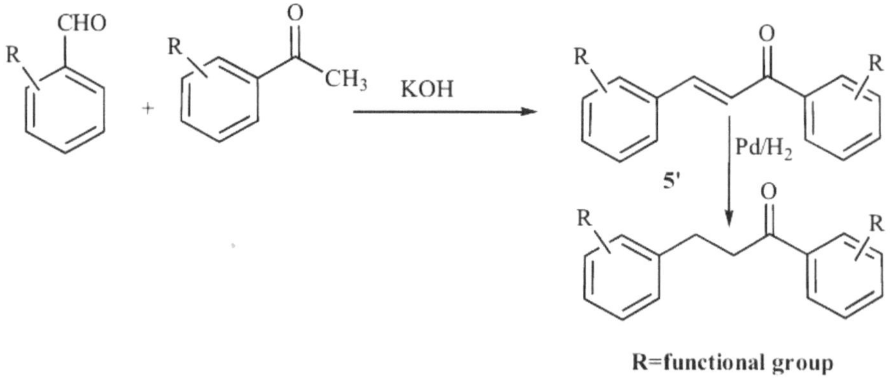

Fig. 4.15 Synthetic layout of dihydrochalcones

4'-O-methyldiplacol were extracted from *Paulownia tomentosa* and showed their antiviral therapeutic action by inhibiting the PL^{pro} of SARS-CoV [61] (Figs. 4.20 and 4.21).

In addition, juglanin also reported as an antiviral agent against SARS-CoV, by blocking the 3a channel of SARS-CoV having 2.3 μM IC_{50} value [62].

Fig. 4.16 Structure of artogomezianone

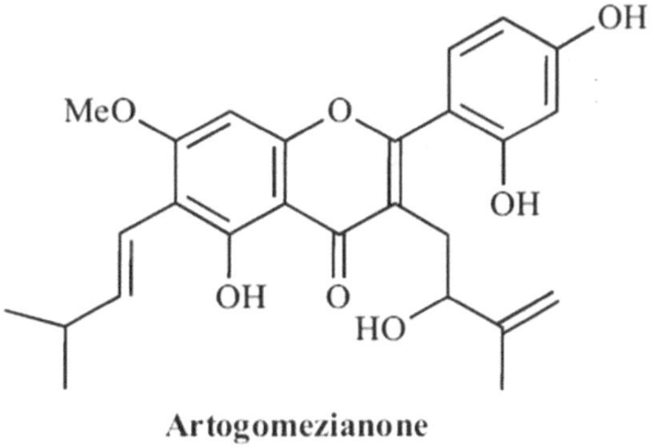

Artogomezianone

Fig. 4.17 Structure of naringin

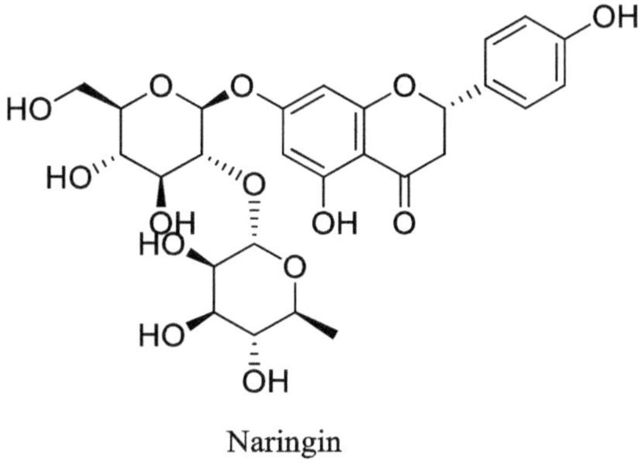

Naringin

Dihydroflavonols

Dihydroflavonols (Fig. 4.22) are categorized by OH substituent at C3 of the flavonone pharmacophore. Flavanonols and their derivatives showed antiviral potential against various strains of viruses like hepatitis B, liver protection, mycotic infection, autoimmune disease, and inflammation disease [27].

Flavonol

Flavonol consists of identical basic structural backbone as present in the flavones having OH functional group at 3-position of the flavone. Quercetin (Fig. 4.23) exhibited the efficient antiviral potential against different viral strains like respiratory syncytial virus (RSV), poliovirus type 1, HSV-I, parainfluenza virus type 3 (Pf-3), etc. Quercetin not only reduced the intracellular replication but also triggered a concentration-dependent decrease in the viral infection [63]. Myricetin (3,5,7,3′,4′,5′-hexa-hydroxyflavone), a plant-derived flavonoid, has natural compound with potent nutraceuticals value and so is used in many foods and beverages.

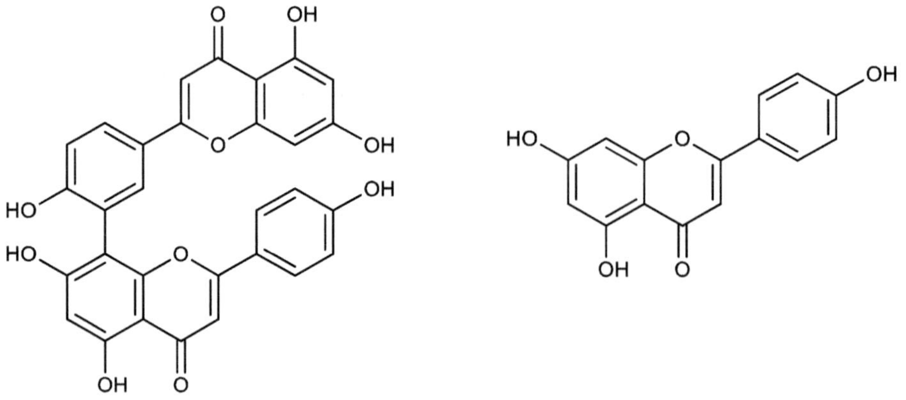

Amentoflavone Apigenin

Fig. 4.18 Structures of amentoflavone and apigenin

Fig. 4.19 Structure of Abyssinone II

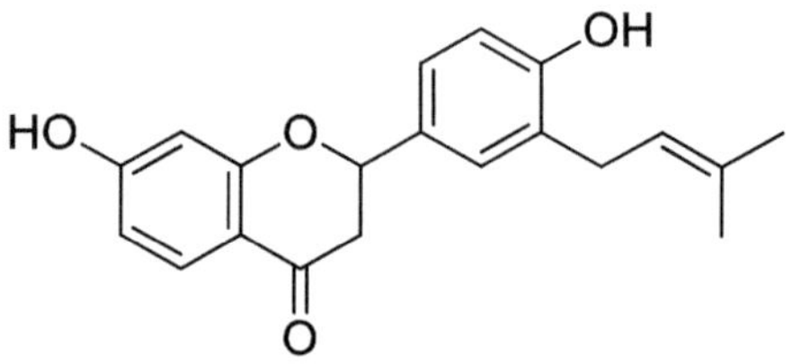

Abyssinone II

Fig. 4.20 Structure of 6-geranyl-4′,-5,7-trihydroxy-3′,5′-dimethoxyflavanone

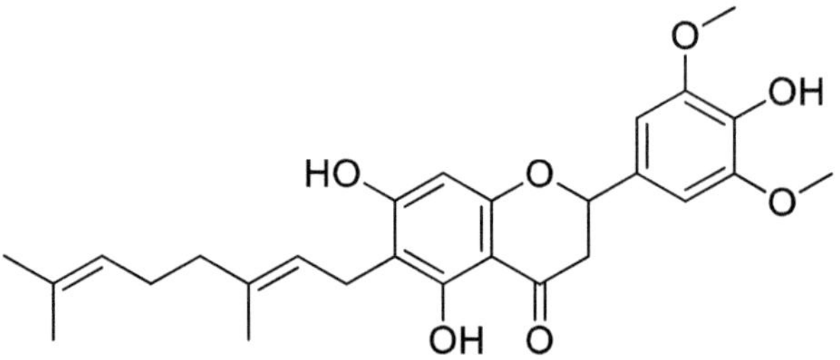

It is also used as preservative in oils and fats to increase the shelf life by acting as antioxidant. This phytochemical showed a wide range of bioactivities like antioxidant, anti-inflammatory, antidiabetic, and anticancer. This compound showed excellent therapeutic results against Parkinson and Alzheimer's disease. Myricetin also exhibited significant antiviral activity against coronavirus, influenza virus, and hepatitis B virus [64].

The natural phytochemicals cinnamtannin B1, procyanidin A2, and procyanidin B1 (Fig. 4.24) were extracted from *Cinnamomi cortex* and reported their antiviral potential against SARS-CoV [65].

Fig. 4.21 Structure of juglanin

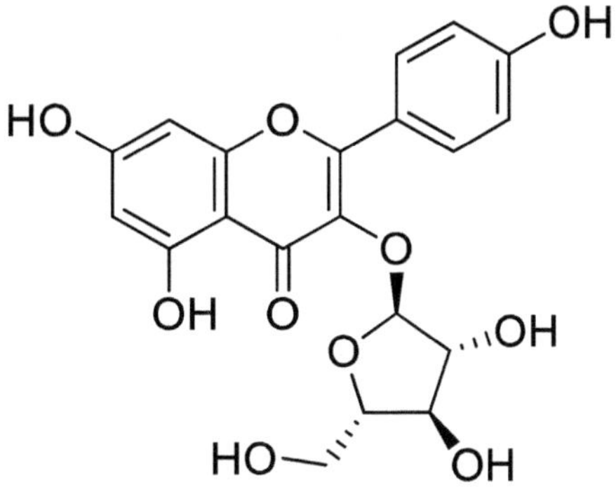

Juglanin

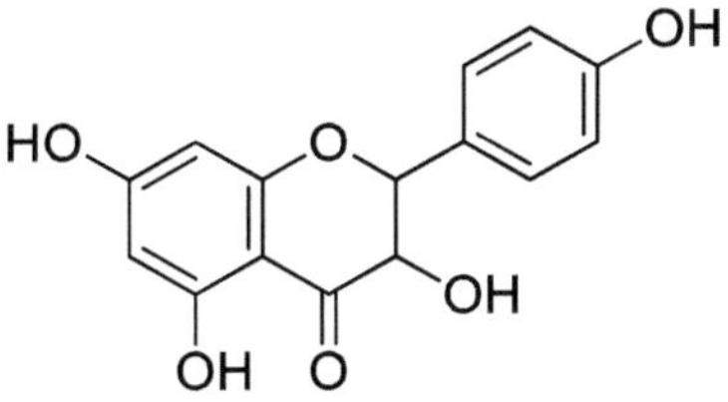

(+)-dihydrokaempferol

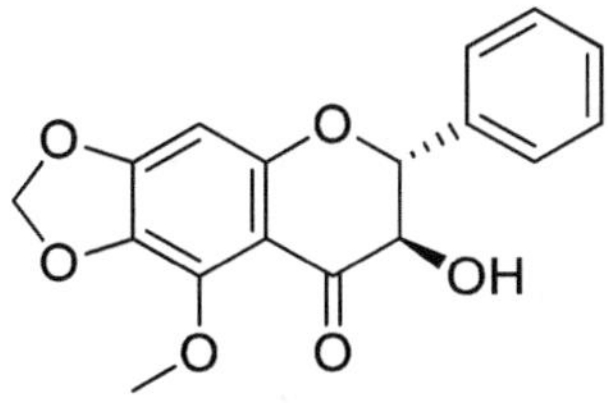

(2R,3R)-3-hydroxy-5-methoxy-6,7- methylenedioxyflavanone

Fig. 4.22 Structures of dihydroflavonol

Isoflavonoids

Isoflavonoids are the derivatives of flavonoids having phenyl group at 3-position instead of 2-position (Fig. 4.25) as a result of migration, with important therapeutic activities.

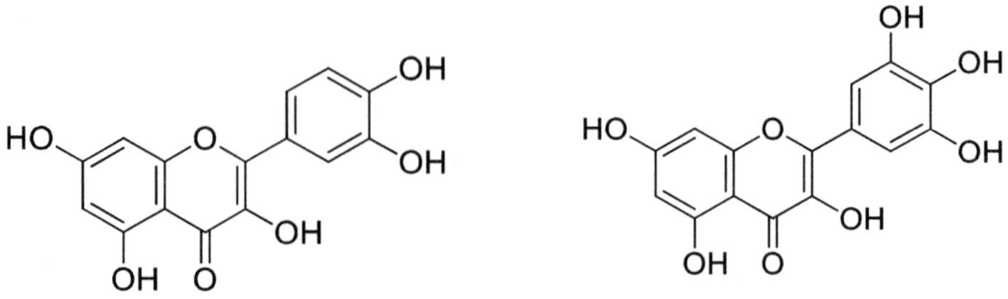

Quercetin Myricetin

Fig. 4.23 Basic structures of quercetin and myricetin

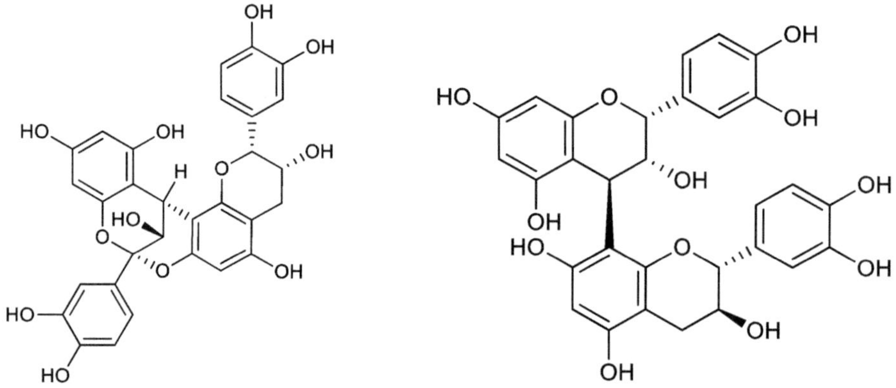

Procyanidin A2 Procyanidin B1

Fig. 4.24 Structure of procyanidin A2 and procyanidin B1

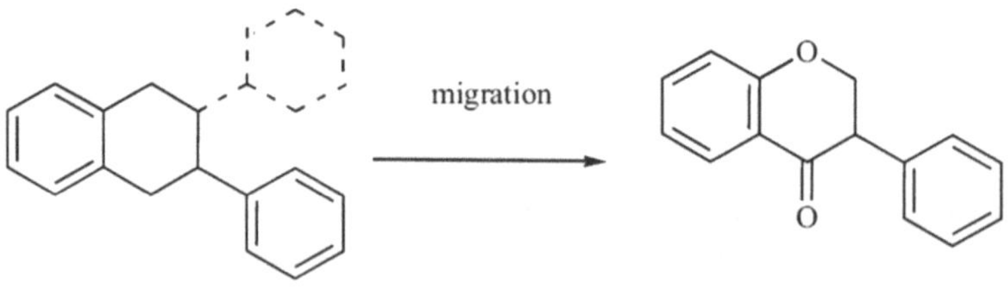

Isoflavone

Fig. 4.25 Structure of isoflavones

Isoflavones

Isoflavones (Fig. 4.25) are mainly isolated from the Leguminosae family. Rotenone (Fig. 4.26) showed excellent antiviral activity against Newcastle disease virus as tested by means of plate and tube assay methods [66].

Fig. 4.26 Structure of rotenone

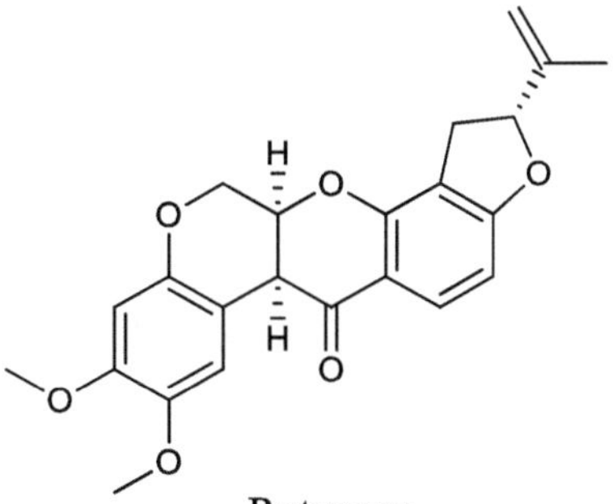

Rotenone

Fig. 4.27 Structure of PMZ-1

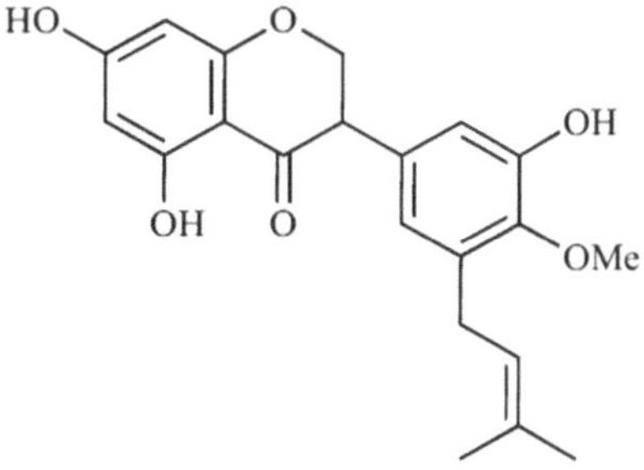

Isoflavanone PMZ-1

Isoflavanones

Isoflavanones are consist of the same basic structural backbone as present in the isoflavones except the presence of the carbonyl moiety at the C4 carbon and also have chiral center at C3. A prenylated isoflavanone, i.e., PMZ-1, was extracted from *Bolusanthus speciosus* (Bolus Harms) and evaluated for its anti-HIV potential with wide therapeutic index (TI > 300) [67] (Fig. 4.27).

Neoflavonoids

Neoflavonoids are the class of flavonoids having aryl group at C_4. Four neoflavonoids, i.e., calophyllolide and inophyllums (Fig. 4.28), were isolated from *Calophyllum inophyllum* and tested for their antiviral potential against HIV-1 RT. The results showed that inophyllums (i.e., P and B) were the most effective with IC_{50} values of 0.130 mM and 0.038 mM, respectively [68].

AnthocyanidinAnthocyanidin (Fig. 4.29) consists of aglycone (anthocyanidine) backbone with glycone sugar moiety, an important class of plant pigments. Their biological potential is based on the coordination of free OH group with metal ions, e.g., Ca^{2+} and Mg^{2+}, in basic conditions.

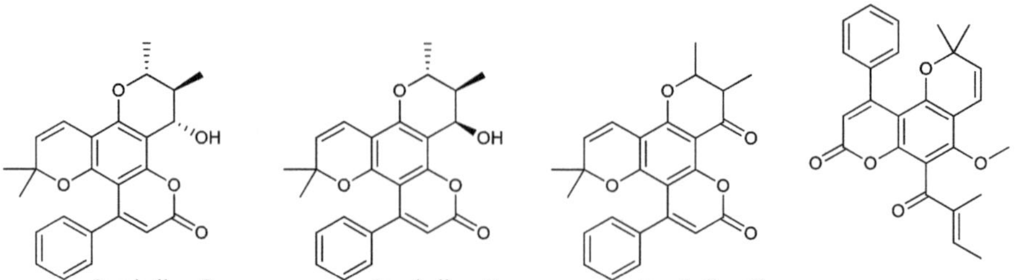

Inophyllum B Inophyllum P Inophyllum C Calophyllolide

Fig. 4.28 Structures of (+) inophyllums B, (+) inophyllums P, (+) inophyllums C, and calophyllolide

Fig. 4.29 Structures of general anthocyanidin

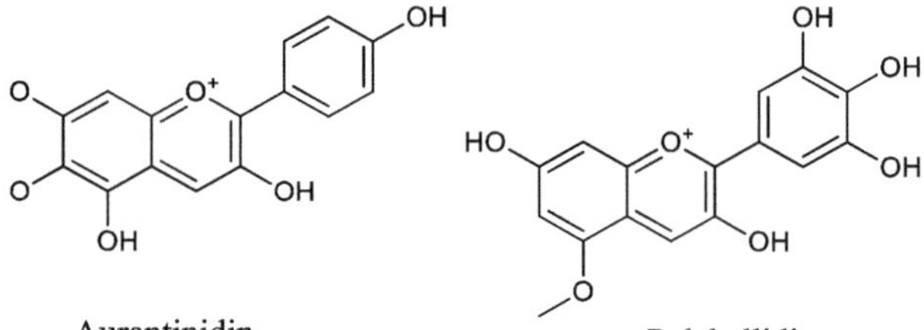

Aurantinidin Pulchellidin

Fig. 4.30 Structure of SP-303

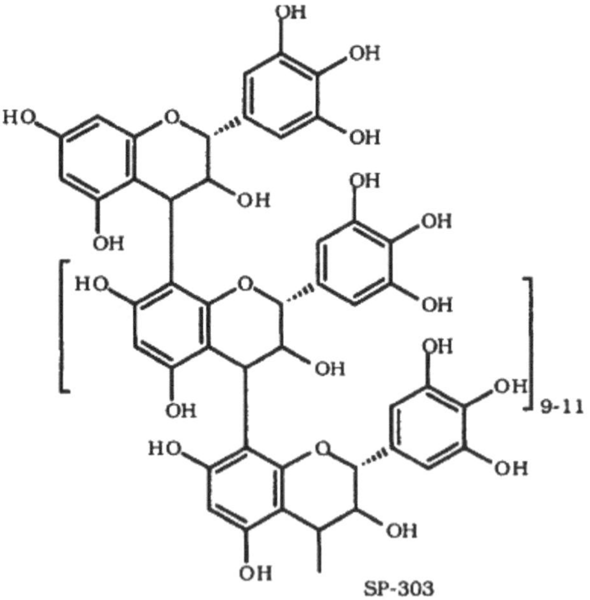

SP-303

SP-303 (Fig. 4.30) is a mixture of oligomeric proanthocyanidins up to Mol. wt. 2100 Daltons. It is extracted from the latex of *Croton lechleri* and evaluated for its in vitro antiviral activity against various strains of RNA, DNA viruses, and HSV.

Virend, a topical formulation of SP-303, was evaluated in phase II clinical trials and used in combination with acyclovir for the treatment of genital herpes. But these trials were stopped later on due to no extra benefits of virend over using acyclovir alone [69].

Terpenoids

Terpenoids are abundantly natural-occurring secondary metabolites, having five-carbon isoprene units as basic skeleton, and classified according to the number of isoprene units present in a molecule.

These are basically classified into:

Monoterpenes (C10), 2 isoprene units with 10 carbon atom basic skeleton
Sesquiterpenes (CI5), 3 isoprene units with 15 carbon atom basic skeleton
Diterpenes (C20), 4 isoprene units with 20 carbon atom basic skeleton
Triterpenes (30), 6 isoprene units with 30 carbon atom basic skeleton
Tetraterpenes (40), 8 isoprene units with 40 carbon atom basic skeleton

Sterols and saponins also classified as terpenoids.

Terpenoids possess diverse class of natural therapeutic phytochemicals. Many of the terpenoids were tested for their antiviral potential against severe acute respiratory syndrome (SARS-CoV) caused by coronavirus, and results showed their excellent antiviral potential (Fig. 4.31) [70].

These bioactive compounds also include abietane-type (diterpenes), labdane-type (i.e., both sesquiterpenes and triterpenes).

The saikosaponins (A, B_2, C, and D, 5–25 µM/L) exhibited potent antiviral activity against human CoV-229E, with EC_{50} values of 13.2, 19.9, 1.7, and 8.6 µM for D, C, B_2, and A, respectively. Saikosaponin B2 showed its antiviral activity by inhibiting the attachment and penetration stages of the virus [71] (Fig. 4.32).

According to the research report, carotenoids (tetraterpenoids, having 40-carbon polyene chain) like β-carotene, lycopene, α-carotene, and zeaxanthin/lutein (Fig. 4.33) increase the death rate during HIV infection [72].

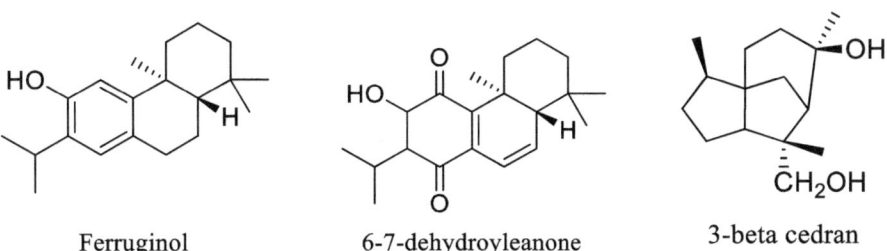

Fig. 4.31 Structures of Ferruginol, 6-7-dehydroyleanone and 3-beta cedran

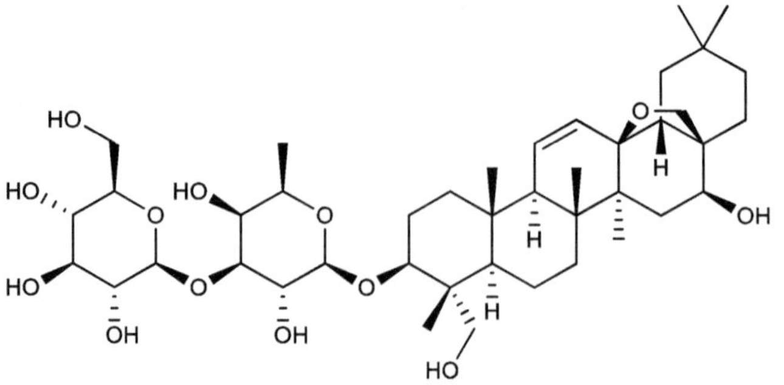

Fig. 4.32 Structures of saikosaponins (A, B$_2$, C, and D)

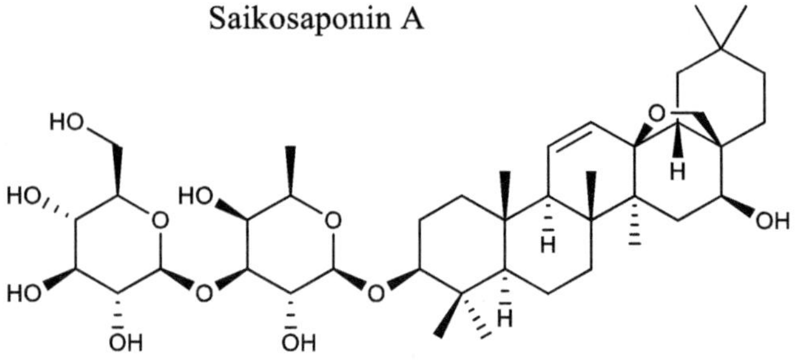

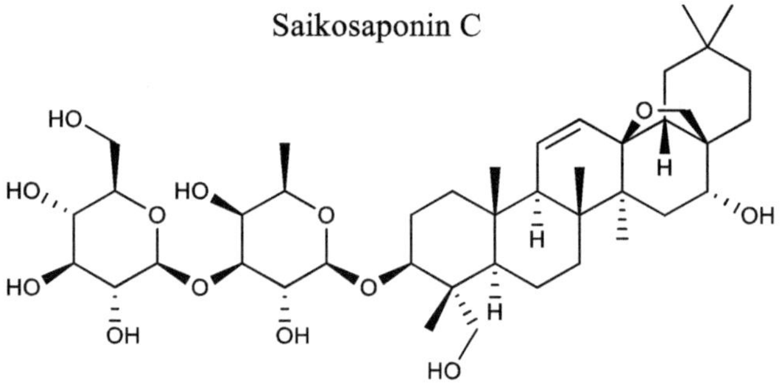

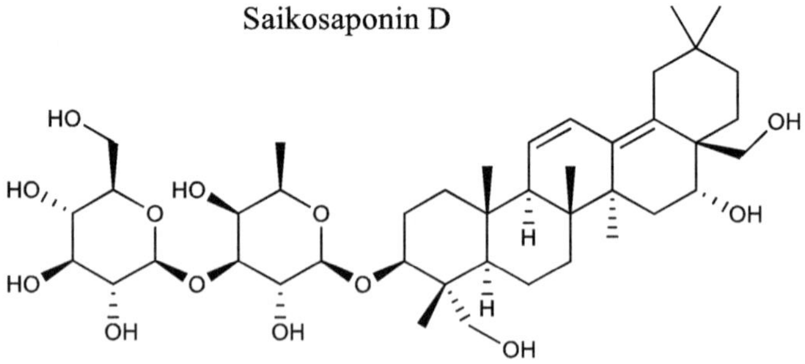

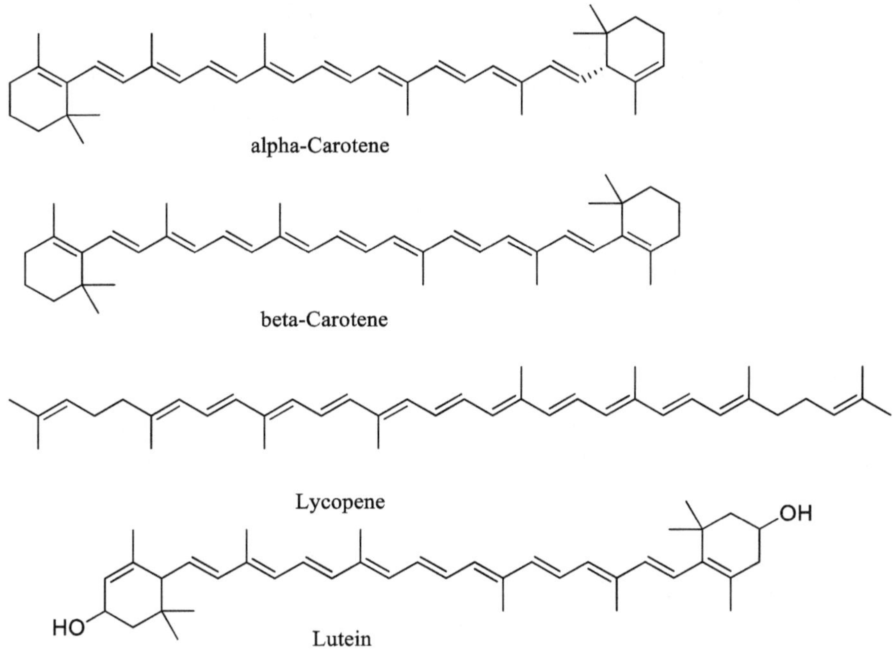

Fig. 4.33 Structures of α-carotene, β-carotene, lycopene, and lutein

Fig. 4.34 Structure of hexahydroxydiphenic acid

Hexahydroxydiphenic acid

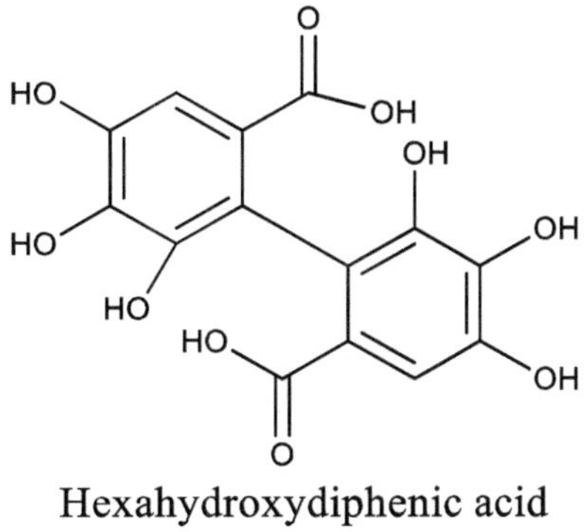

Tannins

Tannins are basically high molecular weight phenolic compounds also containing other functional groups like (carboxyl, e.g., hexahydroxydiphenic acid), which give suitable property to coordinate for strong complex formation [73]. They are classified into two types, i.e., hydrolyzable and non-hydrolyzable (condensed). Hydrolyzable tannins, basically simple phenolic acids, are linked with the sugar moiety, for example, gallic acid. The condensed types have structural resemblance to flavonoids (Fig. 4.34).

Lemon balm (*Melissa officinalis*, Labiatae) is well known for its antiviral potential. This medicinal tannin containing plant is broadly studied. Leaves of lemon balm comprise 5 percent dry weight of tannins, in which caffeic acid is present as a main constituent. A cream having 1 percent dried leave extract of lemon balm has been introduced in Germany for the topical treatment of herpes infection of the skin [74].

Many tannin compounds like pentagalloylglucose, galloyl geraniin, sanguin, genothein B, punicalagin, punicallin, gemin D, etc. were reported for their antiviral potential against different chronic viral infections, i.e., HSV, HIV, and HIV-RT [75–80] (Fig. 4.35).

Seven ellagitannins were extracted from *P. urinaria* (Euphorbiaceae) and *Phyllanthus myrtifolius* and reported their antiviral potential against Epstein-Barr virus DNA polymerase (EBV-DP).

Dieckol, phlorofucofuroeckoln, eckol, and 7-phloroeckol isolated from *Ecklonia cava* displayed their antiviral activity by blocking the viral binding to porcine epidemic cells and reported experimental IC_{50} values of 14.6, 12.2, 18.6, and 22.5 µM, respectively [81] (Fig. 4.36).

Three polyphenolic compounds, i.e., tannic acid, 3-isotheaflavin-3-gallate, and theaflavin-3,3′-digallate (Fig. 4.37), were extracted from black tea and tested for their inhibitory action against SARS-CoV 3CL^{pro} with IC_{50} values of 9.5, 7, and 3 µM, respectively [82].

The hydrolyzable tannins, i.e., punicalagin and chebulagic acid which exhibited broad-spectrum antiviral potential, include RSV infection. These tannins showed their antiviral potential either by inactivating the RSV particles or by blocking the viral entry into the host cell, i.e., binding and fusion. Interestingly, both punicalagin and chebulagic acid were reported as futile against RSV postinfection spread. However, they are still effective against measles virus (MV, paramyxovirus) postinfection spread [83] (Fig. 4.38).

Punicalagin and chebulagic acid also exhibited anti-HSV-1 potential and showed their virucidal activity by acting as a glycosaminoglycan (GAG) competitors, so inhibiting the entry as well as cell-to-cell spread. Both directly targeted the HSV-1 glycoproteins which interact with glycosaminoglycan and inhibit the binding receptors [10].

Vitamins

Vitamin E includes eight isomeric derivatives, i.e., tocopherols and tocotrienols. They are fat-soluble and act as an excellent antioxidant. Vitamin E improves the immune system of the human body (Fig. 4.39). Vitamin E supplementation could be used as therapeutic agent for chronic hepatitis B [84].

Vitamin C also acts an antioxidant and enhances the immune defense against many infectious diseases [85] (Fig. 4.40).

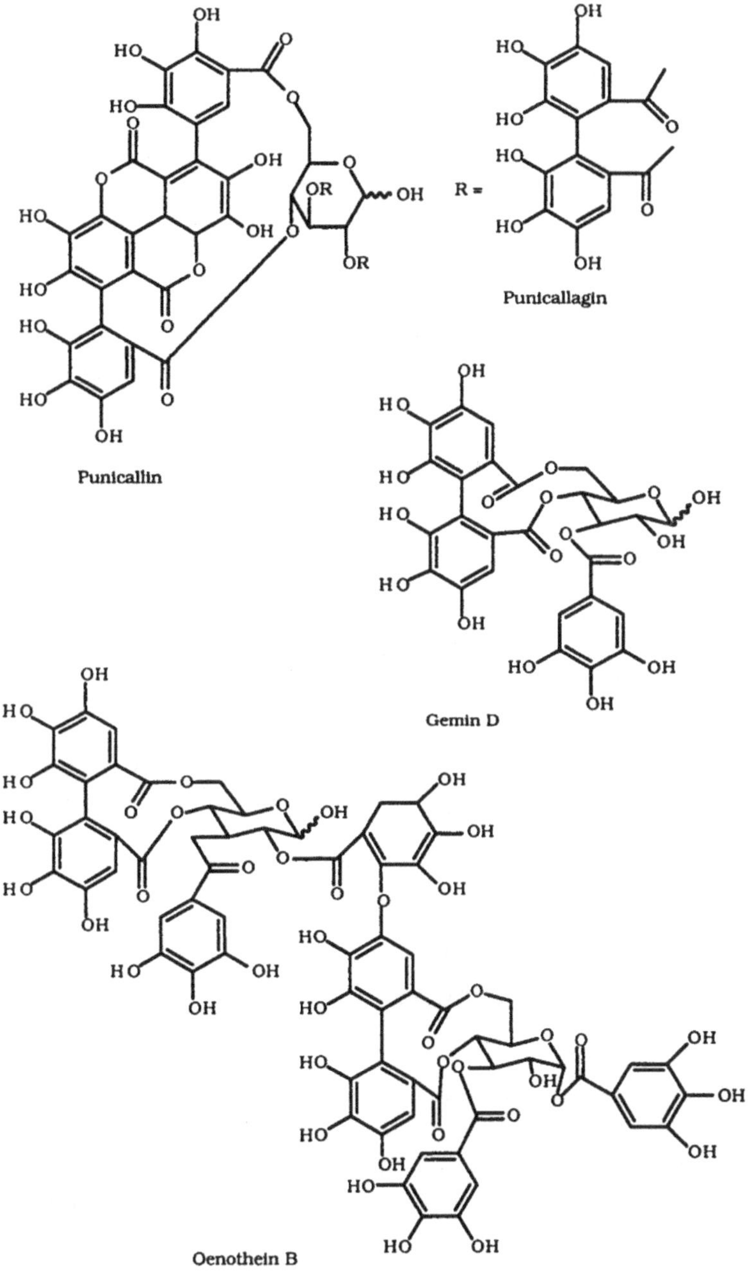

4 Plant-Based Natural Products: Potential Anti-COVID-19 Agents

Fig. 4.35 Structures of punicalagin, gemin D, and genothein B

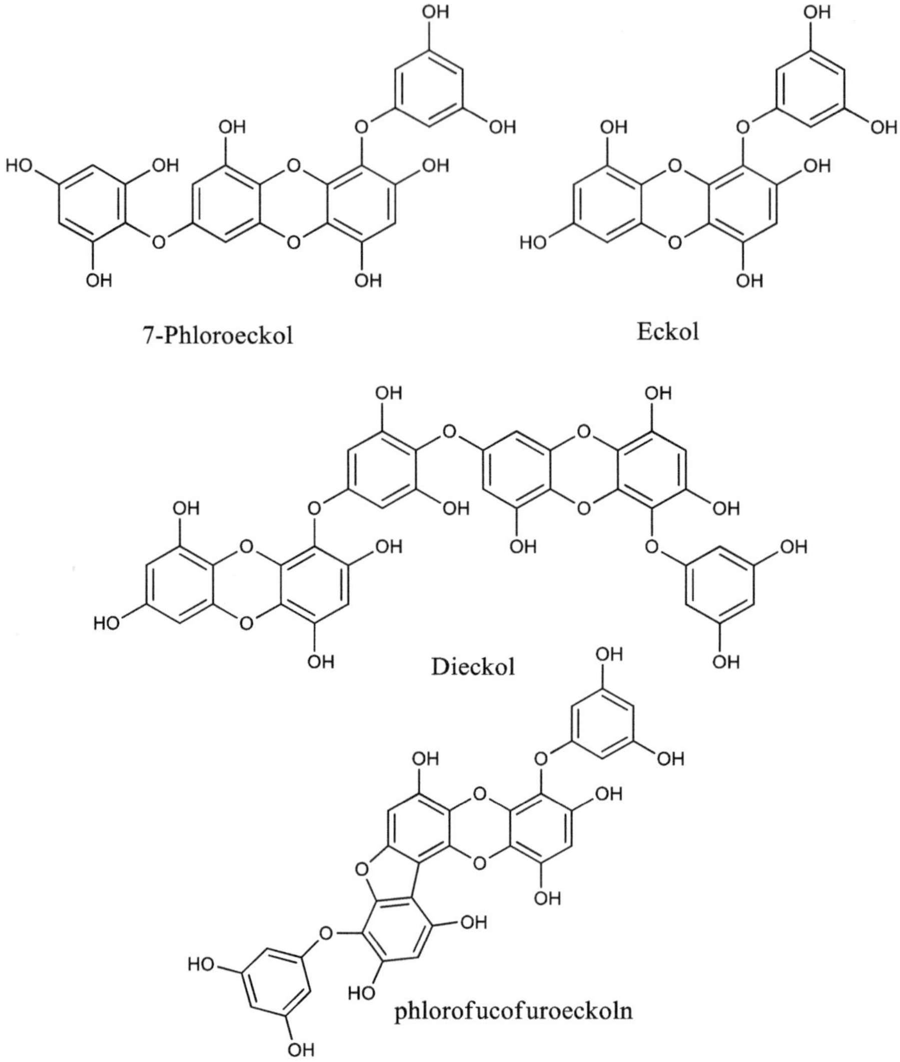

Fig. 4.36 Structures of 7-phloroeckol, eckol, dieckol, and phlorofucofuroeckoln

Vitamin C not only boosts the immune systems but also exerts anticancer, antibacterial, and antiviral activity [86–90]. In a comparative research study, antiviral potential of vitamin C against influenza virus type A, HSV-1, and poliovirus was reported. The decreasing sensitivity order of vitamin C was as follows: influenza virus> HSV-1 > poliovirus [91].

Dehydroascorbic acid (DHA) is an oxidized form of ascorbic acid. It is reported that DHA showed more potent antiviral activity against HSV-1 and influenza virus as compared to vitamin C (ascorbic acid), due to its stronger chemical stability of

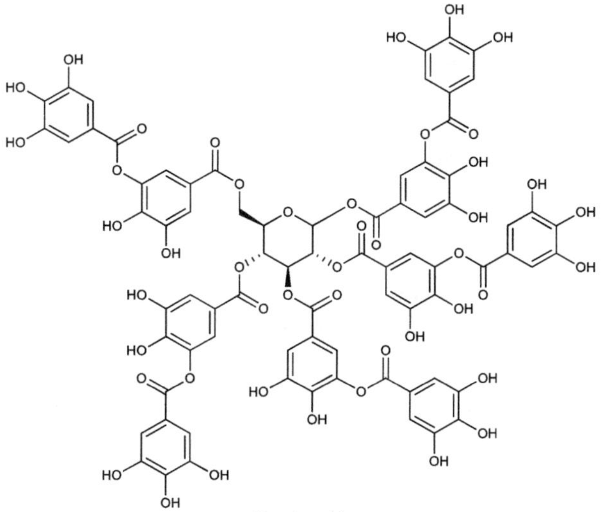

Tannic acid

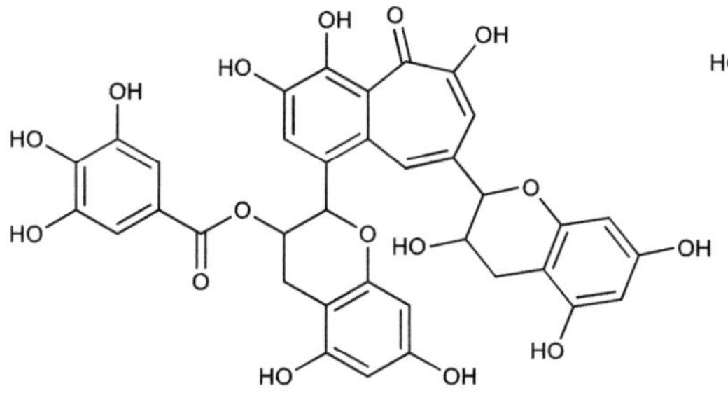

Isotheaflavin 3'-gallate

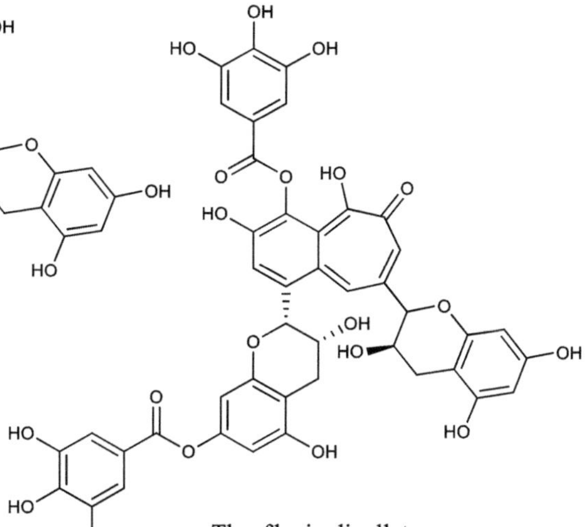

Theaflavin digallate

Fig. 4.37 Structures of tannic acid, 3-isotheaflavin-3-gallate, and theaflavin-3,3′-digallate

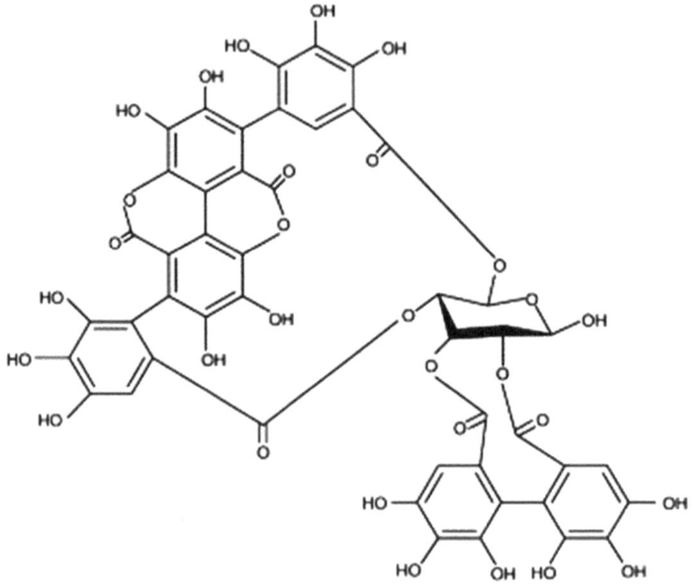

Fig. 4.38 Structure of punicalagin

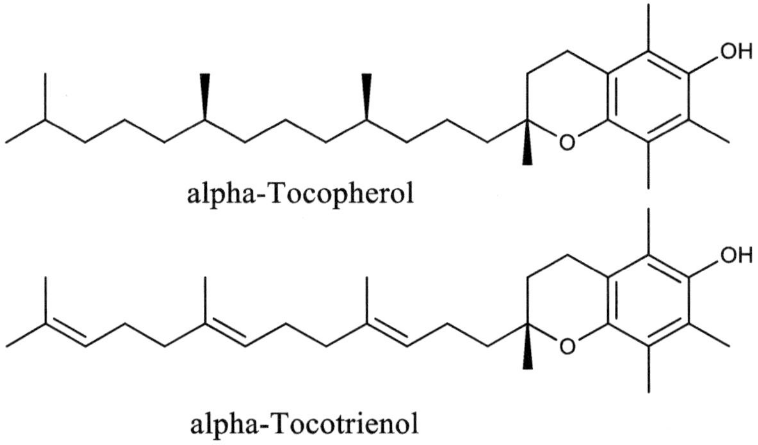

alpha-Tocopherol

alpha-Tocotrienol

Fig. 4.39 Structure of vitamin E (i.e., tocopherol and tocotrienol)

Fig. 4.40 Structure of vitamin C

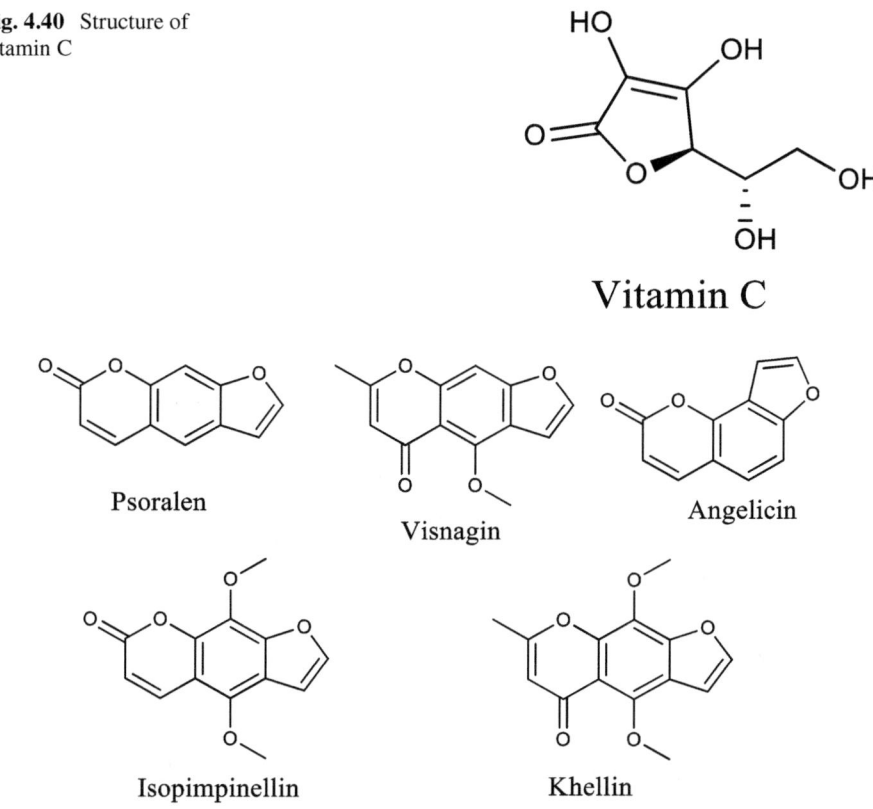

Fig. 4.41 Structures of psoralen, isopimpinellin, khellin, visnagin, and angelicin

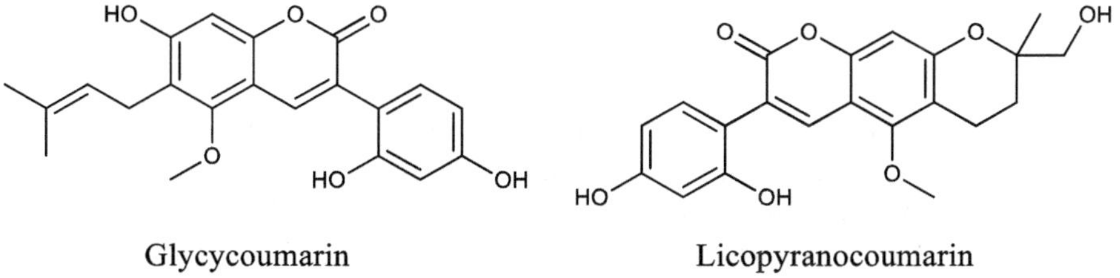

Fig. 4.42 Structures of glycycoumarin and licopyranocoumarin

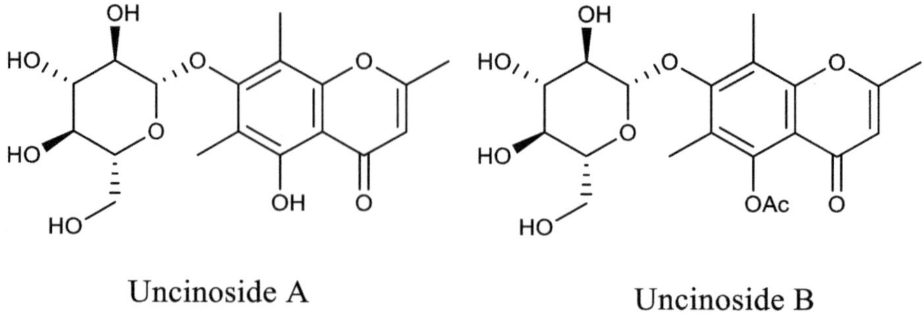

Fig. 4.43 Structures of uncinosides A and B

Table 4.2 Organosulfur phytochemicals isolated from Brassicaceae family and *Allium* [96]

Sr. no	Compounds	Type of compounds	Natural source
1.	Glucobrassicin	Glucosinolates	Cauliflower, Brussels, mustard, cabbage, choy, kale, water garden cress, sprouts, Bok, radish
2.	Allyl sulfides Dithiolethiones	Sulfides	Garlic, broccoli, onion
3.	Sulphoraphanes Phenylethyl Isothiocyanates	Isothiocyanates	Cabbage, kale, cauliflower, radish, Brussels, mustard, water garden cress, bok choy, sprouts

DHA. So results suggested that the antiviral potential of vitamin C is not only of its antioxidant property.

Chromones and Coumarins

Chromones and furanocoumarins are commonly present phytochemicals in various plant families. These natural compounds are abundantly found in families Umbelliferae and Rutaceae. According to research study, Khellin and visnagin were isolated from *Ammi* species (Umbelliferae) and reported their antiviral activities [92]. Psoralen, isopimpinellin, 8-methoxypsoralen, coriandrin, and angelicin were extracted from *Coriandrum sativum* and tested their antiviral potential against HIV, DNA, and RNA viruses and bacteriophages [93] (Figs. 4.41, 4.42, and 4.43).

Two potent anti-HIV coumarins, i.e., glycycoumarin and licopyranocoumarin, were isolated from *Glycyrrhiza Glabra* [94].

According to research study, two chromone glycosides, i.e., uncinosides A and B, were extracted from *Selaginella uncinata* and reported their potent antiviral activity against respiratory syncytial virus (RSV) infection [95].

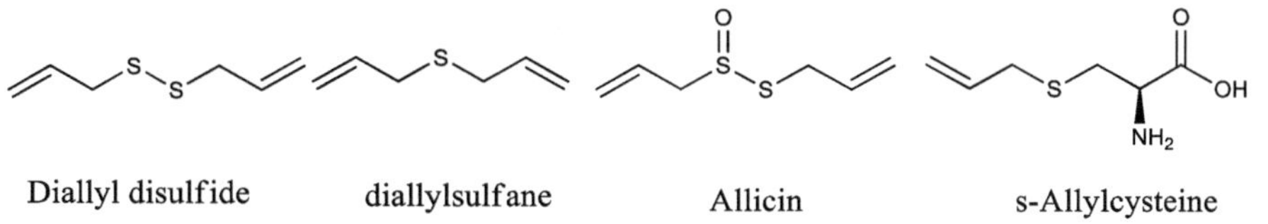

Diallyl disulfide   diallylsulfane   Allicin   s-Allylcysteine

Fig. 4.44 Structures of diallyl disulfide, diallyl sulfide, allicin, and s-allylcysteine

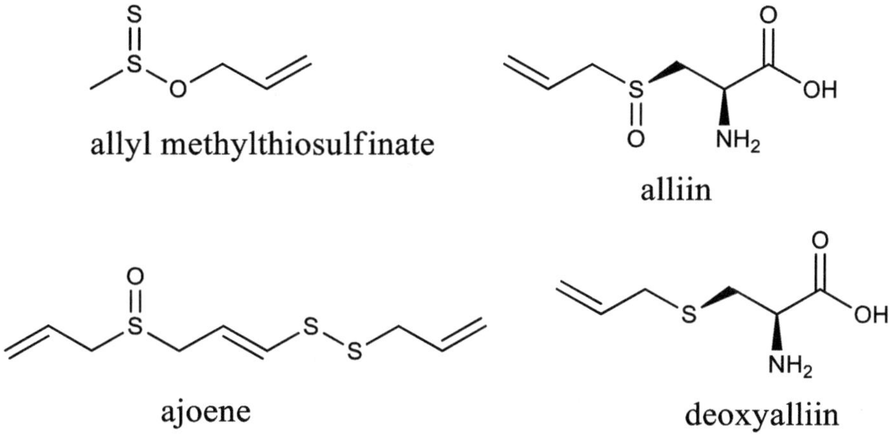

allyl methylthiosulfinate

alliin

ajoene

deoxyalliin

Fig. 4.45 Structures of allyl methyl thiosulfinate, ajoene, deoxyalliin, and alliin

Organosulfur Compounds

Phytochemicals having sulfur atom are mainly isolated from Brassicaceae family and *Allium* (Table 4.2). The organosulfur phytochemicals having pungent odor are chemically unstable. The organosulfur compounds possess excellent antiviral potential [96]. A handsome number of organosulfur compounds showed antiviral activities (Fig. 4.44).

Various unsymmetrical alkyl-aryl disulfides were synthesized and then oxidized into thiolsulfinate [97].

A considerable large number of organosulfur compounds were isolated from fresh garlic extract like diallyl thiosulfinate, diallyl disulfide, diallyl trisulfide, allyl methyl thiosulfinate, alliin, allicin, deoxyalliin, and ajoene. These compounds were then evaluated for their in vitro virucidal potential [98] (Fig. 4.45).

The antiviral mechanistic action of garlic containing phytochemicals against selected strains of viruses, i.e., HSV type 1 and type 2, vesicular stomatitis virus, human rhinovirus type 2, vaccinia virus, and parainfluenza virus type 3, were reported. The results indicated that in vitro virucidal activity depends upon the type of viral envelope and the cytotoxicity may depend upon the cell membrane. Virucidal activity is shown by inhibition of viral penetration or adsorption for non-enveloped

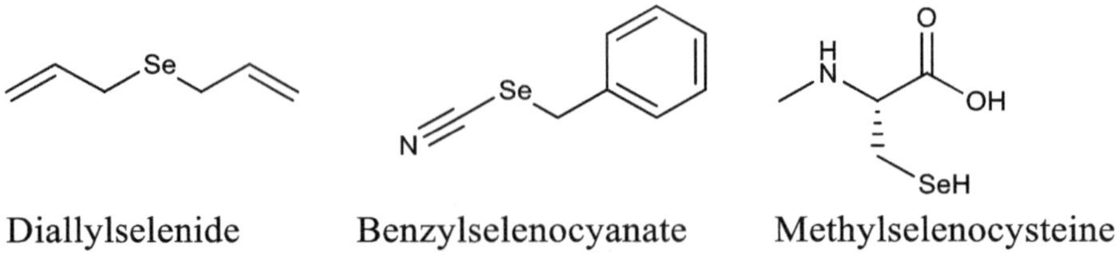

Fig. 4.46 Structures of diallylselenide, benzylselenocyanate, and methylselenocysteine

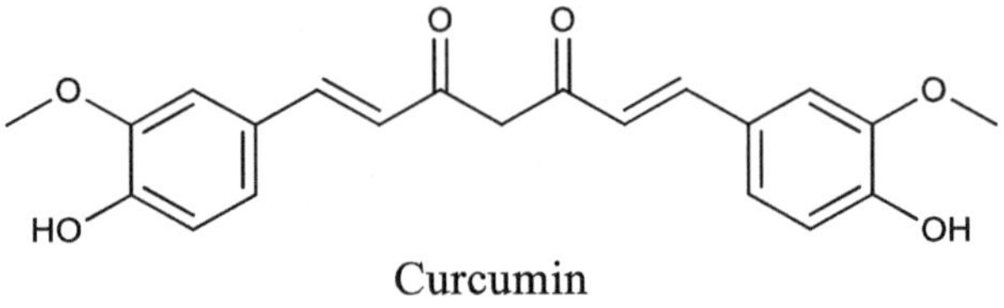

Fig. 4.47 Structure of curcumin

virus. The increasing virucidal activity order was ajoene > allicin > allyl-methyl thiosulfinate.

Selenium Compounds

Selenium compounds also have antiviral potential against different strains of viruses. Various different selenium compounds with effective antiviral activity are shown in Fig. 4.46.

The experimental results of these compounds against different viral infections demonstrate the significance of selenium-based compounds. Antiviral potential of three selenium compounds against *Coxsackievirus* was B5 reported via targeting their replication rate [99, 100]. Selenite inhibited the replication of *Coxsackievirus* B5 more effectively, while selenomethionine and selenite did not show any significant antiviral potential. Ebselen derivatives were synthesized and tested for their in vitro antiviral potential. The results demonstrated that few tested analogs efficiently target the herpes simplex virus type 1 (HSV-1) and encephalomyocarditis virus (EMCV) by inhibiting the cytopathic activity [101].

Miscellaneous Antiviral Phytochemicals

Curcumin and Its Derivative

Curcumin is one of the important constituents of turmeric. Different analogs of curcumin showed significant antiviral activity against HIV-1 integrase [102]. Curcumin also shows anti-HCV potential by inhibiting the replication of HCV,

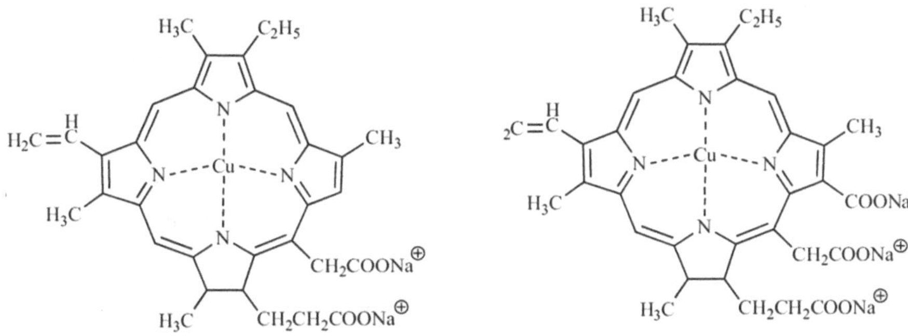

D isodum Copper Chlorin

Trisodum Copper Chlorin

Fig. 4.48 Structures of disodium copper chlorin and trisodium copper chlorin

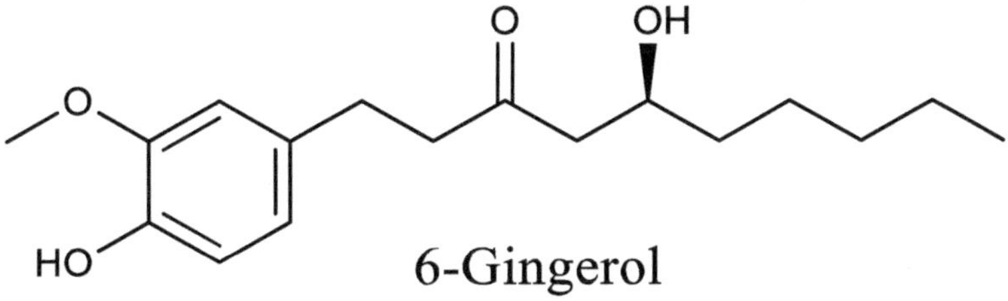

6-Gingerol

Fig. 4.49 Structure of 6-gingerol

targeting the suppression of sterol regulatory element binding protein-1 (SREBP-1) Akt pathway [103] (Fig. 4.47).

Chlorophyllin

Synthetic derivative of chlorophyll, i.e., chlorophyllin (CHLN) (Fig. 4.48), has antimutagenic activity against various environmental pollutants. Chlorophyllin was evaluated for virucidal activity against poliovirus by inhibiting the nuclear fragmentation (NF) in HEp-2-infected cells [104].

Gingerols

Gingerols (Fig. 4.49) have been isolated from ginger and traditionally have been used for the treatment of throat infections and common colds. Gingerols also form an important ingredient of Ayurvedic formulations. Gingerols reported excellent antiviral activity against many strains of viruses [105].

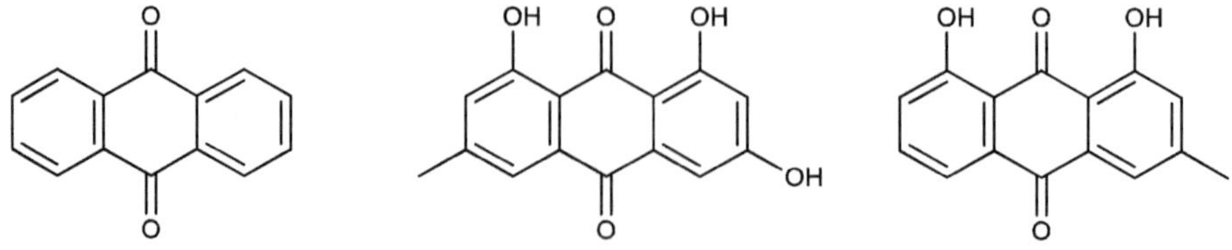

Fig. 4.50 Structures of anthraquinone, emodin, and chrysophanic

Chitin and Chitosan

Chitin, a natural polysaccharide, has partially deacetylated amino sugar *N*-acetylglucosamine basic skeleton, while chitosan is the deacetylated form of chitin. A research study on SCM chitin III (i.e., carboxymethyl chitin having 7.66% degree of sulfation) as its antiviral potential against HSV and Friend murine leukemia helper virus (F-MuLV) was reported [106].

Anthraquinone

Anthraquinone (9, 10-dioxoanthracene Fig. 4.50) is an anthracene derivative. Chrysophanic acid (1,8-dihydroxy-3-methylanthraquinone) is extracted from *Dianella longifolia* and tested for its in vitro antiviral activity against poliovirus types 2 and 3 by inhibiting the viral replication. Emodin was extracted from genus *Polygonum* and *Rheum* and exhibited excellent potential for the treatment of severe acute respiratory syndrome (SARS), which is caused by novel coronavirus (SARS-CoV). A type I membrane-bound protein, i.e., SARS-CoV spike (S) protein, is necessary for the attachment of virus to the angiotensin-converting enzyme 2 (ACE2) receptor of the host cell. Emodin effectively blocked the interaction of S protein with ACE2. It also showed significant inhibitory antiviral action against S protein-pseudotyped retrovirus to Vero E6 cells [107, 108].

Conclusion

In this chapter, we have discussed antiviral agents belonging to the different classes of natural phytochemicals like alkaloids, terpenes, flavonoids, tannins, vitamins, etc. There lies a huge scope in the development of new antiviral drugs due to the dire need. Nature has preserved a huge range of therapeutic natural compounds in the plant body. The coronavirus pandemic has enforced the scientific community to search for new antiviral agents. The work described herein is the sum-up of literature data described in previous few decades for the antiviral natural products and is expected to be useful for researchers working to develop new antiviral products.

References

1. Ball MJ, Lukiw WJ, Kammerman EM, Hill JM (2013) Intracerebral propagation of Alzheimer's disease: Strengthening evidence of a herpes simplex virus etiology. Alzheimers Dementia 9:169–175
2. Hober D, Sane F, Jaidane H, Riedweg K, Goffard A, Desailloud R (2012) Immunology in the clinic review series; focus on type 1 diabetes and viruses: Role of antibodies enhancing the infection with Coxsackie virus Bin the pathogenesis of type 1 diabetes. Clin Exp Immunol 168:47–51
3. Morgan RL, Baack B, Smith BD, Yartel A, Pitasi M, Falck Ytter Y (2013) Eradication of hepatitis C virus infection and the development of hepatocellular carcinoma: A meta analysis of observational studies. Ann Intern Med 158:329–337
4. Christou L (2011) The global burden of bacterial and viral zoonotic infections. Clin Microbiol Infect 17:326–330
5. Cascio A, Bosilkovski M, Rodriguez Morales AJ, Pappas G (2011) The socio ecology of zoonotic infections. Clin Microbiol Infect 17:336–342
6. Grais RF, Strebel P, Mala P, Watson J, Nandy R, Gayer M (2011) Measles vaccination in humanitarian emergencies: a review of recent practice. Confl Heal 5:21
7. International Symposium on Medicinal Plants, April 19–21, 1993, in Philadelphia, U.S.A The Morris Arboretum and the World Health Organization (WHO)
8. Hsu FL, Chang HT, Chang ST (2007) Evaluation of antifungal properties of octyl gallate and its synergy with cinnamaldehyde. Bioresour Technol 98:734–738
9. Serkedjieva JA (1996) A polyphenolic extract from Geranium sanguineum L. Inhibits influenza virus protein expression. Phytother Res 10(5):441–443
10. Cheng HY, Yang CM, Lin TC, Shieh DE, Lin CC (2006) ent Epiafzelechin (4alpha >8) epiafzelechin extracted from Cassia javanica inhibits herpes simplex virus type 2 replication. J Med Microbiol 55:201–206
11. Cheng HY, Yang CM, Lin TC, Lin LT, Chiang LC, Lin CC (2011) Excoecarianin, Isolated from Phyllanthus urinaria Linnea, Inhibits Herpes Simplex Virus Type 2 Infection through Inactivation of Viral Particles. Evid Based Complement Alternat Med 2011:259–103
12. Lin LT, Chen TY, Chung CY, Noyce RS, Grindley TB, McCormick C, Lin TC, Wang GH, Lin CC, Richardson CD (2011) Hydrolyzable tannins (chebulagic acid and punicalagin) target viral glycoprotein-glycosaminoglycan interactions to inhibit herpes simplex virus 1 entry and cell-to-cell spread. J Virol 85:4386–4398
13. Petrera E, Coto CE (2009) Therapeutic effect of meliacine, an antiviral derived from Melia azedarach L., in mice genital herpetic infection. Phytother Res 23:1771–1777
14. Bettega JMR, Teixeira H, Bassani VL, Barardi CRM, Simões CMO (2004) Evaluation of the antiherpetic activity of standardized extracts of Achyrocline satureioides. Phytother Res 18:819–823
15. Hood JL, Jallouk AP, Campbell N, Ratner L, Wickline SA (2013) Cytolytic nanoparticles attenuate HIV 1 infectivity. Antivir Ther 18:95–103
16. Kudo E, Taura M, Matsuda K, Shimamoto M, Kariya R, Goto H, Hattori S, Kimura S, Okada S (2013) Inhibition of HIV-1 replication by a tricyclic coumarin GUT-70 in acutely and chronically infected cells. Bioorg Med Chem Lett 23(3):606–609
17. Lin YM, Flavin MT, Schure R, Chen FC, Sidwell R, Barnard DI, Huffmann JH, Kern ER (1999) Antiviral activities of biflavonoids. Planta Med 65(02):120–125
18. Semple SJ, Nobbs SF, Pyke SM, Reynolds GD, Flower RL (1999) Antiviral flavonoid from Pterocaulon sphacelatum, an Australian Aboriginal medicine. J Ethnopharmacol 68(1–3):283–288
19. Yu YB, Miyashiro H, Nakamura N, Hattori M, Park JC (2007) Effects of triterpenoids and flavonoids isolated from Alnus firma on HIV-1 viral enzymes. Arch Pharm Res 30(7):820

20. Li BQ, Fu T, Dongyan Y, Mikovits JA, Ruscetti FW, Wang JM (2000) Flavonoid baicalin inhibits HIV-1 infection at the level of viral entry. Biochem Biophys Res Commun 276(2):534–538
21. Theisen LL, Muller CP (2012) EPs (R) 7630 (Umckaloabo (R)), an extract from Pelargonium sidoides roots, exerts anti influenza virus activity in vitro and in vivo. Antivir Res 94:147 56
22. He W, Han H, Wang W, Gao B (2011) Anti-influenza virus effect of aqueous extracts from dandelion. Virol J 8:538
23. Ishitsuka H, Ohsawa C, Ohiwa T, Umeda I, Suhara Y (1982) Anti-Picornavirus flavone Ro 09-0179. Antimicrob Agents Chemother 22(4):611–616
24. Malhotra B, Onyilagha JC, Bohm BA, Towers GH, James D, Harborne JB, French CJ (1996) Inhibition of tomato ringspot virus by flavonoids. Phytochemistry 43(6):1271–1276
25. Spedding G, Ratty A, Middleton E Jr (1989) Inhibition of reverse transcriptases by flavonoids. Antivir Res 12(2):99–110
26. Naithani R, Huma LC, Holland LE, Shukla D, McCormick DL, Mehta RG, Moriarty RM (2008) Antiviral activity of phytochemicals: a comprehensive review. Mini-Rev Med Chem 8(11):1106–1133
27. Vlietinck AJ, De Bruyne T, Vanden Berghe DA (1997) Plant substances as antiviral agents. Curr Org Chem 1(4):307–344
28. Watkins T, Resch W, Irlbeck D, Swanstrom R (2003) Selection of high-level resistance to human immunodeficiency virus type 1 protease inhibitors. Antimicrob Agents Chemother 47(2):759–769
29. Farnsworth NR (1966) Biological and phytochemical screening of plants. J Pharm Sci 55(3):225–276
30. Houghton PJ, Woldemariam TZ, Khan AI, Burke A, Mahmood N (1994) Antiviral activity of natural and semi-synthetic chromone alkaloids. Antivir Res 25(3–4):235
31. Zhang CH, Wang YF, Liu XJ, Lu JH, Qian CW, Wan ZY, Yan XG, Zheng HY, Zhang MY, Xiong S, Li JX (2005) Antiviral activity of cepharanthine against severe acute respiratory syndrome coronavirus in vitro. Chin Med J 118(6):493–496
32. Yang CW, Lee YZ, Kang IJ, Barnard DL, Jan JT, Lin D, Huang CW, Yeh TK, Chao YS, Lee SJ (2010) Identification of phenanthroindolizines and phenanthroquinolizidines as novel potent anti-coronaviral agents for porcine enteropathogenic coronavirus transmissible gastroenteritis virus and human severe acute respiratory syndrome coronavirus. Antivir Res 88(2):160–168
33. Shen L, Niu J, Wang C, Huang B, Wang W, Zhu N, Deng Y, Wang H, Ye F, Cen S, Tan W (2019) High-throughput screening and identification of potent broad-spectrum inhibitors of coronaviruses. J Virol 93(12):e00023-19
34. Vanden Berghe DA, Ieven M, Mertens F, Vlietinck AJ, Lammens E (1978) Screening of higher plants for biological activities. II. Antiviral activity. Lloydia 41:463–471
35. Vanden Berghe DA, Vlietinck AJ, Van Hoof L (1986) Plant substances as antiviral agents. Buletin de l Institute Pasteur 84:101–147
36. Ieven M, Vlietinck AJ, Berghe DV, Totte J, Dommisse R, Esmans E, Alderweireldt F (1982) Plant antiviral agents. III. Isolation of alkaloids from Clivia miniate Regel (Amaryl-lidaceae). J Nat Prod 45(5):564–573
37. Montanha JA, Amoros M, Boustie J, Girre L (1995) Anti-herpes virus activity of aporphine alkaloids. Planta Med 61(05):419–424
38. Thomas PR, Nash GB, Dormandy JA (1988) White cell accumulation in dependent legs of patients with venous hypertension: a possible mechanism for trophic changes in the skin. Br Med J (Clin Res Ed) 296(6638):1693–1695
39. Elsohly HN, El-Feraly FS, Joshi AS, Walker LA (1997) Antiviral flavonoids from Alkanna orientalis. Planta Med 63(04):384
40. Loewenstein WR (1979) Junctional intercellular communication and the control of growth. Biochimica et Biophysica Acta (BBA)-Reviews on Cancer 560(1):1–65

41. Gerdin B, Svensjö E (1983) Inhibitory effect of the flavonoid O-(beta-hydroxyethyl)-rutoside on increased microvascular permeability induced by various agents in rat skin. Int J Microcirc Clin Exp 2(1):39–46
42. Li H, Zhou CX, Pan Y, Gao X, Wu X, Bai H, Zhou L, Chen Z, Zhang S, Shi S, Luo J (2005 Nov) Evaluation of antiviral activity of compounds isolated from Ranunculus sieboldii and Ranunculus sceleratus. Planta Med 71(12):1128–1133
43. Likhitwitayawuid K, Supudompol B, Sritularak B, Lipipun V, Rapp K, Schinazi RF (2005) Phenolics with Anti-HSV and Anti-HIV Activities from Artocarpus gomezianus., Mallotus pallidus., and Triphasia trifolia. Pharm Biol (Philadelphia, PA, United States) 43:651–657
44. Deng J, Kelley JA, Barchi JJ, Sanchez T, Dayam R, Pommier Y, Neamati N (2006) Mining the NCI antiviral compounds for HIV-1 integrase inhibitors. Bioorg Med Chem 14:3785–3792
45. Nowakowska Z (2007) A review of anti-infective and anti-inflammatory chalcones. Eur J Med Chem 42(2):125–137
46. Park JY, Yuk HJ, Ryu HW, Lim SH, Kim KS, Park KH, Ryu YB, Lee WS (2017) Evaluation of polyphenols from Broussonetia papyrifera as coronavirus protease inhibitors. J Enzyme Inhibition Med Chem 32(1):504–512
47. Kim DW, Seo KH, Curtis-Long MJ, Oh KY, Oh JW, Cho JK, Lee KH, Park KH (2014) Phenolic phytochemical displaying SARS-CoV papain-like protease inhibition from the seeds of Psoralea corylifolia. J Enzyme Inhibition Med Chem 29(1):59–63
48. Phrutivorapongkul A, Lipipun V, Ruangrungsi N, Kirtikara K, Nishikawa K, Maruyama S, Watanabe T, Ishikawa T (2003) Studies on the chemical constituents of stem bark of Millettia leucantha: isolation of new chalcones with cytotoxic, anti-herpes simplex virus and anti-inflammatory activities. Chem Pharm Bull 51(2):187–190
49. Likhitwitayawuid K, Chaiwiriya S, Sritularak B, Lipipun V (2006) Antiherpetic flavones from the heartwood of Artocarpus gomezianus. Chem Biodivers 3(10):1138–1143
50. Prendergast, P.T.US Patent 2003, Patent No 6555523
51. Santhosh C, Mishra PC (1996) Molecular electrostatic potential mapping and structure-activity relationship for 3-methoxy flavones. Indian J Biochem Biophys 33:458
52. Ryu YB, Jeong HJ, Kim JH, Kim YM, Park JY, Kim D, Lee WS (2010) Biflavonoids from Torreya nucifera displaying SARSCoV3CL (pro) inhibition. Bioorg Med Chem 18:7940–7947
53. Polyak SJ, Morishima C, Shuhart MC, Wang CC, Liu Y, Lee DY (2007) Inhibition of T cell inflammatory cytokines, hepatocyte NF kappaB signaling, and HCV infection by standardized Silymarin. Gastroenterology 132:1925–1936
54. Polyak SJ, Morishima C, Lohmann V, Pal S, Lee DY, Liu Y, Graf TN, Oberlies NH (2010) Identification of hepatoprotective flavonolignans from silymarin. Proc Natl Acad Sci 107(13):5995–5999
55. Ferenci P, Scherzer TM, Kerschner H, Rutter K, Beinhardt S, Hofer H, Schöniger–Hekele M, Holzmann H, Steindl–Munda P (2008) Silibinin is a potent antiviral agent in patients with chronic hepatitis C not responding to pegylated interferon/ribavirin therapy. Gastroenterology 135(5):1561–1567
56. Neumann UP, Biermer M, Eurich D, Neuhaus P, Berg T (2010) Successful prevention of hepatitis C virus (HCV) liver graft reinfection by silibinin mono-therapy. J Hepatol 52:951–952
57. Mariño Z, Crespo G, D'Amato M, Brambilla N, Giacovelli G, Rovati L, Costa J, Navasa M, Forns X (2013) Intravenous silibinin monotherapy shows significant antiviral activity in HCV-infected patients in the peri-transplantation period. J Hepatol 58(3):415–420
58. Huang W, Zhang X, Wang Y, Ye W, Ooi VE, Chung HY, Li Y (2010) Antiviral biflavonoids from Radix wikstroemiae (Liaogewanggen). Chin Med 5(1):23
59. Wang Y, Chen M, Zhang J, Zhang XL, Huang XJ, Wu X, Zhang QW, Li YL, Ye WC (2012) Flavone C-glycosides from the leaves of Lophatherum gracile and their in vitro antiviral activity. Planta Med 78(01):46–51

60. Moriarty R M, Surve B C, Naithani R, Chandersekera S N, Tiwari V, Shukla D. Synthesis and antiviral activity of Abyssinone II analogs., Abstracts of Papers, 233rd ACS National Meeting, Chicago, IL, United States. 2007;25–29
61. Cho JK, Curtis-Long MJ, Lee KH, Kim DW, Ryu HW, Yuk HJ, Park KH (2013) Geranylated flavonoids displaying SARS-CoV papain-like protease inhibition from the fruits of Paulownia tomentosa. Bioorg Med Chem 21(11):3051–3057
62. Schwarz S, Sauter D, Wang K, Zhang R, Sun B, Karioti A, Bilia AR, Efferth T, Schwarz W (2014) Kaempferol derivatives as antiviral drugs against the 3a channel protein of coronavirus. Planta Med 80(02–03):177
63. Kaul TN, Middleton E Jr, Pearay MD, Ogra L (1985) Antiviral Effect of Flavonoids on Human Viruses. J Med Vir 15(1):71–79
64. Twaij HAA, Sayed-Ali HM, AlZohry AM (1988) J Biol Sci Res 19(1):41
65. Zhuang M, Jiang H, Suzuki Y, Li X, Xiao P, Tanaka T, Ling H, Yang B, Saitoh H, Zhang L, Qin C (2009 Apr 1) Procyanidins and butanol extract of Cinnamomi Cortex inhibit SARS-CoV infection. Antivir Res 82(1):73–81
66. Takatsuki A, Tamura G, Arima K (1969) Antiviral and antitumor antibiotics. XIV. Effects of ascochlorin and other respiration inhibitors on multiplication of Newcastle disease virus in cultured cells. Appl Microbiol 17:825
67. Mashava P (1996) PCT Int Appl:38
68. Charles L, Laure F, Raharivelomanana P, Bianchini JP (2005) Sheath liquid interface for the coupling of normal phase liquid chromatography with electrospray mass spectrometry and its application to the analysis of neoflavonoids. J Mass Spectrom 40(1):75–82
69. Shu YZ (1998) Recent natural products based drug development: a pharmaceutical industry perspective. J Nat Prod 61(8):1053–1071
70. Wen CC, Kuo YH, Jan JT, Liang PH, Wang SY, Liu HG, Lee CK, Chang ST, Kuo CJ, Lee SS, Hou CC (2007) Specific plant terpenoids and lignoids possess potent antiviral activities against severe acute respiratory syndrome coronavirus. J Med Chem 50(17):4087–4095
71. Cheng PW, Ng LT, Chiang LC, Lin CC (2006) Antiviral effects of saikosaponins on human coronavirus 229E in vitro. Clin Exp Pharmacol Physiol 33(7):612–616
72. Melikian G, Mmiro F, Ndugwa C, Perry R, Jackson JB, Garrett E, Tielsch J, Semba RD (2001) Relation of vitamin A and carotenoid status to growth failure and mortality among Ugandan infants with human immunodeficiency virus. Nutrition 17(7–8):567–572
73. McCormick JL, McKee TC, Cardellina JH, Boyd MR (1996) HIV inhibitory natural products. 26. Quinoline alkaloids from Euodia roxburghiana. J Nat Prod 59(5):469–471
74. Eich E (1998) Secondary metabolites from plants as antiretroviral agents: promising lead structure for anti-HIV drugs of the future. In: Lawson LD, Bauer R (eds) ACS symposium series 691: Phytomedicines of Europe chemistry and biological activity. American Chemical Society, Washington DC, pp 83–95
75. Takechi M, Tanaka Y (1981) Purification and characterization of antiviral substance from the bud of Syzygium aromatica. Planta Med 42(05):69–74
76. Takechi M, Tanaka Y (1982) Antiviral Substances from The Root of Paeonia species. Planta Med 4:252–253
77. Okuda T, Yoshida T, Hatano T (1989) Ellagitannins as active constituents of medicinal plants. Planta Med 55(02):117–122
78. Takechi M, Tanaka Y, Takehara M, Nonaka GI, Nishioka I (1985) Structure and antiherpetic activity among the tannins. Phytochemistry 24(10):2245–2250
79. Corthout J, Pieters LA, Claeys M, Berghe DV, Vlietinck AJ (1991) Antiviral ellagitannins from Spondias mombin. Phytochemistry 30(4):1129–1130
80. Kakiuchi N, Hattori M, Namba T, Nishizawa M, Yamagishi T, Okuda T (1985) Inhibitory effect of tannins on reverse transcriptase from RNA tumor virus. J Nat Prod 48(4):614–621
81. Kwon HJ, Ryu YB, Kim YM, Song N, Kim CY, Rho MC, Jeong JH, Cho KO, Lee WS, Park SJ (2013 Aug 1) In vitro antiviral activity of phlorotannins isolated from Ecklonia cava

against porcine epidemic diarrhea coronavirus infection and hemagglutination. Bioorg Med Chem 21(15):4706–4713
82. Chen CN, Lin CP, Huang KK, Chen WC, Hsieh HP, Liang PH, Hsu JT (2005) Inhibition of SARS-CoV 3C-like protease activity by theaflavin-3, 3′-digallate (TF3). Evid Based Complement Alternat Med 2(2):209–215
83. Lin LT, Chen TY, Lin SC, Chung CY, Lin TC, Wang GH, Anderson R, Lin CC, Richardson CD (2013) Broad-spectrum antiviral activity of chebulagic acid and punicalagin against viruses that use glycosaminoglycans for entry. BMC Microbiol 13(1):187
84. Andreone P, Fiorino S, Cursaro C, Gramenzi A, Margotti M, Di Giammarino L, Biselli M, Miniero R, Gasbarrini G, Bernardi M (2001) Vitamin E as treatment for chronic hepatitis B: results of a randomized controlled pilot trial. Antivir Res 49(2):75–81
85. Jariwalla R, Harakeh S (1996) In: Harris JR (ed) Subcellular Biochemistry: Ascorbic Acid: Biochem. Biomed. Cell Biol., vol 25. Springer US, PlenumPress New York, p 215–231
86. Hamuy R, Berman B (1998) Treatment of herpes simplex virus infections with topical antiviral agents. Eur J Dermatol 8:310–319
87. Mossad SB (2003) Current and future therapeutic approaches to the common cold. Expert Rev Anti-Infect Ther 1(4):619–626
88. Pauling L (1974) Are recommended daily allowances for vitamin C adequate? Proc Natl Acad Sci 71(11):4442–4446
89. Cameron E (1982) Vitamin C and cancer: an overview. International journal for vitamin and nutrition research. Supplement Internationale Zeitschrift fur Vitamin-und Ernahrungsforschung. Supplement 23:115–127
90. Bub A, Watzl B, Blockhaus M, Briviba K, Liegibel U, Müller H, Pool-Zobel BL, Rechkemmer G (2003) Fruit juice consumption modulates antioxidative status, immune status and DNA damage. J Nutr Biochem 14(2):90–98
91. Kolb H, Kolb-Bachofen V (1992) Nitric oxide: a pathogenetic factor in autoimmunity. Immunol Today 13(5):157–160
92. Hudson JD (1990) Antiviral compounds from plants. CRC Press, Inc., BocaRaton, pp 43–57
93. Ng TB, Huang B, Fong WP, Yeung HW (1997) Anti-human immunodeficiency virus (anti-HIV) natural products with special emphasis on HIV reverse transcriptase inhibitors. Life Sci 61(10):933–949
94. Hatano T, Yasuhara T, Miyamoto K, Okuda T (1988) Anti-human immunodeficiency virus phenolics from licorice. Chem Pharm Bull 36(6):2286–2288
95. Ma LY, Ma SC, Wei F, Lin RC, But PP, Lee SH, Lee SF (2003) Uncinoside A and B, two new antiviral chromone glycosides from Selaginella uncinata. Chem Pharm Bull 51(11):1264–1267
96. Heldreth B, Turos E (2005) Microbiological properties and modes of action of organosulfur-based anti-infectives. Curr Med Chem – Anti-Infect Agents 4:293–315
97. Hunter R, Caira M, Stellenboom N (2005) Thiolsulfinate allicin from garlic: inspiration for a new antimicrobial agent. Nat Prod Mol Ther 1056:234–241
98. Tait S, Salvati AL, Desideri N, Fiore L (2006) Antiviral activity of substituted homoisoflavonoids on enteroviruses. Antivir Res 72(3):252–255
99. Cermelli C, Vinceti M, Scaltriti E, Bazzani E, Beretti F, Vivoli G, Portolani M (2002) Selenite inhibition of Coxsackie virus B5 replication: implications on the etiology of Keshan disease. J Trace Elem Med Biol 16(1):41–46
100. Wójtowicz H, Kloc K, Maliszewska I, Młochowski J, Piętka M, Piasecki E (2004) Azaanalogues of ebselen as antimicrobial and antiviral agents: synthesis and properties. Il Farmaco 59(11):863–868
101. Singh S, Malik BK, Sharma DK (2007) Targeting HIV-1 through molecular modeling and docking studies of CXCR4: leads for therapeutic development. Chem Biol Drug Design 69:191–203

102. La Colla P, Tramontano E, Musiu C, Marongiu ME, Novellino E, Greco G, Massa S, Di Santo R, Costi R, Marino A (1998) In vitro and in vivo antiproliferative activity of IPCAP, a new pyrazole. Antivir Res 7:A57
103. Kim K, Kim KH, Kim HY, Cho HK, Sakamoto N, Cheong J (2010) Curcumin inhibits hepatitis C virus replication via suppressing the Akt-SREBP-1 pathway. FEBS Lett 584(4):707–712
104. Botelho MV, Orlandi JM, de Melo FL, Mantovani MS, Linhares RE, Nozawa C (2004) Chlorophyllin protects HEp-2 cells from nuclear fragmentation induced by poliovirus. Lett Appl Microbiol 39:174–177
105. Chrubasik S, Pittler MH, Roufogalis BD (2005) Zingiberis rhizoma: a comprehensive review on the ginger effect and efficacy profiles. Phytomedicine 12(9):684–701
106. Ishihara C, Yoshimatsu K, Tsuji M, Arikawa J, Saiki I, Tokura S, Azuma I (1993) Antiviral activity of sulfated chitin derivatives against Friend murine leukaemia and herpes simplex type-1 viruses. Vaccine 11(6):670–674
107. Semple SJ, Pyke SM, Reynolds GD, Flower RL (2001) In vitro antiviral activity of the anthraquinone chrysophanic acid against poliovirus. Antivir Res 49(3):169–178
108. Ho TY, Wu SL, Chen JC, Li CC, Hsiang CY (2007) Emodin blocks the SARS coronavirus spike protein and angiotensin-converting enzyme 2 interaction. Antivir Res 74(2):92–101

Chapter 5
Foods as First Defense Against COVID-19

Mahwish and Sarah I. Alothman

Introduction

COVID-19 [1] is a rapidly human-to-human transmissible disease [2, 3] that brings massive cataclysm to the world. The causative agent of the disease named severe acute respiratory syndrome coronavirus 2 (SARS-CoV-2) [4] causes severe complications in infected individuals and a significant number of deaths [5]. The highly susceptible populations to severe COVID-19 complications are immunocompromised [6]. The existence of noncommunicable diseases such as obesity [7], hypertension, cardiovascular diseases, diabetes, chronic lung diseases, chronic kidney diseases, renal diseases, cancer, and others renders the patient immunocompromised and more prone to infections [8].

The origin of all diseases is cellular. If the cells of the body remain healthy, they will make healthy tissues and in return healthy organs and ultimately healthy individuals. So, selection of right foods to nourish these cells is crucial to protect from noxious bodies and reduce complications associated with their invasion. Protection from diseases and healing and recovery from disease depend primarily on the food components we eat. There is no other way to improve health unless proper nourishment is not provided to the cells. Food is essential for sustenance of life because it provides essential moieties needed for normal functioning of the body, in fighting off diseases, and in reduction of severe outcomes/morbidity in case of infections.

Selection of right combination of food from each food group and across all major food groups is a preferred way to ensure appropriate nutrient intake [9]. The consumption of diversified food is important contributing factor particularly for critical

Mahwish (✉)
Institute of Home Sciences, University of Agriculture Faisalabad, Faisalabad, Pakistan

S. I. Alothman
Faculty of Sciences, Biology Department, Princess Nourah bint Abdulrahman University, Riyadh, Saudi Arabia

nutrient homeostasis that is continuously needed in health and diseased state. Human body is actually made from food substances that are consumed. The right food choices provide right information and materials needed for proper body functioning. Our health declines if right food choices are lacking, affecting metabolic processes and leading to different types of maladies.

The nutrients are actually nourishing substances in the form of chemical moieties that are crucial for regulating body functions, mainly growth and development. Certain nutrient intake is so critical that their deficiencies harness major aspects of metabolic pathways and declining of health. In extreme conditions, prolonged and continuous deficiency of particular nutrient hinders many aspects of cell activities resulting in slow accomplishment or even stoppage of metabolic processes, hence declining health status important to maintain in the wake of COVID-19.

Food intake with increased amount of unhealthy substances gives wrong instructions to our bodies that results in malnourished individuals that have more risk of developing noncommunicable diseases and vulnerability to COVID-19 as well. Hence, proper food intake and identification of such food groups that have profound outcomes in minimizing respiratory infections, boosting immunity, and maintaining organ health are crucial during COVID-19 pandemic.

Food choices may differ in different geological regions of the world due to variations in cultural aspects and eating behaviors. Selection of right combination of food irrespective of where we live is vital throughout life. Actually our main emphasis is on the foods rich in nutrients, vitamins, minerals, trace elements, phytochemicals, and antioxidants that are helpful to develop immunity needed to fight viruses.

In present massive upheaval time, complex combinations of food and nutrient intake are such an approach that are more translatable in terms of increasing immunity against infectious diseases and reducing multiple complications associated with COVID-19.

Role of Nutrition and Diet in Prevention and Management of COVID-19

Good nutrition and a healthy well-balanced diet is a pragmatic risk management approach for individuals during COVID-19 pandemic [10]. Diet plays an important role to reduce the susceptibility of developing an infection, minimizes complications during infection, and is continuously needed for normal functioning after recovery from infectious disease like COVID-19. Although no natural remedy or single food has been approved so far to prevent COVID-19 infections [11], still, it is mandatory to attain and maintain good nutritional status for overall health of individual and strengthening of immune system to fight against viruses. The importance of healthy diet due to its beneficial effects in past viral infections is of worth consideration in relation to COVID-19 pandemic. Different dietary patterns are adopted by individuals around the globe. The most famous are Western pattern diet and Mediterranean diet.

The Western pattern diet is generally characterized by high intakes of prepackaged foods, processed meat, red meat, butter, fried foods, sweets, candies, conventionally raised high-fat dairy and other animal products, highly refined grains, potatoes, corn, high-sugar drinks, and high-fructose corn syrup and low intakes of whole grains, grass-fed animal products, vegetables, fruits, nuts, seeds, and fish [12]. This dietary pattern is developed due to modern rapid industrial revolution in past decades.

The major features of Mediterranean diet include regular and high consumption of unrefined cereals, vegetables, fruits, legumes, olive oil, and reasonable amount of fish and dairy products (mostly yogurt and cheese) and minimum non-fish meat product consumption [12, 13]. Olive oil has been studied as a potential health factor for reducing all-cause mortality and the risk of chronic diseases lacking in other diets.

Some countries hit by COVID-19 are worsening [14]. The communities in these countries have now started to reexamine their nutrition status and dietary patterns. Mediterranean diet is important at this pandemic time for overall health maintenance. So, Mediterranean diet should be preferred over other diets to avoid COVID-19 symptoms and severe outcomes. Traditional Mediterranean diet intake has beneficial impact in reducing inflammation [15], diabetes, cancer, coronary artery disease, and cardiovascular disease [16, 17] and helps in minimizing vulnerability to COVID-19. Obesity, considered as mother of all diseases, is also substantially reduced by adherence to Mediterranean-type diet [18]. Mediterranean diet with least involvement of processing procedures, naturally low sodium contents, and abundant intake of fruit and vegetables [19] is imperative approach for controlling diseases and improving health.

The diet plays a pivotal role in enhancing host reaction to noxious and infectious agents and may be potentially beneficial to minimize the incidence or severity of COVID-19. The diet rich in various nutrients, phytochemicals, and antioxidants is mandatory for immunomodulation and immunocompetence, whereas deficiency in any nutritional aspect has direct link to high host vulnerability to viral infections and severe clinical outcomes during disease progression [20]. The dietary guidelines declared by WHO during present worldwide outbreak of COVID-19 also stress to adopt well-balanced diet comprised of suitable proportion of all major food groups to maintain and develop strong immunity to avoid chances of acquiring chronic diseases and infections [21].

The nutritional substances intake and absorption is considerably affected as age progresses leading to functional abnormalities and decline in immune system potency to fight off lethal pathogens in old-age people known as immunosenescence [22]. Hence, particular focus must be given to the nutritional aspects and dietary patterns of elderly communities at this time, as the elderly populations are not only vulnerable to develop chronic diseases, but they are extremely predisposed risk group possessing malnourishment and high chances of developing infections and COVID-19 [23, 24]. In current pandemic era, with enormous elderly cases, age itself is thought to be a major risk factor for COVID-19 [24]. The level of vitamin

D, folate, calcium, vitamin C, zinc, and many other vital bioactive molecules become deficient with age [25] due to several reasons including poor mental health, improper dietary approaches with enormous nutritional deficiencies, adverse socio-economic circumstances, and a host of number of multifactorial issues [26]. Malnourishment, particularly undernutrition, in older age can exasperate proper functioning of cells in the immune system and resultantly increase the risk of developing infectious diseases [27].

The old-age individuals can be protected from chronic inflammation that is the major cause of onset of noncommunicable diseases and infections, by a regular intake of balanced diet that constitutes all the necessary macronutrients, micronutrients, probiotics, and prebiotics that are predominantly essential to restore immune cell function [28].

In addition, a healthy balanced diet is also valuable for the general population. Utilization of a diet comprised of abundant nutritional substances with anti-inflammatory properties on continuous basis by unaffected individuals reduces the chances of infliction of disease on the one hand and prevention from severe outcomes due to comorbidities and COVID-19 on the other hand. It is noteworthy from past viral infections that good nutritional status reduces the chances of severe complexities during the course of disease [29] urging to advocate this strategy against COVID-19. The prevalence of COVID-19 outbreak around the globe emphasizes upon maintaining good nutritional status of individuals [30]. In fact, a large number of nutrients from dietary sources have a potentially enhancing immunomodulatory and anti-inflammatory role against NCDs [31–33]; thus, nutrient-rich diet is critically focal point against COVID-19.

Nutrition might play a profound role in present challenging task of maintaining health of infected as well as unaffected individuals due to safe and ease in application. Intake of all the necessary dietary components and metabolites and adherence to healthy dietary patterns affect the systemic markers related to function of cells in immune system. However, complex interaction between immunology and nutrition is not yet fully comprehended, but in this crucial period of massive devastation due to SARS-CoV-2, attaining reference nutrient intakes or recommended daily allowance must be devised particularly for those nutrients that play a dynamic role in proper functioning of immune system. The efficient immune system supports the restraint of viral entry and its spread and minimizes severe complications associated with such viral infections.

The increased risk of infection and related complexities are directly related to nutritional deficiencies, so monitoring of nutritional status of an infected patient is crucial to avoid severity of infection of COVID-19 [34]. Nutrition of intensive care unit patients must also be considered [35] due to rapid decline of nutrients in their critical state. Due to potential role and beneficial clinical outcomes of nutritional interventions, nutritional status of infected individuals must not be overlooked [36].

Variations in dietary habits in different ethnicities can be considered as contributory factor for differences in COVID-19 cases and death rates in different geographical regions of the world [14]. In fact, lower infection rate and mortality might be due to consuming certain foods [14]. This is due to the reason that the transmembrane angiotensin-converting enzyme (ACE2) facilitates the entry of SARS-CoV-2 into the cells and levels of ACE2 are directly related to food substances and dietary patterns. For example, broccoli protein hydrolysate showed prominent ACE inhibitory activity and subsequent hypotensive effects [37], whereas diet with high saturated fat increases the level of ACE [38]. Many other nutrients and dietary approaches affect ACE levels, with some foods having enhancing activities and others having ACE inhibitory activities [39]. Furthermore, the level of ACE in the blood is highly sensitive to intake of food [40]. Hence, it is vital to maintain a healthy nutritional status.

Currently, there is scarcity of knowledge about reliable COVID diet that reduces the incidence and mortality due to COVID-19. However, a healthy nutritional status has significance at this time because the SARS-CoV-2 is affecting populations continuously and the battle against COVID-19 seems to persist for longer period from that predicted initially. SARS-CoV-2 rapidly infects millions of people around the globe and resultantly badly affects all aspects of life. Due to lifestyle changes, like social isolation and restrictions in movements, noncommunicable diseases may increase to alarming level that is already a major problem for global health systems [41]. To maintain robust immune system at this difficult time, maintaining a healthy diet and proper nutrition is an exclusive way to minimize the crises [16].

As SARS-CoV-2 severely affects lungs and respiratory system, nutritional advices to minimize damages and other infections of lungs in COVID-19 patients are of worth consideration [24]. As virus may affect other organs as well, optimum nutritional status is essential for protection against viral infections [42].

Prioritizing Nutrition Interventions During COVID-19

Grains

Grains including wheat, rice, barley, oats, corn, rye, etc., are recognized as an important component of a healthy diet [43] because of rich source of macronutrients, micronutrients, bioactive compounds, and dietary fibers. Many of these dietary components contribute in wide-ranging physiological functions in the body [44]. The consumption of whole grain in higher proportion results in better health outcomes [45] in present COVID-19 pandemic crises. Intake of whole grains should be preferred because grains with minimal processing maintain their nutritive values, while refining of grains results in reduction of dietary fibers and numerous bioactive compounds. Although cereal grains are a major source of energy and designated as a staple food in many countries, surprisingly little attention was given to the quality of cereal grains to specify their use in refined form or as a whole grain form in dietary recommendations.

Whole grains are complete grains with bran and germ as well; therefore, they still remain naturally rich in valuable chemical constituents. Whole grains are defined as comprised of grain with intact, flaked, cracked, or ground caryopsis whose major components, the germ, endosperm, and bran, are present in natural proportion as in the intact grain [46]. The grains are resorvoir of carbohydrates and provide reasonable proportion of proteins and lipids and also rich source of important vitamins and minerals. Whole grains are also valuable source of bioactive molecules and dietary fibers, which increase the interest to use them in the development of high-value food products with enormous healthy outcomes [47]. Recently, whole grains become more important in dietary guidelines due to the discovery of their effective role in prebiotic activities for gut microbiota, important for immunity and well-being of the host [48, 49]. Moreover, ample quantity of whole grain consumption is beneficial in lowering risk of multi-morbidities [50]. The nutritional components of grains may vary due to differences in agroclimatic conditions and cultivar selection, but despite variations, they play a fundamental role in controlling lipid levels in blood, glucose-insulin homeostasis, and gastrointestinal health [51].

The inclusion of grains is recommended in many globally recognized dietary guidelines [52] to promote health and reduce the chances of developing lifestyle-related diseases, predominantly obesity, diabetes, cancer, and cardiovascular diseases [48, 53–55], and to treat some chronic inflammatory diseases with positive outcomes [56, 57]. Despite the well-documented health benefits of whole grains, whole grain intake continues to fall below recommended levels [58], indicating the importance and potential impact of promoting increased consumption of whole grains across the globe during this pandemic crisis. Certain dietary guidelines continuously focus to increase the consumption of whole grain components in the diet from past years [59]. The recommended amount for daily consumption of whole grain, in these guidelines, is at least 75 grams for adults [60]. However, in many countries, the recommended level is far less and requires reconsideration. To optimize the nutritional requirements, the World Health Organization recommended to consume 180 grams of whole grain per day on continuous basis during pandemic [21].

Furthermore, in some countries, the dietary guidelines are vague with no advice to choose whole grains or refined grains that may result in less positive outcomes associated with grain consumption as both quantity and quality are compromised in these guidelines. At present, the intake of whole grain is far behind the recommended levels in most of the countries; hence, a comprehensive plan is needed to increase the whole grain consumption that has potential health benefits essential to cope with COVID-19.

Vegetables and Fruits

Fruits and vegetables are among the most important foods needed for normal growth, physiological functioning, metabolic activities, and boosting immunity against infectious diseases. The regular consumption of fresh fruits and vegetables is generally linked with immunomodulation and rapid generation of immune responses to defeat noxious invading microorganisms. Fruits and vegetables actually supply indispensable macronutrients, micronutrients, and phytonutrients that are necessary for health promotion. These nutrients substantively increase antioxidant activity and control the anti-inflammatory markers and are actively involved in recuperation from chronic diseases. The most important nutrients provided by many fruits and vegetables include vitamin C, vitamin E, beta-carotene, and polyphenolic compounds that are all involved in immunomodulation [42].

Vitamin C is present in abundance in oranges, lemons, grapefruit, tangerine, kiwi, strawberries, mangoes, broccoli, red peppers, and other fruits and vegetables. Vegetable oil is a good source of vitamin E. Beta-carotene sources include green leafy vegetables, carrots, and sweet potatoes. Quercetin, a bioflavonoid found in a variety of fruits and vegetables, has prospective capacity to be used as antiviral agent and minimize severe outcomes in case of virally spread infectious diseases particularly respiratory infections including COVID-19 by maintaining antioxidant profile of individuals and protecting lungs from damages.

Many micronutrients and phytonutrients present in fruits and vegetables are actively involved in immunomodulation by increasing the number of T-cell subsets, potentiate response of lymphocyte to mitogen, stimulate natural killer cell activity, increase production of interleukin-2, and generate specific response against viral infections like influenza [61].

In addition, several positive effects have a direct link with regular consumption of fruits and vegetables. The known beneficial effects include high intake of flavonol essential for cardiovascular health, potential rise in nitric oxide species that paradoxically reduce blood pressure by vasodilation, concomitant loss in weight, and decrease in the number of inflammatory markers [62, 63].

The ingestion of vegetables with high antioxidant potential acts on insulin intolerance and may help in reducing complexities and severe outcomes of COVID-19. In many countries, COVID-19-related symptoms recovered rapidly due to intake of fermented vegetables [64]. Most of the vegetables possess sufficient antioxidant activity that play effective role against the development of noncommunicable diseases, particularly cardiovascular diseases [65] or diabetes [66].

The consumption of fruits and vegetables on daily basis with high fiber contents and huge reservoirs of many vitamins and minerals, together with abundant protective phytonutrients and low caloric density, helps in reducing multiple inflammatory markers and adhesion molecules [67, 68]. The healthy lifestyle and adherence to diet are remarkably influenced by gender [17, 69], but the beneficial effects of intake of fruit and vegetables were analogous in both in men and women.

Among phytochemicals or secondary plant metabolites that possess immense pharmacological properties are polyphenols that are obtained mainly from vegetables. Despite protective effects against several degenerative diseases, polyphenol intake in many countries is very low [70] and poses great concern in the wake of COVID-19. Polyphenols are important in viral infections as they influence the adhesion and penetration of viruses into the host cell and replication of viruses and suppress neuraminidase and hemagglutinin activity and viral hemagglutination, extensively modifying transcription factors and cellular signaling pathways [71].

Many phytochemicals in foods are fundamentally involved in establishing interactions with transcription factors, especially nuclear factor erythroid 2-related factor 2 (Nrf2) [72]. Various foods regulate transcription factors, particularly Nrf2, in relation to potent antioxidant activity [73]. Cucurbitacin B, a natural compound that exists in edible plants (bitter melons, cucumbers, pumpkins, and zucchini), has hypoglycemic and anti-inflammatory role by activating Nrf2 [74, 75]. Recently, a strong link was found between Nrf2 and the protection against COVID-19 and related severe complications by intake of fermented vegetables [64]. The transcription factors, noticeably Nrf2, can be activated by a vast array of natural compounds having vegetable origin such as vitamin D, resveratrol, sulforaphane, curcumin, and many others [76]. Sulforaphane, a potent activator of Nfr2, is present in high amount in cruciferous vegetables [77]. The most important cruciferous vegetables possessing potent anti-oxidative and chemopreventive role against infectious diseases are broccoli [37, 78, 79] and cauliflower [80]. In addition, European countries with large consumption of cucumber and head cabbage exhibit reduction in mortality rates due to COVID-19 [64]. Increase consumption of some of the vegetables results in decline of risk factors associated with COVID-19 and casualties [64].

Increase in consumption of fruits is also a pleiotropic factor in reducing COVID-19 infections and mortality rates [81]. The fruit intake enhances the supply and functionality of many vitamins and minerals that have supportive role in the establishment of physical barrier for microbes and activation of antimicrobial proteins [82]. Some other vital functions are also played by micronutrients including supportive role in growth parameters, differentiation, and chemotaxis of innate cells; increased phagocytosis and ingested activity of macrophages and neutrophils; and speedup recovery from acute and chronic inflammation by managing cytokine production and antioxidant activity [82].

The beneficial impacts of biologically active substances of fruits are due to modulation of immunoregulatory defense, induction of antiviral immune response, stimulation of apoptosis, and genetic or epigenetic regulation [83]. Hence, fruit and vegetable intake must be increased during COVID-19 with four servings of fruits and five servings of vegetables as suggested in WHO guideline to consume in daily diet to cope with the adverse outcome of COVID-19 [21].

Fruits and vegetables that boost immune system to fight SARS-CoV-2 include beetroot, eggplant, turnip, cauliflower, green beans, carrot, pumpkin, parsley, celery, red onion, radish, pepper, potatoes, tomatoes, watermelon, apricot, strawberry, peach, cherry, avocado, berries, dates, oranges, and others [84].

Nuts

Nuts are highly nutritious, shell-covered fruit with bevy of macronutrients, micronutrients, and phytonutrients. They are considered as natural functional foods with important components of a healthy diet and can be used in health promotion due to their ease inclusion in the usual diet. Tree nuts (almond, walnut, cashew, pistachio, hazelnut, Brazil nut, pecan, pine nut, and macadamia) and groundnuts (peanuts) are highly valuable plant-based diet rich in bioactive molecules.

Frequent consumption of nuts favorably supports innocuous plasma lipid profiles [85] and endothelial functions [86] and reduces the risk factors associated with type 2 diabetes [87], chronic heart diseases [88, 89], cancer [90], and inflammation [86] and may be beneficial in reducing severe complications of COVID-19 patients.

Nuts and their coproducts such as hard shell or rind, skin or testa, hull or husks, and leafy cover and leaves are valuable as they provide important phytochemicals that play a role in reducing oxidative stress by neutralizing free radicals and boosting free radical scavenging activities and promoting antimutagenic, anticarcinogenic, and antiproliferative properties. Therefore, addition of phytochemicals from these coproducts is a cheapest way to increase natural antioxidants in diet. Moreover, nuts are energy-dense foods with the presence of essential fats and proteins; their consumption on regular basis has positive impact on body weight and obesity [91, 92]. Hence, a moderate consumption of nuts is valuable to manage weight [92], as upper and lower respiratory tract infections have severe outcomes for obese patients [93].

In addition to good source of several vitamins and minerals, nuts are densely packed with essential amino acids, fat-soluble bioactive molecules, soluble fibers, unsaturated fatty acids, and numerous phytochemicals that play a contributing role in health promotion and curtailing the risk of chronic diseases which are the main cause of morbidity and mortality across the globe.

Nuts contain numerous types of antioxidants. For example, almonds contain a variety of flavonoids such as flavonols, flavonones, catechins, and their glycoside and aglycone forms, while walnuts possess a wide range of polyphenols typically non-flavonoid ellagitannins and tocopherols. Cashew nuts have abundant quantity of alkylphenols. Several flavonoids and appreciable amount of resveratrols are present in pistachios and peanuts. Macadamia nuts have properties that lower oxidative stress, cholesterol, and inflammation. Peanuts are high in vitamin B_3, B_6, and magnesium. They may improve circulation in the brain and enhance cognitive function. Nuts are a rich source of vitamins and minerals necessary for normal immune functions.

In present COVID-19 pandemic and global health emergency situation, addition of nuts in daily nutrition is helpful in reducing incidence and severity of infection.

Legumes

Legumes play an important role in human nutrition; especially, it occupies a major part in the diet of people with low income in most of the developing countries. Legumes are a great source of slow-releasing carbohydrates and all types of proteins and hence are designated as poor man's meat [94]. Legumes are fiber-rich plants that have several health implications and many therapeutic benefits [94]. The presence of minerals like zinc, iron, and potassium in legumes is important in many physiological processes and strengthening immunity [95]. Legumes also have considerable proportion of lectins and peptides that dynamically stimulate anti-inflammatory responses [96]. Several components in legumes including lectins possess antiviral effects that have potent contribution in the prevention of infections [97, 98]. The infection due to SARS-CoV-2 causes acute respiratory distress syndrome that is involved in the release of pro-inflammatory cytokines like TNF-α and IL-6 [99]. The dietary components in legumes assist in prevention, reduction in complications, and faster relief from COVID-19 by inhibiting the activities of TNF-α and IL-6 [100]. The increase in the consumption of legumes and beans reduces the mortality rates by COVID-19 [81]. In addition, legumes with enormous functional components such as hydroxychloroquine are in spotlight to be a remedy for treating COVID-19 patients [101]. Fava beans are a rich source of hydroxychloroquine, and their consumption can be helpful against viral pathogenesis due to antiviral properties without possible side effects. It is also evident that legumes that are low in glycemic index, low in fat, and a source of vegetarian proteins are helpful in controlling chronic diseases like hypertension, type 2 diabetes, and cardiovascular disease [102], which all are common comorbidities for COVID-19.

Meat

Meat is nutrient-dense food and is considered valuable as its intake provides complete protein with all essential amino acids. In addition, it is a major dietary source of most common shortfall nutrients like zinc, iron, selenium, and B vitamins, specially B_{12}. Meat also possesses many valuable bioactive molecules such as carnosine, anserine, creatine, taurine, and hydroxyproline that are helpful in the killing of pathogenic parasites, bacteria, fungi, and viruses (including coronaviruses like SARS-CoV-2) [103]. These compounds also have a pivotal role against chronic and aging-associated diseases which serve as the most common comorbidities with worse health outcomes during a SARS-CoV-2 pestilence [24]. In addition, fish meat is a rich source of polyunsaturated fatty acids (omega-3 fatty acids) that effectively boost functionality of immune system. The dietary intake of polyunsaturated fatty acids has a crucial role, during a pandemic [104], in the reduction of microbial load due to their ability to inactivate enveloped viruses such as the SARS-CoV-2 virus [104, 105].

Consumption of meat is associated with a large number of health benefits due to the presence of beneficial compounds, but detail research is needed to assess accurate dosage, possible negative outcomes, or any other metabolic abnormalities to declare meat as a functional food to provide resilience and desirable immunity against COVID-19. WHO recommends eating 160 g of protein with recommendation of red meat to consume one to two times per week and poultry two to three times per week [21].

Milk and Milk Products

Milk and milk products are a major part of well-balanced diet. They have absolutely good natural balance of carbohydrates, high-quality proteins, and fats and may be considered itself as a complete meal. Milk is excellent source of many essential vitamins (vitamins A and B12) and minerals calcium, magnesium, zinc, potassium, phosphorous, and selenium. These vitamins and minerals regulate and maintain a healthy immune system needed against SARS-CoV-2 virus [42]. Despite immunomodulatory role of all minerals that constrained in milk and milk products, zinc is considered to be more valuable in maintaining immune system homeostasis because its deficiency adversely affects functioning of innate and adaptive immune cells [106]. Furthermore, high amount of zinc induces high host responses against viruses, retards replication of viruses by preventing viral membrane fusion, and improves clinical symptomatology [107].

A valuable milk product is yogurt that is actively involved in augmentation of natural killer cell activities and plays a profound role in the reduction of respiratory infections [108] suggesting its auspicious utilization in moderately infected COVID-19 patients.

Cheese is an important milk product, rich in natural yeast, which increases the number of useful microorganisms that convert undigested plant-based fiber contents into chemicals to be absorbed in the blood stream and strengthens immunity [109] necessary to maintain during an infection.

Herbs

Herbs, combination of herbs in specific manner (herbal formulae), and their extracts are important for the treatment of viral diseases like COVID-19. Several herbs are valuable: garlic, black cumin, and liquorice herbal extract boost immunity [6]; *Pelargonium sidoides* effectively inhibit the replication of respiratory viruses [110]; ginseng roots are used to reduce symptoms associated with viral respiratory diseases such as those linked to influenza strains [111]; and *Astragalus membranaceus* is beneficial in the treatment of common cold and upper respiratory infections [112]. Many extracts and bioactive compounds of herbs exhibit antiviral activity against the influenza virus such as Korean red ginseng [113], ginger, garlic [114], tea tree, and eucalyptus [115] and can be helpful against SARS-CoV-2.

The controlling of H1N1 and SARS influenza in the past with Chinese herbal formulae in the high-risk population group provides an insight that herbal formulae are helpful against viral infections and may employ as an alternative approach for the prevention of COVID-19 [112].

Traditional Chinese medicine (TCM) based on herbal formulae may play a critical role in the prevention and treatment of COVID-19, bringing new hope to control COVID-19 [116]. The presence of bioactive molecules in different herbs and traditional Chinese medicines (e.g., flavonoids, phenolic compounds, quercetin, and kaempferol) has inhibitory role in the enzymatic activity of SARS 3-chymotrypsin-like protease (3CLpro). This enzyme actively plays a role in replication process of SARS-CoV and is thus thought to be a potential treatment option for SARS-CoV-2-related infections and supportive care agent for COVID-19 patients. The implications of herbal formula in relation to specific symptoms and pattern identification were revised, and these guidelines are helpful in increasing the supportive care for COVID-19 patients [117].

TCM based on a combination of various herbs helps in reducing overall symptoms of SARS-CoV-2 pneumonia patients. Many decoctions including sheganmahuang decoction, gancaoganjiang decoction, qingfei paidu decoction (QPD), qingfei touxie fuzheng recipe, etc., have been suggested to be prescribed for COVID-19 patients. QPD, which consisted of *Gypsum Fibrosum, Ephedrae Herba, Armeniacae Semen Amarum, Glycyrrhizae Radix et Rhizoma Praeparata* cum *Melle, Atractylodis Macrocephalae Rhizoma, Alismatis Rhizoma, Bupleuri Radix, Polyporus, Cinnamomi Ramulus, Asteris Radix et Rhizoma, Pinelliae Rhizoma Praepratum* cum *Zingibere et Alumine, Poria, Zingiberis Rhizoma Recens, Scutellariae Radix, Belamcandae Rhizoma, Asari Radix et Rhizoma, Aurantii Fructus Immaturus, Farfarae Flos, Dioscoreae Rhizoma, Pogostemonis Herba, and Citri Reticulatae Pericarpium*, has been used extensively as general prescription for treating COVID-19 in China [118].

The use of QPD for the treatment of COVID-19 patients is important as it provides absolute cure in some patients, improves the symptoms in others, or in some cases stabilizes the symptoms without aggravation [119]. The effectiveness in cure rate and overall regulatory effect of QPD against COVID-19 is due to the presence of multicomponents and multi-targeted approaches which include lung protection and inhibit the replication of SARS-CoV-2 virus by acting on multiple ribosomal proteins and others. COVID-19 is associated with strong responses by immune system and inflammatory storm [120]. The use of QPD impedes these responses by immune system and eliminates excessive inflammation by maintaining pathways related to immune and cytokines [121].

Certain drugs like ergosterol, shionone, and patchouli alcohol obtained from herbs had shown reasonable anti-COVID-19 effects that provided new molecules necessary for the development of new drugs [122].

Several herbs such as *Astragalus membranaceus, Ganoderma lucidum, Aloe vera, Angelica gigas, Scutellaria baicalensis,* and *Panax ginseng* exhibited immunomodulatory properties that include lymphocyte activation, increased activity of natural killer cells, regulation of cytokines, and rise of macrophage actions [123].

Herbal formulae to use as medicines rely on accurate symptom analysis and pattern identifications for the treatment of COVID-19 patients [117]. Different pattern identifications and herbal formulae were suggested for mild stage, moderate stage, severe stage, and recovery stage in the Chinese and Hong Kong guidelines. In the frequency analysis of herbs, in Chinese guidelines, few herbs like *Ephedrae Herba*, *Gypsum Fibrosum, Armeniacae Semen Amarum*, and *Glycyrrhizae Radix et Rhizoma* have highest frequency of usage than the others. Moreover, for the treatment of pediatric COVID-19 patients, different herbal formulae prepared from different combinations of 56 herbs and pattern identifications were established in traditional medicines in China [124].

Early intervention of patients infected with COVID-19 with Chinese herbal medicines was found to be an excellent way to improve the cure rate, delay progression of disease, reduce the course of disease, and decrease death rates. These herbal medicines not only control the viral activities but are involved in the regulation of immune system, generate immune-related responses, reduce inflammation, and promote simultaneous repairing of affected body parts [116].

Recently, Lianhuaqingwen, a Chinese herbal formulation, effectively suppressed the replication of SARS-CoV-2, reduced production of pro-inflammatory cytokines, and changed the SARS-CoV-2 morphology [125].

Interestingly, hundreds of herbs and herbal formulae have been found effective for prevention and treatment of infections caused by viruses [126]. However, prospective clinical trials with rigorous research are required to ascertain the potential preventive effect of these herbs and herbal formulae [112] (Fig. 5.1).

Fig. 5.1 Immunity booster foods

Role of Vitamins in COVID-19

Vitamin A

Vitamin A is a group of fat-soluble compounds such as retinol, retinoic acid, and carotene. These compounds promote cell growth and development, protect epithelial and mucosal integrity within the body, regulate immune functions, and are important factors to lower the susceptibility to infectious agents and diseases [127]. Vitamin A supplementation of this important vitamin enhances the humeral immunity in pediatric patients following vaccination of influenza [128]. Vitamin A has a supportive role in enhancing immune functions including promotion of antibody production, apoptosis, cytokine expression, mucins and keratins, and lymphopoiesis and increases the functional stability of monocytes, natural killer cells, neutrophils, T-cells, and B-cells [129, 130]. During respiratory diseases, vitamin A plays important role in reducing the severe complications and deaths [131]. An important derivative of vitamin A is isotretinoin that mediates downregulation of the ACE2 receptor. Since ACE2 is host cellular protein to which SARS-CoV-2 virus attaches to facilitate its entry into host cells and ultimately in respiratory tract of victim, this profound role of vitamin A reduces the chances of viral attachment and hence decreases the susceptibility of individuals to contract COVID-19 [132].

However, deficiency of vitamin A is a predisposed risk factor for viral infections with high severity due to impaired immune responses as was happened in individuals affected by measles virus, respiratory syncytial virus, and influenza virus [128, 133].

Foods that are high in vitamin A include eggs, liver, carrots, spinach, kale, broccoli, lettuce, sweet potato, mango, apricot, and others.

Vitamin B

B vitamins are a complex of water-soluble vitamins and work as important component of coenzymes needed in vital metabolic reactions. The interactions of B vitamins with immune cells mediate anti-inflammatory responses and create hindrance in many pathophysiological pathways [134]. The deficiency of B vitamin may delay the host response to infections due to weak immune functionality [135]. Each B vitamin also performs wide-ranging roles in the body.

For example, in human plasma products, the titer of MERS-CoV is effectively reduced due to vitamin B_2 and UV light [34] and hence may be helpful against SARS-CoV-2 as well.

Vitamin B_6 has a pivotal role in immune function as antibodies and cytokines build up from amino acids and B_6 acts as coenzyme in their metabolism. Vitamin B_6 deficiency impairs growth, antibody production, lymphocyte mitogenic response, lymphocyte maturation, and T-cell activities [136]. Moreover, in COVID-19 critically ill patients, vitamin B_6 supplementation increases total lymphocyte cells (T-helper cells and T-suppressor cells) [137]. It is well evident fact that vitamin B_6 intake is inversely related to inflammation status of individuals [138].

Vitamin B_{12} also acts as immunomodulator for cellular immunity especially in relation to cytotoxic cells, e.g., natural killer cells, CD8+, and T-lymphocytes [139]. Vitamin B_{12} also plays a supportive role for maintaining a healthy gut microbiome which is important for functional aspects of both innate and adaptive immune systems [140, 141]. Hence, this could be helpful in prevention of excessive immune response [142] specially needed for COVID-19 patients with microbiota dysbiosis that may lead to severe complications [143]. Foods rich in vitamin B_6 are bananas, avocados, pineapple, beans, spinach, potatoes, etc.

A diet rich in vitamin B_{12}, vitamin D, and magnesium given to COVID-19-affected old-age men and women within 1 day after hospitalization results in better health outcomes and reduces their chances to require oxygen therapy and intensive care support [144]. The major dietary sources of vitamin B_{12} are fish, eggs, milk, meat, shellfish, and others [145].

In COVID-19 patients, high vitamin B_{12} intake significantly lowers the rate of inflammation [144]. Therefore, B vitamins could be a basic choice to reduce the infection rate and severe complications in COVID-19 patients [34].

Vitamin C

Vitamin C (ascorbic acid), a water-soluble vitamin, is a vital nutrient and potent antioxidant. Vitamin C functions as enzymatic cofactor in many crucial physiological reactions that mediate essential biological functions including absorption of iron, collagen synthesis, production of hormones, and increased immune responses [146]. Vitamin C is important in growth, development, and repairing of all the tissues in the body [147]. Vitamin C also exhibits immunomodulatory effects in respiratory infection [148]. Prevention from lower respiratory tract infection has a direct relationship with vitamin C intake [147, 149].

The individual with deficiency of vitamin C can be benefitted by vitamin supplementation [150]. The supplementation of vitamin C on regular basis has modest but consistent effect in shortening the infection period of common cold [151]. Moreover, administration of vitamin C in high doses before or after the appearance of flu symptoms reduces severity of illness [152]. The risk of common cold decreases by vitamin C supplementation without any adverse effects [153]. Vitamin C boosts antiviral immune responses against the influenza A virus by increasing the production of interferon-α/β [146].

As vitamin C has effective role in respiratory tract infections and antiviral immune responses, vitamin C intake is crucial particularly in micronutrient-deficient individuals with high risk of COVID-19, although vitamin C supplementation to prevent and treat acute respiratory diseases is inconclusive [149, 150].

Adequate vitamin C consumption in diet may be beneficial in COVID-19 prevention or prognosis. The dietary recommendation may be achieved through natural foods such as citrus fruits, kiwi, broccoli, and other products [101].

Vitamin D

Vitamin D is an essential micronutrient needed to maintain bone and musculoskeletal health and many other functions in the body. Recent epidemiological data support the importance of vitamin D in COVID-19 severity and complications [154]. Vitamin D acts as modulator for both innate and adaptive immunities [155] through cytokines and regulation of cell signaling pathways [156]. Vitamin D is crucial in functionality of immune cells and plays a critical role during viral infections by modulating inflammatory responses [157]. The presence of vitamin D receptors on T and B immune cells helps in proliferation, inhibition, and differentiation of these cells [106]. In experimental models of lipopolysaccharide-induced inflammation, vitamin D is directly linked with lower concentrations of pro-inflammatory cytokine interleukin- 6 [158, 159]. Vitamin D also reduces lipopolysaccharide-induced lung injury in mice by blocking the Ang-2-Tie-2 and renin-angiotensin pathways [160] that are highly relevant to SARS-CoV-2 pathogenicity.

A sufficient vitamin D serum level is linked to a switch from pro- to anti-inflammatory profiles in older adults [154]. This impact on the regulation of inflammation is of particular importance in older adults, the obese, and those with chronic conditions as they may already be pre-set for higher inflammatory responses if exposed to COVID-19. A heightened immune response in vitamin D-deficient people may therefore increase the potential for cytokine storm and consequent acute respiratory distress syndrome [120].

Lower level of vitamin D has a direct link with higher respiratory infection rates, and the effect was more pronounced in individuals having underlying lung conditions [161]. Case-control studies have also declared that increased risk of infection is due to low level of vitamin D [162], and supplementation with vitamin D seems to help reduce both symptoms and antibiotic use [163]. Vitamin D supplementation reduces the risk of acute respiratory infections [164], while a higher blood vitamin D status has been associated with a small reduction in risk of pneumonia [165]. Vitamin D deficiencies probably increase the risk of upper respiratory viral infections, although the size of this effect is small.

The humeral immunity increases in pediatric patients by vitamin D supplementation after influenza vaccination [128]. Older age, obesity, being male, and having preexisting chronic conditions are risk factors for vitamin D deficiency [166], all of which increase vulnerability to COVID-19 and related complications [167, 168].

Importantly, it is evident that there is a worldwide association between northern latitude and increased COVID-19 mortality [169]. Recent reports have indicated that those residing at higher latitudes or with darker skin pigmentation may be particularly affected by COVID-19. These individuals are also at higher risk of obesity, preexisting chronic disease, and vitamin D deficiency [170, 171]. While there could be various explanations for this, it supports the assumption that sunlight exposure and hence vitamin D status could be impacting on COVID-19 severity.

Several mechanisms are proposed to establish a role between vitamin D deficiency and increased risk of viral infections, but all underlying mechanisms are still inconclusive [83]. The potential mechanisms include the modulation of immunoregulatory defense, induction of autophagy and apoptosis, antiviral immune induction, and genetic or epigenetic regulations [83]. Furthermore, vitamin D reduces the risk of viral infections as it can promote the binding of SARS-CoV-2 cell entry receptor ACE2 to angiotensin II receptor type 1 and hence reduces the attachment of virus particles to ACE2 and inhibits the viral entry into the cell [172].

Furthermore, vitamin D enhances the activity of certain peptides like defensins and cathelicidins that inhibit the viral replication and increase the concentrations of anti-inflammatory cytokines and decrease the level of pro-inflammatory cytokines that cause inflammation-related pneumonia [173]. The effectiveness of vitamin D in reducing COVID-19 risks is confirmed by fatality rates with comorbidity, chronic disease, and age, all reported with low level of vitamin D.

In pandemic restrictions, with less outdoor activities and less exposure to sun, vitamin D production in the skin is reduced considerably. Adequate vitamin D level is associated with reduction of developing chronic diseases such as diabetes, cardiovascular disease, hypertension, and cancers that are high-risk group for severe respiratory infections during COVID-19 [174]. Due to limited sun exposure, vitamin D intake may be increased from dietary sources. Foods rich in vitamin D are egg yolks, mushrooms, cod liver oil, sardines, tuna, salmon, and meat [175]. Besides, vitamin D supplementation is also important in strengthening the resilience against COVID-19 [176].

Vitamin E

Vitamin E, a fat-soluble vitamin, is actually a group of four tocopherols and four tocotrienols. Vitamin E is vital for maintaining the overall health and immunity. The role of vitamin E as powerful antioxidant helps in protection from many viral infections by modulating immune functions of hosts [177]. Vitamin E deficiency is associated with impaired humoral and cell-mediated immunity [178]. The major role of vitamin E is preserving immune responses. The vitamin E deficiency is more pronounced in older people, and vitamin E supplementation to old-age people improves their humoral and cell-mediated immunity.

Vitamin E increases the activity of natural killer cells modulating concentration of nitric oxide [179]. Administration of vitamin E increases humoral (B cells) immune responses and antibody actions [180]. Vitamin E is also involved in the regulation of maturation of dendritic cells and related functions [181], which are essential to orchestrate immune response for intertwining adaptive and innate immune systems [182]. Vitamin E is beneficial in improving immune synapse formation in naïve T-cells and generating T-cell activation signals [180]. Hence, rec-

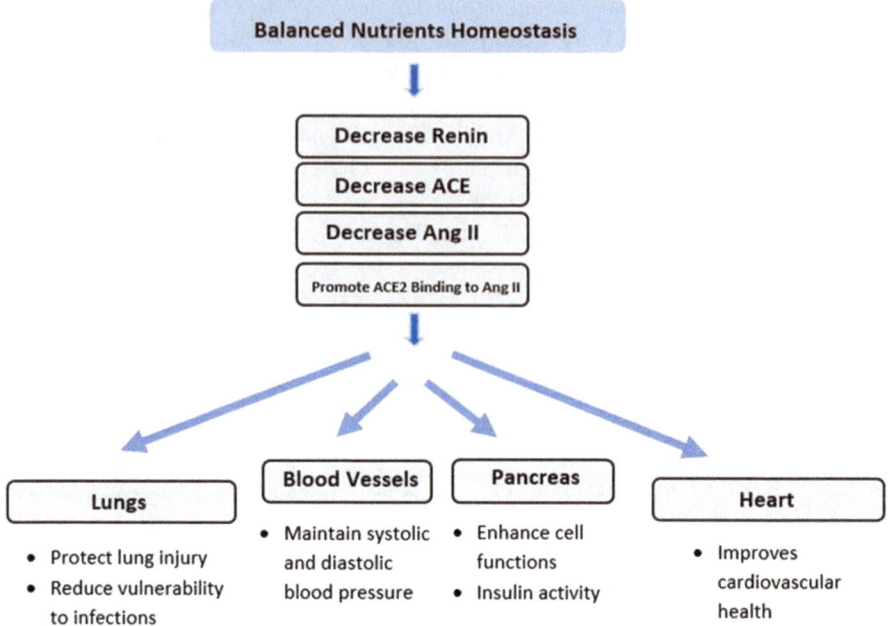

Fig. 5.2 Nutrient homeostasis in relation to organ protection and functionality

ommendation of vitamin E as a beneficial nutrient against COVID-19 infection and related disorders is mandatory [34].

The major foods rich in vitamin E are spinach, broccoli, nuts, particularly soaked almonds, peanut butter, hazelnuts, sunflower seeds, and vegetable oils (soybean, sunflower, corn, wheat germ, and walnut), which should be consumed on a daily basis to get a recommended dose of vitamin E (Fig. 5.2).

Role of Minerals in COVID-19

Iron

Iron is essential nutrient for both humans and pathogenic microbes, and complex system evolves for its accurate acquisition and utilization in the body. The innate immune system is particularly involved to restrict iron availability to invading microbes [183]. Many aspects related to severe competition for iron between hosts and pathogens are yet elusive [184].

Iron has overwhelming role in minimizing viral infections [185] and respiratory infections [186]. Iron helps to fight off infections by regulating proliferation and maturation of T-lymphocyte immune cell and cytokine production [187, 188].

The deficiency of iron results in impaired host immunity, while iron overload in the body increases the oxidative stress and deleterious viral mutations [189]. Iron deficiency is a major risk factor for developing periodic respiratory infections [190]. In many developing countries, the deficiency of iron is more prevalent [191] and is associated with infection [192].

The best-absorbed form of iron (heme) is found in meat, poultry, and fish, whereas nonheme iron is found in leafy green vegetables, seeds of legumes, fruits, and dairy products [193].

Selenium

Selenium is a trace element with diversified pleiotropic effects, ranging from powerful antioxidant activities to stimulatory role in many anti-inflammatory processes [194]. Selenium acts as cofactor in many enzyme regulations. The primary source of selenium for human is food, and, surprisingly, only five molecules named selenocysteine, selenoneine, selenite, selenate, and selenomethionine constitute the bioavailability of selenium via food intake [195].

Selenium acts in dose-dependent manner to increase the activity of macrophages and to enhance the immunoglobulin production. Selenium appears to enhance the cytolysis of natural killer cells [196]. Most of the biological activities of selenium are carried out by its incorporation in the form of rare amino acid selenocysteine into an essential group of proteins, selenoproteins [197]. Selenium as part of several selenoproteins, especially thioredoxin reductases and glutathione peroxidases, has a crucial role to maintain host defense system against viral infections through its redox homeostatic contributions, powerful antioxidant activities, and redox signaling [197], regulating natural killer cell activity and leukocyte functions [198]. Selenium is actively involved in proliferation of T-lymphocyte and specifically influences the humoral system to enhance immunoglobulin production [187, 198].

Selenium intake in high amount is related to antiviral effects [194], while low levels of selenium result in poor immune functions, onset of infections, and increased risk of fatality. Selenium deficiency is primarily related to increase the pathogenicity of many viral infections [199]. Selenium deficiency may result in alterations of viral genome [199, 200] such that a mild strain becomes highly virulent form [199]. Furthermore, production of pro-inflammatory chemokines increases under pathogenic conditions in selenium-deficient state [199]. Selenium status has direct influence in regulation of numerous immune responses and related mechanisms; therefore, selenium supplementation may have effective role against diseases. In deficient state, the selenium supplements are a treatment option against viral diseases [201] including COVID-19. Selenium supplementation has shown beneficial effects after influenza vaccination in developing immunity [202].

The selenium deficiencies are widespread in many individuals around the globe and can be supplemented by proper diet and selenium-enriched food. The recommendations for selenium intake on daily basis range between 50 µg and 70 µg [197]. Food sources of selenium include meat, cereals, milk, fishes, asparagus, mushrooms, garlic, giblets, bread, nuts, and seafoods [203–205].

Zinc

Zinc, a nutritionally essential mineral, has vital contributions in growth, development, synthesis of DNA, proliferation of cells [206], enzymatic functions, and the regulations of many transcription factors in the human body [34, 42]. It is also involved in the production of immune cells, facilitation of cell signaling pathways, and regulation of responses related to innate and adaptive immune system [207, 208]. Zinc is also crucial for normal functioning of cells related to nonspecific immunity such as neutrophils and natural killer cells. Zinc is a basic structural component of hundreds of zinc-finger transcription regulators of immune cells [209].

Zinc is actively involved in mediating antiviral effects by improving immune responses and suppressing viral replications. Zinc has promising role in improving symptoms and pathologies associated with acute and chronic viral infections [107]. Zinc blocks the viral replications by inhibiting the function of RNA-dependent RNA polymerase, a core enzyme involved in replication of positive-sense, single-stranded RNA viruses [210] like SARS-CoV-2.

The deficiency of zinc is surprisingly common in modern lifestyle [107] and hence may impart severe outcomes during COVID-19 outbreak. The deficiency of this vital micronutrient is associated with impaired antiviral immunity [107] by restricting certain T-lymphocyte functions such as production and stimulation of Th1 cytokines [211]. In addition, macrophage function is adversely affected by zinc deficiency through the dysregulation in secretion of cytokines, phagocytosis, and intracellular killing activity [211].

Zinc intake in diet on regular basis is necessary as some viruses have evolved mechanisms to modify zinc homeostasis that favor viral genome in rapid replication and their persistence [212]. Zinc supplementation in high doses enhances the immunity in virally infected patients [213]. Zinc supplementation also has beneficial effects in reducing the symptoms and duration of common cold infection [214]. Therefore, zinc intake is important as a protective measure against COVID-19 to reduce the severe outcomes in patient and strengthen host's immunity against infection.

A daily intake of up to 50 mg of zinc may provide resilience against SARS-CoV-2 [209]. The most common food sources of zinc are poultry, beans, lentils, sesame seeds, pumpkin seeds, nuts, red meat, and shellfish.

Magnesium

Magnesium is a very important mineral that acts as cofactor in many biochemical pathways. Magnesium is favorably involved in the synthesis of vitamin D and acts as cofactor in many enzymatic reactions involved in metabolism of vitamin D [215]. Magnesium is bound to adenosine triphosphate (ATP) for its biological activation to release energy which is critically important as our cells cannot function properly without this energy.

Magnesium keeps the immune system strong by regulating many immune functions including macrophage response to lymphokines, immune cell adherence, immunoglobulin synthesis, immunoglobulin lymphocyte binding, antibody-dependent cytolysis, and helper T-cell and B-cell adherence [216]. Magnesium by blocking calcium channels acts as bronchodilator and vasodilator [217] and is helpful during lung infections.

Magnesium is also involved in the activation of immune responses against viral infections [218]. Magnesium maintains the level of hemoglobin in our blood which provides oxygen to all parts of body from the lungs, which is helpful during COVID-19 infection since respiratory system is badly affected by SARS-CoV-2.

Magnesium homeostasis is vital to strengthen resistance against infections in old-age people as they have compromised immune system [219]. Magnesium deficiency results in impaired immunological processes. The deficiency of this mineral readily arises as it may be lost during sweating in hot countries and during strenuous physical activities; it cannot be synthesized by the human body and needs to be replaced via the diet [220]. Foods rich in magnesium are dark chocolate, black beans, avocados, and whole grains.

Role of Omega-3 Polyunsaturated Fatty Acids

The omega-3 polyunsaturated fatty acids (PUFAs) are essential nutrients with predominant role in promoting anti-inflammatory and pro-inflammatory effects [221]. Long-chain omega-3 PUFAs play an essential role in the regulation of adaptive immune responses and decrement in inflammation [222]. Omega-3 PUFAs are precursors of important mediators like resolvins and protectins for resolution of inflammation [222]. These mediators (resolvins and protectins) trigger anti-inflammatory responses through oxygenated metabolites such as oxylipins [223].

Many PUFAs exhibited potent antiviral properties including omega-3 PUFAs [224]. An important omega-3 PUFA-derived lipid mediator, protectin D1, could significantly attenuate replication of influenza virus by RNA export machinery [225]. Therefore, omega-3 PUFAs including protectin D1, with consideration of novel antiviral drug, may have prospective benefits against SARS-CoV-2 [99].

Human body is unable to make them omega-3 PUFAs and can only be consumed by intake of dietary components. The food sources of omega-3 PUFAs are fatty fish, seafood (tuna, salmon, sardines), nuts, flaxseed, and plant oils (canola oil, soybean oil, and others) [226].

Nutrients for Immune System and Resilience in COVID-19

The conjoint denominator reflecting the importance of nutrition and dietary recommendations to minimize the virus prognosis during COVID-19 is based on relationship between dietary components and immunity [227]. Diet has a critical role in the regulation of many intricate pathways related to immune system to defend the body and reduce vulnerability to infections. The nutrients and combination of nutrient affect the immune system to generate a complex array of protective mechanisms like modifications in the synthesis of signaling molecules, cell activation, and regulation of appropriate gene expression [228].

Many vitamins and trace minerals play a detrimental role in enhancing immune functions [136]. Several nutraceuticals also regulate immune responses [229] necessary for normal functioning and fight off infections.

There is a complex interaction between consumption of nutrients and regulation of immune functions [230]. A well-balanced diet is valuable as excessive increase in certain macronutrients may cause deficiency of any one or more micronutrients as occurs in obesity [231], which leads to impaired immune responses including phagocyte function, cytokine production, secretory antibody response, antibody affinity, cell-mediated immunity, and the complement system, thus increasing exposure to viral infections [232]. Increase in dietary intake of nutrients or nutrient supplementation has positive effects in enhancing immunity against viral infections. Deficiency of micronutrient malnutrition may result in weak immune responses [20].

The functions of immune system decline in old-age people due to dysregulation of innate immunity with passage of time, high levels of chronic inflammation, and altered development of T-cell and B-cell [233]. The impaired immune system in older individuals increases the risk of contracting SARS-CoV-2, and severe outcomes after infection result in vital organ failure and increased chances of deaths. Nutrition has impact on multitude of immune responses and modulation of immune homeostasis. Hence, nutrient-dense diet during surveillance and response phase of immune system provide huge advantages to reduce severity of infection [234] when fighting viruses like SARS-CoV-2 (Fig. 5.3).

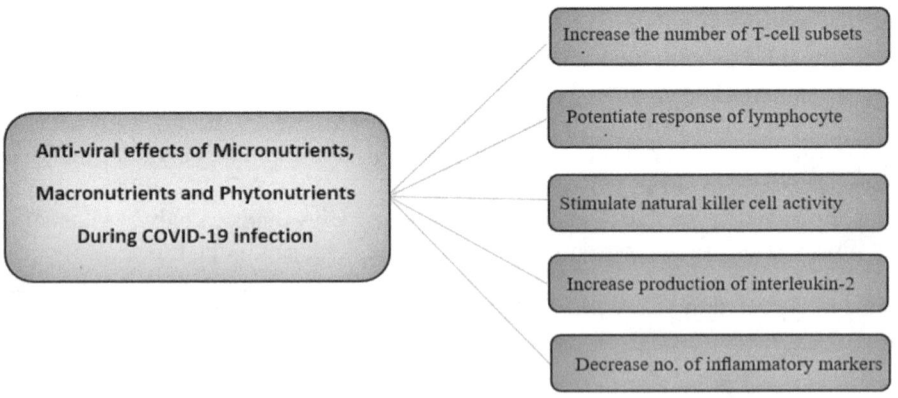

Fig. 5.3 Role of nutrients to enhance resilience and immunity

Nutrition in Lowering Stress and Depression in COVID-19

In present global spread of SARS-CoV-2, stress is a pervasive dilemma. Life becomes stagnant and quandary due to imposed lockdown after the declaration of global health emergency, which may instigate mood deterioration and provoke stress. This persistent and severe stress condition may affect the nervous system and result in alterations in brain homeostatic functions, hence increasing the risk of onset of depression [235]. Stress critically affects the development and functioning of the body. The COVID-19 pandemic, with radical alterations in every norm of life, can be classified as a stressful event with constant fear of infliction of disease, feelings of helplessness, falling sick, or dying, and stigma [236] further boosts the stress, and, generally, such stressful events greatly affect the nutrition [237]. Stress may cause hypophagia or hyperphagia, and binge eating may apply that results in a significant change in weight [238, 239]. Stay-at-home orders, smart working, and limitations of outdoor activities could result in increase in stress and boredom. Boredom markedly influences the dietary patterns and food choices. In most of the cases, boredom is associated with excessive intake of energy and consumption of relatively more quantities of carbohydrates, proteins, and fats [240]. Moreover, continuous hearing or reading news about lockdown, abrupt increase in infection rate, and death during pandemic can further increase the stress. Consequently, these stressed people start overeating and mostly rely on sugar-rich foods to compensate stress condition [241]. The intense desire of consuming a specific food is known as food craving, which is a multidimensional approach including behavioral (seeking food), cognitive (thoughts about food), emotional (intense desire to eat), and physiological (salivation) attributes [242].

SARS-CoV-2 badly affects the health system and global economy. Economic crises and high unemployment rates further exploit the stress condition and also induce anxiety, irritability, restlessness, and sleep disturbances. Stress increases the risk of developing infectious diseases, mainly viral infections [243]. In healthy people, stress suppresses the functions of immune system by shifting Th1 cytokines toward Th2 cytokine and reduces natural killer cell activity. Moreover, stress inflicts impaired response to vaccines, and decreased CD4+ to CD8+ ratio and low antibody titer occur in stress [244]. Hence, stress induces upper respiratory tract infections and compromised immune responses [245].

During stress, foods that contain and promote the serotonin and melatonin synthesis are important as stress can deplete serotonin levels in the brain [246] which can be restored by the intake of carbohydrate-rich diet. Sleep disturbance further worsens the stress state and increases the consumption of foods, thus provoking dangerous vicious cycle. Milk, salmon, poultry, and eggs contain sleep-inducing amino acid tryptophan, which is a precursor of serotonin and melatonin. Moreover, tryptophan regulates the satiety and caloric intake via serotonin to reduce the intake of carbohydrates/fats and inhibits the activity of neuropeptide Y which is the most powerful hypothalamic orexigen peptides that stimulate food intake [247]. Polyphenols from grapeseed, cayenne, and cardamom possess stress-reducing activity [248]. Therefore, to cope with stress condition, healthy food choices may be preferred to sensibly avoid malnutrition and chronic diseases.

Food Choices for Diabetic Patients During COVID-19

Diabetes is associated with an increased risk for infectivity and prognosis of COVID-19 as more than two-thirds of those who died of SARS-CoV-2 in Italy had diabetes [249]. Moreover, due to the wide fluctuations in blood glucose levels or presence of diabetes complications, COVID-19 can be more difficult to treat in diabetic patients.

Individuals with diabetes and COVID-19 infection experience uncontrolled acute hyperglycemic condition as ACE2 receptors are also present in pancreas that facilitate the entry of SARS-CoV-2 into the pancreatic islets and result in dysfunction of acute beta-cells with severe hyperglycemic status [250]. In COVID-19 patients with diabetes, uncontrolled hyperglycemic state remains a serious concern for health authorities as after taking recommended strategies to manage patients' blood glucose level, fatality rates still increase in number [251]. In these circumstances, attention must be given to nutrition besides medication and other preventive measures [252].

A varied and well-balanced diet given to diabetic patients on a regular basis is important to maintain blood glucose level and proper functionality of immune system; priority should be given to those foods that have low glycemic index, limit intake of starchy or sugary foods and high-fat diet, and prefer lean protein variety. Diet rich in vitamin D or vitamin D supplements should be suggested in COVID-19 diabetic patients to maintain glycemic status [253].

The dietary constituents with anti-inflammatory properties may have beneficial role during hyper-inflammation, such as cytokine storm during COVID-19. A healthy diet should consist of protein, fiber, vitamin D-rich food, fresh fruit and vegetables, and limited saturated fats [253].

Food Choices for Cardiovascular Patients During COVID-19

Cardiovascular patients are badly affected during COVID-19 pandemic progression. Some patients after being affected with SARS-CoV-2 remain asymptomatic; cardiac symptoms are considered as the first clinical manifestation in these patients [254, 255]. Such patients have an adverse prognosis. The ACE2 receptor required for the entry of SARS-CoV-2 into the host cell is expressed on vascular endothelial cells and myocytes [256, 257], so the possibility of direct involvement of cardiac cells by the virus cannot be ignored. Moreover, SARS-CoV-2 elevates the blood pressure due to vasoconstriction and increased activity of aldosterone and is associated with potential downregulation effect of ACE2. COVID-19 infection may increase dysregulation of blood pressure that results in increased vulnerability to cardiovascular complications [253].

To overcome cardiac complications associated with COVID-19, a combination of vitamins C and E may be a beneficial antioxidant therapy focusing on the importance of dietary components in infectious state.

A healthy diet is one of the best weapons to fight cardiovascular disease. A diet rich in high saturated and trans fat (Western diet) increases the risk of developing cardiovascular disease, while vitamins, minerals, and antioxidant-rich food (Mediterranean diet and DASH diet) are vascular protective. Adherence to Mediterranean diet provides adiponectin, a protein hormone, involved in fatty acid metabolism that helps in improving cardiovascular functionality [258]. Mediterranean diet is naturally low in sodium, rich in fruits and vegetables, and low in fatty substances that is directly linked with reducing incidence, prevalence, and mortality from hypertension, cardiovascular disease, and coronary artery disease. Moreover, whole grain oats and rye might be more beneficial than whole grain wheat in relation to cardiovascular disease [259].

Food Choices for Cancer Patients During COVID-19

Cancer patients are generally in more immunocompromised state due to regular anticancer, malignancy treatments and related issues and hence are more vulnerable to COVID-19 infection. In most of the cases, cancer patients may be prone to severe outcomes than other individuals after SARS-CoV-2 infection [260].

Several stratagems have been proposed to avoid severe infections of COVID-19 in cancer patients, such as stronger personal protection provisions, intentionally postponed adjuvant chemotherapy or elective surgery on a patient-by-patient basis, and more intensive surveillance or treatment [260]. Furthermore, healthy dietary habits, omega-3-rich diets, probiotic use, and vitamin D supplementation, as well as obesity prevention, are likely to be the most efficacious preventive approaches to control hyper-inflammation and impaired immune responses and decrease the severity of inflammatory diseases [261].

Food Choices for Patients with Digestive Disorders During COVID-19

COVID-19 patients with preexisting digestive disorders also face difficulties after infection due to improper intake of nutrients, their absorption, and assimilation in cells. The gastrointestinal symptoms such as vomiting and diarrhea may also appear after infection by SARS-CoV-2 that may highlight possible connection between gastrointestinal tract and the lung. This supposition further strengthens the finding of SARS-CoV-2 RNA from the stool specimen of COVID-19 patient [262].

In many laboratory-confirmed COVID-19 patients in China, gastrointestinal symptoms like nausea or vomiting, or both, and diarrhea were reported in many patients depicting abnormal functioning of digestive tract in these patients. Autopsy studies are necessary to understand and manage digestive issues related to COVID-19. Segmental dilatation and stenosis in small intestine were shown in the autopsy of one COVID-19 patient [263].

There is certainly a beneficial role of gut microbiota during COVID-19 infection as some guidelines focus on the use of probiotics to retain the balance of intestinal microbiome. The gut microbiota has paramount importance in the reduction of enteritis and ventilator-associated pneumonia and plays a role to minimize certain side effects associated with the use of antibiotics for control of early replications by influenza virus in epithelial cells of lungs [264].

Gut microbiota diversity plays a putative role in human health, and subtle balance of these microbiota alters in old age that causes severe implications for health. Elderly patients are badly affected by COVID-19 which might be due to dysbiosis of gut microbiota in these patients. Furthermore, gut microbiota diversity and stability also play a key role in myriad of metabolic processes and immunoregulatory functions. So, improvement in gut microbiome profile may be vital to reduce the impact of COVID-19 infection by increasing immunity. Personalized nutrition and supplementation are prophylactic ways to stabilize microbiota profile in immunocompromised and older patients having digestive issues. The foods that stabilize microbiome status such as fermented foods may be increased in diet. Dietary polyphenols also stimulate the growth of several beneficial microbes in the gut [265]. Maintaining composition of gut microbiota through proper nutrition and balance diet may have a valid scope as new therapeutic option and adjuvant therapeutic choices to reduce incidence and severity of COVID-19.

Food Choices for Pulmonary Disease Patients During COVID-19

COVID-19 infection causes atypical pneumonia in most of the cases that leads to acute respiratory distress syndrome (ARDS) in certain patients. This infection also modifies the secretion of upper and lower respiratory tract products including pulmonary surfactant and mucin. The production of mucin is upregulated during infection, with the function to prevent invading microbes from binding to epithelial cells and causing infection in these cells [266].

Dietary constituents regulate the inflammatory responses and play a critical role to minimize lung oxidative stress [267] by improving patients' outcomes during COVID-19. In particular, certain bioactive molecules such as polyphenols, flavonoids, and vitamin C actively influence inflammatory mediators and immune modulators and protect the lungs from infections. Antioxidants play a pivotal role in protecting lung tissues by reducing inflammation [268]. Flavonoids are also actively involved in reducing lung injury during infection [269]. During progression of an infection, vitamin C level is depleted continuously; vitamin C supplements in this situation attenuate the infection. So, vitamin C infusion may be a pragmatic treatment approach in seriously ill COVID-19 patients [270]. Omega-3 PUFAs are beneficial in upregulating the host's specific and nonspecific immune responses [271].

Malnutrition May Exacerbate COVID-19

Malnutrition is a condition that arises due to either not enough of one or more nutrient intake or otherwise excessive intake in diet. It may be due to imbalance in calories, carbohydrates, proteins, fats, vitamins, and mineral intake. Malnutrition increases the risk of susceptibility and more serious complications of COVID-19. Malnutrition may also involve morbidity, delayed recovery, and fatality in COVID-19 pandemic. Malnourished host leads to increased pathogenicity of infectious agent.

In patients with diagnosis of COVID-19, screening for malnutrition is important with valid nutrition screening tools [272] as nutritional deficiencies lead to impaired immune functions and increase the susceptibility, duration, and severity of disease. A diet deficient in nutritional antioxidants can alter a normal avirulent type into a virulent by modifying viral genetic characteristics. These mutants with altered genome have now the ability to infect healthy host with adequate nutritional status [29].

Evaluation of nutritional status in all infected individuals is necessary before hospital admission [273] and starting treatment [274]. In COVID-19 patients, nausea, vomiting, and diarrhea are common symptoms that result in nutrient deficiency due to poor absorption and assimilation of nutrients [275]; thus, good nutritional status is vital for survival against infections. Individuals with inadequate nutritional status should give prompt nutritional support [276] because immune response is weakened during the course of infectious disease and demand for several nutrients increases. It is particularly important to increase oral intake of protein.

Patients with high risk of malnutrition should maintain sufficient protein and calorie intake [273]. In addition, continuous supply of nutrients is important due to their wide-ranging antioxidants and anti-inflammatory actions [34]. Thus, the nutritional status should be urgently managed in COVID-19 patients. Maintaining proper nutritional status is mandatory for patients at risk of nutritional deficiency to reduce the severe complications and other adverse effects who might acquire COVID-19 in the future.

Nutrition and Post-COVID Recovery

COVID-19 infection may lead to acute respiratory distress syndrome and multiorgan failure [8]. People who have officially declared recovered from COVID-19, feeling fully normal by a patient may be a long process. These people are now highly enfeebled, and good nutrition and healthy diet are essential for full recovery. The World Health Organization (WHO) estimates that it can take 6 weeks or longer for someone to fully recover if they were in critical state during COVID-19.

After recovery from disease, patients still experience certain clinical symptoms like fatigue and dyspnea which may result in inadequate nutritional intake. However, it is crucial to minimize the gap of energy intake and energy expenditures during disease, and nutrient-rich diet after recovery balances all the depleted nutrients.

Critically ill COVID-19 patients who may be admitted to ICUs for days have resulted in severe depletion in muscle mass, strength, and resilience that leads to impaired immune system. After recovery, these patients need adequate nutrition like enough protein to repair their bodies and more vitamins and minerals to help smooth functions of metabolism. Such a recovery diet would not only help to get strength back but also boost immune system.

Future Perspectives

There is a strong risk that COVID-19 will endure and continue to be a threat in the long term. The COVID-19 pandemic devastatingly affects millions of people across the globe and strains the existing healthcare systems. The pandemic causes fear among healthy people, infected individuals, physicians, and health system workers but meanwhile provides new insights for developing innovative approaches and strategies.

The prolonged restricted conditions during pandemic lead to change in lifestyle, particularly eating habits and behaviors, and advocate for planned support efforts to provide nutrition during future pandemic crises. People with impaired immune system, NCDs, and deficient in essential nutrients must provide necessary guidelines, information, and nutritional support to modify their eating patterns, hence developing human communities with strong resilience and immune functions that defend the man from future pathogens. Moreover, false claims and exaggeration about health benefits of different nutrients should be avoided for better health outcomes.

Conclusion

The COVID-19 pandemic is a significant threat to human life worldwide. As there is no known effective cure or treatment for COVID-19, proper food intake and good nutritional status is an advantage for people at risk for severe COVID-19. Nutritional intervention through a healthy balanced diet comprised of whole grains, vegetables and fruits, nuts, legumes, dairy products, meat, and herbs is needed to boost immunity and reduce the incidence of infection and severity. Several vitamins and minerals are also important for immunomodulation and attenuate complications. In addition, foods that boost immunity, reduce stress, and prevent malnutrition are important to minimize the crises and critical outcomes. Selection of right foods is perspective way to manage comorbidities, associated risk factors, and severe complications during COVID-19. After recovery, good nutritional status is still mandatory to reinstate the nutritional losses incurred during disease progression.

References

1. World Health Organization. WHO Director-General's remarks at the media briefing on 2019-nCoV on 11 February 2020. World Health Organization, Geneva. Available via https://www.who.int/dg/speeches/detail/who-director-general-s-remarks-at-the-media-briefing-on-2019-ncov-on-11-february-2020. Accessed 2020
2. Li Q et al (2020) Early transmission dynamics in Wuhan, China, of novel coronavirus-infected pneumonia. N Engl J Med 382(13):1199–1207
3. Rothe C et al (2020) Transmission of 2019-nCoV infection from an asymptomatic contact in Germany. N Engl J Med 382(10):970–971
4. Gorbalenya A et al (2020) Coronaviridae Study Group of the International Committee on Taxonomy of Viruses. The species severe acute respiratory syndrome-related coronavirus: classifying 2019-nCoV and naming it SARS-CoV-2. Nat Microbiol 2020:03–04
5. Wang D et al (2020) Clinical characteristics of 138 hospitalized patients with 2019 novel coronavirus-infected pneumonia in Wuhan, China. JAMA 323(11):1061–1069
6. Arshad MS, et al. Coronavirus Disease (COVID-19) and immunity booster green foods: a mini Review. Food Sci Nutr 8:3971–3976
7. Dietz W, Santos-Burgoa C (2020) Obesity and its Implications for COVID-19 Mortality. Obesity (Silver Spring) 28(6):1005
8. Zhou F et al (2020) Clinical course and risk factors for mortality of adult inpatients with COVID-19 in Wuhan, China: a retrospective cohort study. Lancet 395(10229):1054–1062
9. Food and Agriculture Organization of the United Nations (March 2020) Maintaining a healthy diet during the COVID-19 pandemic 2020:27
10. Gasmi A et al (2020) Individual risk management strategy and potential therapeutic options for the COVID-19 pandemic. Clin Immunol 215:108409
11. World Health Organization (March 2020) Off-label use of medicines for COVID-19. Scientific brief 31:2020
12. Bloomfield HE et al (2015) VA evidence-based synthesis program reports. In: Benefits and harms of the mediterranean diet compared to other diets. Department of Veterans Affairs (US, Washington (DC)
13. Martinez-Gonzalez MA, Bes-Rastrollo M (2014) Dietary patterns, Mediterranean diet, and cardiovascular disease. Curr Opin Lipidol 25(1):20–26
14. Bousquet J et al (2020) Is diet partly responsible for differences in COVID-19 death rates between and within countries? Clin Transl Allergy 10:16
15. Chrysohoou C et al (2004) Adherence to the Mediterranean diet attenuates inflammation and coagulation process in healthy adults: The ATTICA Study. J Am Coll Cardiol 44(1):152–158
16. Mattioli AV et al (2020) Quarantine during COVID-19 outbreak: changes in diet and physical activity increase the risk of cardiovascular disease. *Nutr Metab Cardiovasc Dis* 30(9):1409–1417
17. Mattioli AV et al (2019) Cardiovascular prevention in women: a narrative review from the Italian Society of Cardiology working groups on 'Cardiovascular Prevention, Hypertension and peripheral circulation' and on 'Women Disease'. J Cardiovasc Med (Hagerstown) 20(9):575–583

18. Bendall CL et al (2018) Central obesity and the Mediterranean diet: a systematic review of intervention trials. Crit Rev Food Sci Nutr 58(18):3070–3084
19. Magriplis E et al (2020) Presence of hypertension is reduced by mediterranean diet adherence in all individuals with a more pronounced effect in the obese: the Hellenic National Nutrition and Health Survey (HNNHS). Nutrients 12(3), 853:1–15
20. Bhaskaram P (2002) Micronutrient malnutrition, infection, and immunity: an overview. Nutr Rev 60(5 Pt 2):S40–S45
21. World Health Organization. Nutrition advice for adults during the COVID-19 outbreak. 2020; Available from: http://www.emro.who.int/nutrition/nutrition-infocus/nutrition-advice-for-adults-during-the-covid-19-outbreak.html
22. Pae M, Meydani SN, Wu D (2012) The role of nutrition in enhancing immunity in aging. Aging Dis 3(1):91–129
23. Covid C et al (2020) Preliminary estimates of the prevalence of selected underlying health conditions among patients with coronavirus disease 2019—United States, February 12–March 28, 2020. Morb Mortal Wkly Rep 69(13):382
24. Wu C et al (2020) Risk factors associated with acute respiratory distress syndrome and death in patients with coronavirus disease 2019 Pneumonia in Wuhan, China. JAMA Intern Med 180(7):1–11
25. Power SE et al (2014) Food and nutrient intake of Irish community-dwelling elderly subjects: who is at nutritional risk? J Nutr Health Aging 18(6):561–572
26. Volkert D et al (2020) Joint action malnutrition in the elderly (MaNuEL) knowledge hub: summary of project findings. Eur Geriatr Med 11(1):169–177
27. Haase H, Rink L (2009) The immune system and the impact of zinc during aging. Immun Ageing 6:9
28. Gammoh NZ, Rink L (2017) Zinc in infection and inflammation. Nutrients 9(6), 624:1–25
29. Beck MA, Handy J, Levander OA (2004) Host nutritional status: the neglected virulence factor. Trends Microbiol 12(9):417–423
30. World Health Organization. Information note on COVID-19 and NCDs. 2020; Available from: https://www.who.int/internal-publications-detail/covid-19-and-ncds
31. Georgousopoulou EN et al (2016) Anti-inflammatory diet and 10-year (2002-2012) cardiovascular disease incidence: The ATTICA study. Int J Cardiol 222:473–478
32. de Boer A et al (2017) The effect of dietary components on inflammatory lung diseases - a literature review. Int J Food Sci Nutr 68(7):771–787
33. Phillips CM et al (2019) Dietary inflammatory index and non-communicable disease risk: a narrative review. Nutrients 11(8), 1873:1–32
34. Zhang L, Liu Y (2020) Potential interventions for novel coronavirus in China: a systematic review. J Med Virol 92(5):479–490
35. Barazzoni R et al (2020) ESPEN expert statements and practical guidance for nutritional management of individuals with SARS-CoV-2 infection. Clin Nutr 39(6):1631–1638
36. Caccialanza R et al (2020) Early nutritional supplementation in non-critically ill patients hospitalized for the 2019 novel coronavirus disease (COVID-19): rationale and feasibility of a shared pragmatic protocol. Nutrition 74:110835
37. Dang Y et al (2019) In vitro and in vivo studies on the angiotensin-converting enzyme inhibitory activity peptides isolated from broccoli protein hydrolysate. J Agric Food Chem 67(24):6757–6764
38. Schüler R et al (2017) High-saturated-fat diet increases circulating angiotensin-converting enzyme, which is enhanced by the rs4343 polymorphism defining persons at risk of nutrient-dependent increases of blood pressure. J Am Heart Assoc 6(1):1–20
39. Fan H, Liao W, Wu J (2019) Molecular interactions, bioavailability, and cellular mechanisms of angiotensin-converting enzyme inhibitory peptides. J Food Biochem 43(1):e12572
40. Tejpal S et al (2020) Angiotensin converting enzyme (ACE): a marker for personalized feedback on dieting. Nutrients 12(3), 660:1–13

41. Melaku YA et al (2019) Burden and trend of diet-related non-communicable diseases in Australia and comparison with 34 OECD countries, 1990–2015: findings from the Global Burden of Disease Study 2015. Eur J Nutr 58(3):1299–1313
42. Calder PC et al (2020) Optimal Nutritional Status for a Well-Functioning Immune System Is an Important Factor to Protect against Viral Infections. Nutrients 12(4), 1181:1–10
43. World Health Organization. Healthy diet. 2019; Available from: https://www.who.int/news-room/fact-sheets/detail/healthy-diet
44. Fardet A (2010) New hypotheses for the health-protective mechanisms of whole-grain cereals: what is beyond fibre? Nutr Res Rev 23(1):65–134
45. Jawhara M et al (2019) Biomarkers of whole-grain and cereal-fiber intake in human studies: a systematic review of the available evidence and perspectives. Nutrients 11(12), 2994:1–32
46. The American Association of Cereal Chemists. The American Association of Cereal Chemists—AACCI Definitions of Whole Grain/Sprouted Grain/Whole Grain Product; Available from: http://www.cerealsgrains.org/initiatives/definitions/Pages/WholeGrain.aspx
47. McKevith B (2004) Nutritional aspects of cereals. Nutr Bull 29(2):111–142
48. Gong L et al (2018) Whole cereal grains and potential health effects: involvement of the gut microbiota. Food Res Int 103:84–102
49. Harris S et al (2019) Comparative prebiotic activity of mixtures of cereal grain polysaccharides. AMB Express 9(1):203
50. Ruel G et al (2014) Association between nutrition and the evolution of multimorbidity: the importance of fruits and vegetables and whole grain products. Clin Nutr 33(3):513–520
51. Ye EQ et al (2012) Greater whole-grain intake is associated with lower risk of type 2 diabetes, cardiovascular disease, and weight gain. J Nutr 142(7):1304–1313
52. Slavin JL et al (2013) Whole grains: definition, dietary recommendations, and health benefits. Cereal Foods World 58(4):191–198
53. Seal CJ, Brownlee IA (2015) Whole-grain foods and chronic disease: evidence from epidemiological and intervention studies. Proc Nutr Soc 74(3):313–319
54. Kyrø C et al (2018) Higher whole-grain intake is associated with lower risk of type 2 diabetes among middle-aged men and women: the Danish diet, cancer, and health cohort. J Nutr 148(9):1434–1444
55. Aune D et al (2016) Whole grain consumption and risk of cardiovascular disease, cancer, and all cause and cause specific mortality: systematic review and dose-response meta-analysis of prospective studies. BMJ 353:i2716
56. Andersen V et al (2017) A proposal for a study on treatment selection and lifestyle recommendations in chronic inflammatory diseases: a Danish multidisciplinary collaboration on prognostic factors and personalised medicine. Nutrients 9(5), 499:1–16
57. Awika JM, Rose DJ, Simsek S (2018) Complementary effects of cereal and pulse polyphenols and dietary fiber on chronic inflammation and gut health. Food Funct 9(3):1389–1409
58. Micha R et al (2015) Global, regional and national consumption of major food groups in 1990 and 2010: a systematic analysis including 266 country-specific nutrition surveys worldwide. BMJ Open 5(9):e008705
59. Ministry of Environment and Food of Denmark. The Official Dietary Guidelines. 2018; Available from: https://www.foedevarestyrelsen.dk/Nyheder/Aktuelt/Sider/Nyheder_2017/Rugbr%C3%B8d_til_aftensmad_er_ogs%C3%A5_sund_fornuft.aspx#
60. Foedevarestyrelsen, Rugbrød til aftensmad er også sund fornuft. 2017
61. Chandra RK (1992) Effect of vitamin and trace-element supplementation on immune responses and infection in elderly subjects. Lancet 340(8828):1124–1127
62. Głąbska D et al (2020) Fruit and vegetable intake and mental health in adults: a systematic review. Nutrients 12(1), 115:1–34
63. Mattioli AV et al (2018) Fruit and vegetables in hypertensive women with asymptomatic peripheral arterial disease. Clin Nutr ESPEN 27:110–112
64. Fonseca S et al (2020) Association between consumption of fermented vegetables and COVID-19 mortality at a country level in Europe. In: medRxiv

65. Tang GY et al (2017) Effects of vegetables on cardiovascular diseases and related mechanisms. Nutrients 9(8), 857:1–25
66. Alkhatib A et al (2017) Functional foods and lifestyle approaches for diabetes prevention and management. Nutrients 9(12), 1310:1–18
67. Gao X, Bermudez OI, Tucker KL (2004) Plasma C-reactive protein and homocysteine concentrations are related to frequent fruit and vegetable intake in Hispanic and non-Hispanic white elders. J Nutr 134(4):913–918
68. Esmaillzadeh A et al (2006) Fruit and vegetable intakes, C-reactive protein, and the metabolic syndrome. Am J Clin Nutr 84(6):1489–1497
69. Sciomer S et al (2019) Prevention of cardiovascular risk factors in women: the lifestyle paradox and stereotypes we need to defeat. Eur J Prev Cardiol 26(6):609–610
70. Zamora-Ros R et al (2016) Dietary polyphenol intake in Europe: the European Prospective Investigation into Cancer and Nutrition (EPIC) study. Eur J Nutr 55(4):1359–1375
71. Bahramsoltani R et al (2016) The preventive and therapeutic potential of natural polyphenols on influenza. Expert Rev Anti-Infect Ther 14(1):57–80
72. Kaulmann A, Bohn T (2014) Carotenoids, inflammation, and oxidative stress--implications of cellular signaling pathways and relation to chronic disease prevention. Nutr Res 34(11):907–929
73. Pall ML, Levine S (2015) Nrf2, a master regulator of detoxification and also antioxidant, anti-inflammatory and other cytoprotective mechanisms, is raised by health promoting factors. Sheng Li Xue Bao 67(1):1–18
74. Kim M et al (2015) Cucurbitacin B inhibits immunomodulatory function and the inflammatory response in macrophages. Immunopharmacol Immunotoxicol 37(5):473–480
75. Kim KH et al (2018) Cucurbitacin B induces hypoglycemic effect in diabetic mice by regulation of AMP-activated protein kinase alpha and glucagon-like peptide-1 via bitter taste receptor signaling. Front Pharmacol 9:1071
76. Jiménez-Osorio AS, González-Reyes S, Pedraza-Chaverri J (2015) Natural Nrf2 activators in diabetes. Clin Chim Acta 448:182–192
77. Bai Y et al (2015) Sulforaphane Protects against Cardiovascular Disease via Nrf2 Activation. Oxid Med Cell Longev 2015:407580
78. Fahey JW et al (2019) Bioavailability of sulforaphane following ingestion of glucoraphanin-rich broccoli sprout and seed extracts with active myrosinase: a pilot study of the effects of proton pump inhibitor administration. Nutrients 11(7), 1489:1–16
79. Vanduchova A, Anzenbacher P, Anzenbacherova E (2019) Isothiocyanate from broccoli, sulforaphane, and its properties. J Med Food 22(2):121–126
80. Baenas N et al (2019) Influence of cooking methods on glucosinolates and isothiocyanates content in novel cruciferous foods. Foods 8(7), 257:1–9
81. Abdulah DM, Hassan A (2020) Relation of dietary factors with infection and mortality rates of COVID-19 across the world. J Nutr Health Aging 24(9):1011–1018
82. Gombart AF, Pierre A, Maggini S (2020) A review of micronutrients and the immune system-working in harmony to reduce the risk of infection. Nutrients 12(1), 236:1–41
83. Teymoori-Rad M et al (2019) The interplay between vitamin D and viral infections. Rev Med Virol 29(2):e2032
84. Raghu G, Radha S (2020) Increasing fruits and vegetable consumption to protect against viral diseases. Available at SSRN: https://ssrn.com/abstract=3547168
85. Mukuddem-Petersen J, Oosthuizen W, Jerling JC (2005) A systematic review of the effects of nuts on blood lipid profiles in humans. J Nutr 135(9):2082–2089
86. Salas-Salvadó J et al (2008) The effect of nuts on inflammation. Asia Pac J Clin Nutr 17(Suppl 1):333–336
87. Luo C et al (2014) Nut consumption and risk of type 2 diabetes, cardiovascular disease, and all-cause mortality: a systematic review and meta-analysis. Am J Clin Nutr 100(1):256–269
88. Ros E, Mataix J (2006) Fatty acid composition of nuts--implications for cardiovascular health. Br J Nutr 96(Suppl 2):S29–S35

89. Salas-Salvadó J et al (2006) Dietary fibre, nuts and cardiovascular diseases. Br J Nutr 96(Suppl 2):S46–S51
90. González CA, Salas-Salvadó J (2006) The potential of nuts in the prevention of cancer. Br J Nutr 96(Suppl 2):S87–S94
91. Sabate J (2007) Nut consumption and change in weight: the weight of the evidence. Br J Nutr 98(3):456–457
92. Mattes RD (2008) The energetics of nut consumption. Asia Pac J Clin Nutr 17(Suppl 1):337–339
93. Maccioni L et al (2018) Obesity and risk of respiratory tract infections: results of an infection-diary based cohort study. BMC Public Health 18(1):271
94. Tharanathan R, Mahadevamma S (2003) Grain legumes—a boon to human nutrition. Trends Food Sci Technol 14(12):507–518
95. Iqbal A et al (2006) Nutritional quality of important food legumes. Food Chem 97(2):331–335
96. Zhu F, Du B, Xu B (2018) Anti-inflammatory effects of phytochemicals from fruits, vegetables, and food legumes: a review. Crit Rev Food Sci Nutr 58(8):1260–1270
97. Kitayaporn D et al (1992) Haemoglobin-E in the presence of oxidative substances from fava bean may be protective against Plasmodium falciparum malaria. Trans R Soc Trop Med Hyg 86(3):240–244
98. Uematsu J et al (2012) Legume lectins inhibit human parainfluenza virus type 2 infection by interfering with the entry. Viruses 4(7):1104–1115
99. Messina G et al (2020) Functional role of dietary intervention to improve the outcome of COVID-19: a hypothesis of work. Int J Mol Sci 21(9):3104
100. Hosseinpour-Niazi S et al (2015) Non-soya legume-based therapeutic lifestyle change diet reduces inflammatory status in diabetic patients: a randomised cross-over clinical trial. Br J Nutr 114(2):213–219
101. Kalantar-Zadeh K, Moore LW (2020) Impact of nutrition and diet on COVID-19 infection and implications for kidney health and kidney disease management. J Ren Nutr 30(3):179–181
102. Becerra-Tomás N et al (2018) Legume consumption is inversely associated with type 2 diabetes incidence in adults: A prospective assessment from the PREDIMED study. Clin Nutr 37(3):906–913
103. Polamarasetti P, Martirosyan D (2020) Nutrition planning during the COVID-19 pandemic for aging immunity. Bioactive Compounds Health Dis 3(7):109–123
104. Das UN (2020) Can bioactive lipids inactivate coronavirus (COVID-19)? Arch Med Res 51(3):282–286
105. Kris-Etherton PM, Harris WS, Appel LJ (2002) Fish consumption, fish oil, omega-3 fatty acids, and cardiovascular disease. Circulation 106(21):2747–2757
106. Wu D et al (2019) Nutritional modulation of immune function: analysis of evidence, mechanisms, and clinical relevance. Front Immunol 9:3160
107. Read SA et al (2019) The role of zinc in antiviral immunity. Adv Nutr 10(4):696–710
108. Makino S et al (2010) Reducing the risk of infection in the elderly by dietary intake of yoghurt fermented with Lactobacillus delbrueckii ssp. bulgaricus OLL1073R-1. Br J Nutr 104(7):998–1006
109. Conlon MA, Bird AR (2014) The impact of diet and lifestyle on gut microbiota and human health. Nutrients 7(1):17–44
110. Kolodziej H (2011) Antimicrobial, antiviral and immunomodulatory activity studies of pelargonium sidoides (EPs(®) 7630) in the context of health promotion. Pharmaceuticals (Basel) 4(10):1295–1314
111. Im K, Kim J, Min H (2016) Ginseng, the natural effectual antiviral: protective effects of Korean Red Ginseng against viral infection. J Ginseng Res 40(4):309–314
112. Luo H et al (2020) Can Chinese medicine be used for prevention of corona virus disease 2019 (COVID-19)? A review of historical classics, research evidence and current prevention programs. Chin J Integr Med 26(4):243–250

113. Quan FS et al (2007) Ginseng and Salviae herbs play a role as immune activators and modulate immune responses during influenza virus infection. Vaccine 25(2):272–282
114. Rasool A et al (2017) Anti-avian influenza virus H9N2 activity of aqueous extracts of Zingiber officinalis (Ginger) and Allium sativum (Garlic) in chick embryos. Pak J Pharm Sci 30(4):1341–1344
115. Usachev EV et al (2013) Antiviral activity of tea tree and eucalyptus oil aerosol and vapour. J Aerosol Sci 59:22–30
116. Ren J-l, Zhang A-H, Wang X-J (2020) Traditional Chinese medicine for COVID-19 treatment. Pharmacol Res 155:104743
117. Ang L et al (2020) Herbal medicine and pattern identification for treating COVID-19: a rapid review of guidelines. Integr Med Res 9(2):1–14
118. National Health Commission of the People's Republic of China. Guideline on diagnosis and treatment of COVID-19 (Trial 6th edition). 2020; Available from: http://www.nhc.gov.cn/xcs/zhengcwj/202002/8334a8326dd94d329df351d7da8aefc2.shtml
119. Publicity Department of the People's Republic of China. 2020.; Available from: http://www.nhc.gov.cn/xcs/fkdt/202002/f12a62d10c2a48c6895cedf2faea6e1f.shtml
120. Xu Z et al (2020) Pathological findings of COVID-19 associated with acute respiratory distress syndrome. Lancet Respir Med 8(4):420–422
121. Zhao J et al (2020) Investigating the mechanism of Qing-Fei-Pai-Du-Tang for the treatment of Novel Coronavirus Pneumonia by network. Pharmacology. Chin Tradit Herbal Drugs. Available at http://kns.cnki.net/kcms/detail/12.1108.R.20200216.2044.002.html
122. Wu H et al (2020) Preliminary exploration of the mechanism of Qingfei Paidu decoction against novel coronavirus pneumonia based on network pharmacology and molecular docking technology. Acta Pharm Sin 55(3):374–383
123. Tan BK, Vanitha J (2004) Immunomodulatory and antimicrobial effects of some traditional Chinese medicinal herbs: a review. Curr Med Chem 11(11):1423–1430
124. Ang L et al (2020) Herbal medicine for treatment of children diagnosed with COVID-19: a review of guidelines. Complement Ther Clin Pract 39:1–4
125. Runfeng L et al (2020) Lianhuaqingwen exerts anti-viral and anti-inflammatory activity against novel coronavirus (SARS-CoV-2). Pharmacol Res 156:104761
126. Mousa HA-L (2017) Prevention and treatment of influenza, influenza-like illness, and common cold by herbal, complementary, and natural therapies. J Evid-Based Complement Alternat Med 22(1):166–174
127. Huang Z et al (2018) Role of vitamin A in the immune system. J Clin Med 7(9), 258:1–16
128. Patel N et al (2019) Baseline serum vitamin A and D levels determine benefit of oral vitamin A&D supplements to humoral immune responses following pediatric influenza vaccination. Viruses 11(10), 907:1–21
129. Jee J et al (2013) Effects of dietary vitamin A content on antibody responses of feedlot calves inoculated intramuscularly with an inactivated bovine coronavirus vaccine. Am J Vet Res 74(10):1353–1362
130. Kańtoch M et al (2002) Importance of vitamin A deficiency in pathology and immunology of viral infections. Rocz Panstw Zakl Hig 53(4):385–392
131. Glasziou PP, Mackerras DE (1993) Vitamin A supplementation in infectious diseases: a meta-analysis. BMJ 306(6874):366–370
132. Sinha S, et al (2020) Systematic cell line-based identification of drugs modifying ACE2 expression. Preprints. https://doi.org/10.20944/preprints202003.0446.v1
133. McGill JL et al (2019) Vitamin A deficiency impairs the immune response to intranasal vaccination and RSV infection in neonatal calves. Sci Rep 9(1):15957
134. Spinas E et al (2015) Crosstalk between vitamin B and immunity. J Biol Regul Homeost Agents 29(2):283–288
135. Yoshii K et al (2019) Metabolism of dietary and microbial vitamin B family in the regulation of host immunity. Front Nutr 6:48

136. Wintergerst ES, Maggini S, Hornig DH (2007) Contribution of selected vitamins and trace elements to immune function. Ann Nutr Metab 51(4):301–323
137. Cheng CH et al (2006) Vitamin B6 supplementation increases immune responses in critically ill patients. Eur J Clin Nutr 60(10):1207–1213
138. Morris MS et al (2010) Vitamin B-6 intake is inversely related to, and the requirement is affected by, inflammation status. J Nutr 140(1):103–110
139. Tamura J et al (1999) Immunomodulation by vitamin B12: augmentation of CD8+ T lymphocytes and natural killer (NK) cell activity in vitamin B12-deficient patients by methyl-B12 treatment. Clin Exp Immunol 116(1):28–32
140. Degnan PH, Taga ME, Goodman AL (2014) Vitamin B12 as a modulator of gut microbial ecology. Cell Metab 20(5):769–778
141. Negi S et al (2019) Potential role of gut microbiota in induction and regulation of innate immune memory. Front Immunol 10:2441
142. Dhar D, Mohanty A (2020) Gut microbiota and Covid-19- possible link and implications. Virus Res 285:198018
143. Zuo T et al (2020) Alterations in gut microbiota of patients with COVID-19 during time of hospitalization. Gastroenterology 159:944–955
144. Tan CW et al (2020) A cohort study to evaluate the effect of combination Vitamin D, Magnesium and Vitamin B12 (DMB) on progression to severe outcome in older COVID-19 patients. In: medRxiv
145. Watanabe F (2007) Vitamin B12 sources and bioavailability. Exp Biol Med (Maywood) 232(10):1266–1274
146. Kim Y et al (2013) Vitamin C is an essential factor on the anti-viral immune responses through the production of interferon-α/β at the initial stage of Influenza A Virus (H3N2) infection. Immune Netw 13(2):70–74
147. Carr AC, Maggini S (2017) Vitamin C and immune function. Nutrients 9(11):1211
148. Manning J et al (2013) Vitamin C promotes maturation of T-cells. Antioxid Redox Signal 19(17):2054–2067
149. Hemilä H (2017) Vitamin C and infections. Nutrients 9(4):339
150. Van Driel ML et al (2019) Oral vitamin C supplements to prevent and treat acute upper respiratory tract infections. Cochrane Database Syst Rev (3):1–10
151. Hemilä H, Chalker E (2013) Vitamin C for preventing and treating the common cold. Cochrane Database Syst Rev (1):1
152. Colunga Biancatelli RML, Berrill M, Marik PE (2020) The antiviral properties of vitamin C. Taylor & Francis
153. Maggini S, Beveridge S, Suter M (2012) A combination of high-dose vitamin C plus zinc for the common cold. J Int Med Res 40(1):28–42
154. Laird E et al (2014) Vitamin D deficiency is associated with inflammation in older Irish adults. J Clin Endocrinol Metabol 99(5):1807–1815
155. Rondanelli M et al (2018) Self-care for common colds: the pivotal role of vitamin D, vitamin C, zinc, and echinacea in three main immune interactive clusters (physical barriers, innate and adaptive immunity) involved during an episode of common colds—practical advice on dosages and on the time to take these nutrients/botanicals in order to prevent or treat common colds. Evid Based Complement Alternat Med 2018:1–36
156. Di Rosa M et al (2011) Vitamin D3: a helpful immuno-modulator. Immunology 134(2):123–139
157. Vanherwegen A-S, Gysemans C, Mathieu C (2017) Regulation of immune function by vitamin D and its use in diseases of immunity. Endocrinol Metab Clin 46(4):1061–1094
158. Zhang Y et al (2012) Vitamin D inhibits monocyte/macrophage proinflammatory cytokine production by targeting MAPK phosphatase-1. J Immunol 188(5):2127–2135
159. McGonagle D et al (2020) Interleukin-6 use in COVID-19 pneumonia related macrophage activation syndrome. Autoimmun Rev 19(6):102537

160. Kong J et al (2013) VDR attenuates acute lung injury by blocking Ang-2-Tie-2 pathway and renin-angiotensin system. Mol Endocrinol 27(12):2116–2125
161. Ginde AA, Mansbach JM, Camargo CA (2009) Association between serum 25-hydroxyvitamin D level and upper respiratory tract infection in the Third National Health and Nutrition Examination Survey. Arch Intern Med 169(4):384–390
162. Jolliffe DA, Griffiths CJ, Martineau AR (2013) Vitamin D in the prevention of acute respiratory infection: systematic review of clinical studies. J Steroid Biochem Mol Biol 136:321–329
163. Bergman P et al (2012) Vitamin D3 supplementation in patients with frequent respiratory tract infections: a randomised and double-blind intervention study. BMJ Open 2(6):1–10
164. Martineau AR et al (2017) Vitamin D supplementation to prevent acute respiratory tract infections: systematic review and meta-analysis of individual participant data. BMJ 356:1–14
165. Zhou Y-F, Luo B-A, Qin L-L (2019) The association between vitamin D deficiency and community-acquired pneumonia: a meta-analysis of observational studies. Medicine 98(38):1–7
166. Laird E, Kenny R (2020) Vitamin D deficiency in Ireland: implications for COVID-19. Results from the Irish longitudinal study on ageing (TILDA). TILDA. Available at: https://tilda.tcd.ie/publications/reports/pdf/Report_Covid19VitaminD.pdf. Accessed 2020
167. Jia X, et al (2020) Two things about COVID-19 might need attention. Preprints. https://doi.org/10.20944/preprints202002.0315.v1
168. Thornton J (2020) Don't forget chronic lung and immune conditions during covid-19, says WHO. Br Med J Publ Group 368:1
169. Braiman, M., Latitude Dependence of the COVID-19 Mortality Rate—A Possible Relationship to Vitamin D Deficiency? Available at SSRN 3561958, 2020
170. Farrar MD et al (2011) Recommended summer sunlight exposure amounts fail to produce sufficient vitamin D status in UK adults of South Asian origin. Am J Clin Nutr 94(5):1219–1224
171. Laird E et al (2018) A high prevalence of vitamin D deficiency observed in the Dublin South East Asian population. Proc Nutr Soc 77(OCE3):1
172. Glaab EO, M (2020) The role of spike-ACE2 interaction in pulmonary blood pressure regulation. FAIRDOM Hub. Available at: https://fairdomhub.org/models/7092020
173. Grant WB et al (2020) Evidence that vitamin D supplementation could reduce risk of influenza and COVID-19 infections and deaths. Nutrients 12(4):988
174. Muscogiuri G et al (2017) Vitamin D and chronic diseases: the current state of the art. Arch Toxicol 91(1):97–107
175. Crowe FL et al (2011) Plasma concentrations of 25-hydroxyvitamin D in meat eaters, fish eaters, vegetarians and vegans: results from the EPIC–Oxford study. Public Health Nutr 14(2):340–346
176. Wang L-s et al (2020) A review of the 2019 Novel Coronavirus (COVID-19) based on current evidence. Int J Antimicrob Agents 2020:105948
177. Moriguchi S, Muraga M (2000) Vitamin E and immunity, Vitam Horm 59:305–336
178. Han SN, Meydani SN (2006) Impact of vitamin E on immune function and its clinical implications. Expert Rev Clin Immunol 2(4):561–567
179. Stiff A et al (2018) Nitric oxide production by myeloid-derived suppressor cells plays a role in impairing Fc receptor–mediated natural killer cell function. Clin Cancer Res 24(8):1891–1904
180. Beharka A et al (2000) Long-term dietary antioxidant supplementation reduces production of selected inflammatory mediators by murine macrophages. Nutr Res 20(2):281–296
181. Lee GY, Han SN (2018) The role of vitamin E in immunity. Nutrients 10(11):1614
182. Xuan NT et al (2016) Klotho sensitive regulation of dendritic cell functions by vitamin E. Biol Res 49(1):1–9
183. Cassat JE, Skaar EP (2013) Iron in infection and immunity. Cell Host Microbe 13(5):509–519
184. Ganz T, Nemeth E (2015) Iron homeostasis in host defence and inflammation. Nat Rev Immunol 15(8):500–510

185. Drakesmith H, Prentice A (2008) Viral infection and iron metabolism. Nat Rev Microbiol 6(7):541–552
186. Ali MK et al (2017) Role of iron in the pathogenesis of respiratory disease. Int J Biochem Cell Biol 88:181–195
187. Alpert PT (2017) The role of vitamins and minerals on the immune system. Home Health Care Manag Pract 29(3):199–202
188. Maggini S, Pierre A, Calder PC (2018) Immune function and micronutrient requirements change over the life course. Nutrients 10(10):1531
189. Wessling-Resnick M (2018) Crossing the iron gate: why and how transferrin receptors mediate viral entry. Annu Rev Nutr 38:431–458
190. Jayaweera JAAS, Reyes M, Joseph A (2019) Childhood iron deficiency anemia leads to recurrent respiratory tract infections and gastroenteritis. Sci Rep 9(1):1–8
191. Bhutta ZA (2007) Iron and zinc deficiency in children in developing countries. Br Med J Publ Group 334:104
192. Shaw JG, Friedman JF (2011) Iron deficiency anemia: focus on infectious diseases in lesser developed countries. Anemia 2011
193. Fraga CG (2005) Relevance, essentiality and toxicity of trace elements in human health. Mol Asp Med 26(4–5):235–244
194. Rayman MP (2012) Selenium and human health. Lancet 379(9822):1256–1268
195. Vindry C, Ohlmann T, Chavatte L (2018) Selenium metabolism, regulation, and sex differences in mammals. In: Selenium. Springer, pp 89–107
196. Zwolak I, Zaporowska H (2012) Selenium interactions and toxicity: a review. Cell Biol Toxicol 28(1):31–46
197. Guillin OM et al (2019) Selenium, selenoproteins and viral infection. Nutrients 11(9):2101
198. Saeed F et al (2016) Studying the impact of nutritional immunology underlying the modulation of immune responses by nutritional compounds–a review. Food Agric Immunol 27(2):205–229
199. Beck MA et al (2001) Selenium deficiency increases the pathology of an influenza virus infection. FASEB J 15(8):1481–1483
200. Beck MA, Levander OA, Handy J (2003) Selenium deficiency and viral infection. J Nutr 133(5):1463S–1467S
201. Harthill M (2011) Micronutrient selenium deficiency influences evolution of some viral infectious diseases. Biol Trace Elem Res 143(3):1325–1336
202. Ivory K et al (2017) Selenium supplementation has beneficial and detrimental effects on immunity to influenza vaccine in older adults. Clin Nutr 36(2):407–415
203. Fraczek A, Pasternak K (2013) Selenium in medicine and treatment. J Elem 18(1):145–163
204. Kieliszek M, Błażejak S (2013) Selenium: significance, and outlook for supplementation. Nutrition 29(5):713–718
205. Kieliszek M, Błażejak S (2016) Current knowledge on the importance of selenium in food for living organisms: a review. Molecules 21(5):609
206. Fuhrman, J., Immunity benefits of zinc as we age. 2020
207. Wessels I, Maywald M, Rink L (2017) Zinc as a gatekeeper of immune function. Nutrients 9(12):1286
208. Maywald M, Wessels I, Rink L (2017) Zinc signals and immunity. Int J Mol Sci 18(10):2222
209. Razzaque, M., COVID-19 pandemic: can maintaining optimal zinc balance enhance host resistance? 2020
210. Te Velthuis AJ et al (2010) Zn2+ inhibits coronavirus and arterivirus RNA polymerase activity in vitro and zinc ionophores block the replication of these viruses in cell culture. PLoS Pathog 6(11):e1001176
211. Shankar AH, Prasad AS (1998) Zinc and immune function: the biological basis of altered resistance to infection. Am J Clin Nutr 68(2):447S–463S
212. Lazarczyk M, Favre M (2008) Role of Zn2+ ions in host-virus interactions. J Virol 82(23):11486–11494

213. Iovino L et al (2018) High-dose zinc oral supplementation after stem cell transplantation causes an increase of TRECs and CD4+ naive lymphocytes and prevents TTV reactivation. Leuk Res 70:20–24
214. Johnstone J et al (2012) Zinc for the treatment of the common cold: a systematic review and meta-analysis of randomized controlled trials. CMAJ 184(10):E551–E561
215. Uwitonze AM, Razzaque MS (2018) Role of magnesium in vitamin D activation and function. J Am Osteopath Assoc 118(3):181–189
216. Liang R-y et al (2012) Magnesium affects the cytokine secretion of CD4+ T lymphocytes in acute asthma. J Asthma 49(10):1012–1015
217. Houston M (2011) The role of magnesium in hypertension and cardiovascular disease. J Clin Hypertens 13(11):843–847
218. Chaigne-Delalande B et al (2013) Mg2+ regulates cytotoxic functions of NK and CD8 T cells in chronic EBV infection through NKG2D. Science 341(6142):186–191
219. Tam M et al (2003) Possible roles of magnesium on the immune system. Eur J Clin Nutr 57(10):1193–1197
220. Wishart K (2017) Increased micronutrient requirements during physiologically demanding situations: review of the current evidence. Vitamin Miner 6:1–16
221. Sansbury BE, Spite M (2016) Resolution of acute inflammation and the role of resolvins in immunity, thrombosis, and vascular biology. Circ Res 119(1):113–130
222. Cai C et al (2018) Macrophage-derived extracellular vesicles induce long-lasting immunity against hepatitis C virus which is blunted by polyunsaturated fatty acids. Front Immunol 9:723
223. Serhan CN, Levy BD (2018) Resolvins in inflammation: emergence of the pro-resolving superfamily of mediators. J Clin Invest 128(7):2657–2669
224. Leu G-Z, Lin T-Y, Hsu JT (2004) Anti-HCV activities of selective polyunsaturated fatty acids. Biochem Biophys Res Commun 318(1):275–280
225. Morita M et al (2013) The lipid mediator protectin D1 inhibits influenza virus replication and improves severe influenza. Cell 153(1):112–125
226. Covington M (2004) Omega-3 fatty acids. Am Fam Physician 70(1):133–140
227. Naja F, Hamadeh R (2020) Nutrition amid the COVID-19 pandemic: a multi-level framework for action. Eur J Clin Nutr 74:1117–1121
228. Valdés-Ramos R et al (2010) Diet, exercise and gut mucosal immunity. Proc Nutr Soc 69(4):644–650
229. McCarty MF, DiNicolantonio JJ (2020) Nutraceuticals have potential for boosting the type 1 interferon response to RNA viruses including influenza and coronavirus. Prog Cardiovasc Dis 63(3):383–385
230. Childs CE, Calder PC, Miles EA (2019) Diet and immune function. Multidisciplinary Digital Publishing Institute
231. García OP, Long KZ, Rosado JL (2009) Impact of micronutrient deficiencies on obesity. Nutr Rev 67(10):559–572
232. Thurnham DI (1997) Micronutrients and immune function: some recent developments. J Clin Pathol 50(11):887–891
233. Pinti M et al (2016) Aging of the immune system: focus on inflammation and vaccination. Eur J Immunol 46(10):2286–2301
234. Percival SS (2011) Nutrition and immunity: balancing diet and immune function. Nutr Today 46(1):12–17
235. Takeda E et al (2004) Stress control and human nutrition. J Med Investig 51(3, 4):139–145
236. Jeong H et al (2016) Mental health status of people isolated due to Middle East Respiratory Syndrome. Epidemiol Health 38:1–7
237. Yau YH, Potenza MN (2013) Stress and eating behaviors. Minerva Endocrinol 38(3):255
238. Razzoli M, Bartolomucci A (2016) The dichotomous effect of chronic stress on obesity. Trends Endocrinol Metab 27(7):504–515

239. Maniscalco JW, Rinaman L (2017) Interoceptive modulation of neuroendocrine, emotional, and hypophagic responses to stress. Physiol Behav 176:195–206
240. Moynihan AB et al (2015) Eaten up by boredom: Consuming food to escape awareness of the bored self. Front Psychol 6:369
241. Yılmaz C, Gökmen V (2020) Neuroactive compounds in foods: occurrence, mechanism and potential health effects. Food Res Int 128:108744
242. Rodríguez-Martín BC, Meule A (2015) Food craving: new contributions on its assessment, moderators, and consequences. Front Psychol 6:21
243. Hamer M et al (2019) Psychological distress and infectious disease mortality in the general population. Brain Behav Immun 76:280–283
244. Hayward SE et al (2020) A systematic review of the impact of psychosocial factors on immunity: Implications for enhancing BCG response against tuberculosis. SSM-Population Health 10:100522
245. Lehman H, Ballow M (2014) Immune compromise due to metabolic disorders: malnutrition, obesity, stress, and inborn errors of metabolism. In: Stiehm's immune deficiencies. Elsevier, pp 823–834
246. van den Buuse M, Hale MW (2019) Serotonin in stress. In: Stress: physiology, biochemistry, and pathology. Elsevier, pp 115–123
247. Peuhkuri K, Sihvola N, Korpela R (2012) Diet promotes sleep duration and quality. Nutr Res 32(5):309–319
248. Hashizume S (2010) Stress-reducing activity of various food ingredients. J Int Soc Life Inform Sci 28(1):148–152
249. Remuzzi A, Remuzzi G (2020) COVID-19 and Italy: what next? Lancet 395:1225–1228
250. Yang J-K et al (2010) Binding of SARS coronavirus to its receptor damages islets and causes acute diabetes. Acta Diabetol 47(3):193–199
251. Yang J et al (2006) Plasma glucose levels and diabetes are independent predictors for mortality and morbidity in patients with SARS. Diabet Med 23(6):623–628
252. Abdi A et al (2020) Diabetes and COVID-19: a systematic review on the current evidences. Diabetes Res Clin Pract 166:108347
253. Cuschieri S, Grech S (2020) COVID-19 and diabetes: the why, the what and the how. J Diabetes Complicat 34(9):107637
254. Stefanini GG et al (2020) ST-elevation myocardial infarction in patients with COVID-19: clinical and angiographic outcomes. Circulation 141(25):2113–2116
255. Deng Q et al (2020) Suspected myocardial injury in patients with COVID-19: evidence from front-line clinical observation in Wuhan, China. Int J Cardiol 311:116–121
256. Mendoza-Torres E et al (2015) ACE2 and vasoactive peptides: novel players in cardiovascular/renal remodeling and hypertension. Ther Adv Cardiovasc Dis 9(4):217–237
257. Gallagher PE, Ferrario CM, Tallant EA (2008) Regulation of ACE2 in cardiac myocytes and fibroblasts. Am J Phys Heart Circ Phys 295(6):H2373–H2379
258. Fragopoulou E et al (2010) The association between adherence to the Mediterranean diet and adiponectin levels among healthy adults: the ATTICA study. J Nutr Biochem 21(4):285–289
259. Helnæs A et al (2016) Intake of whole grains is associated with lower risk of myocardial infarction: the Danish Diet, Cancer and Health Cohort. Am J Clin Nutr 103(4):999–1007
260. Liang W et al (2020) Cancer patients in SARS-CoV-2 infection: a nationwide analysis in China. Lancet Oncol 21(3):335–337
261. Garófolo A, Qiao L, Maia-Lemos PdS (2020) Approach to nutrition in cancer patients in the context of the coronavirus disease 2019 (COVID-19) pandemic: perspectives. Nutr Cancer 2020:1–9
262. Holshue ML et al (2020) First case of 2019 novel coronavirus in the United States. N Engl J Med 382:929–936
263. Liu Q, Wang R, Qu G (2020) Macroscopic autopsy findings in a patient with COVID-19. J Forensic Med 36:1–3

264. Bradley KC et al (2019) Microbiota-driven tonic interferon signals in lung stromal cells protect from influenza virus infection. Cell Rep 28(1):245–256. e4
265. Valdes AM et al (2018) Role of the gut microbiota in nutrition and health. BMJ 361:36–44
266. Kreda SM, Davis CW, Rose MC (2012) CFTR, mucins, and mucus obstruction in cystic fibrosis. Cold Spring Harb Perspect Med 2(9):a009589
267. Alvarado A, Arce I (2016) Antioxidants in respiratory diseases: Basic science research and therapeutic alternatives. Clin Res Trials 3(1):1–11
268. Galvão AM et al (2011) Antioxidant supplementation for the treatment of acute lung injury: a meta-analysis. Revista Brasileira de terapia intensiva 23(1):41–48
269. Ling L-j et al (2020) Flavonoids from Houttuynia cordata attenuate H1N1-induced acute lung injury in mice via inhibition of influenza virus and Toll-like receptor signalling. Phytomedicine 67:153150
270. Carr AC (2020) A new clinical trial to test high-dose vitamin C in patients with COVID-19. Crit Care 24(1):1–2
271. Sharma S et al (2013) Dietary supplementation with omega-3 polyunsaturated fatty acids ameliorates acute pneumonia induced by Klebsiella pneumoniae in BALB/c mice. Can J Microbiol 59(7):503–510
272. Reber E et al (2019) Nutritional risk screening and assessment. J Clin Med 8(7):1065
273. Jin Y-H et al (2020) A rapid advice guideline for the diagnosis and treatment of 2019 novel coronavirus (2019-nCoV) infected pneumonia (standard version). Mil Med Res 7(1):4
274. Short KR, Kedzierska K, van de Sandt CE (2018) Back to the future: lessons learned from the 1918 influenza pandemic. Front Cell Infect Microbiol 8:343
275. Chen N et al (2020) Epidemiological and clinical characteristics of 99 cases of 2019 novel coronavirus pneumonia in Wuhan, China: a descriptive study. Lancet 395(10223):507–513
276. Schuetz P et al (2019) Individualised nutritional support in medical inpatients at nutritional risk: a randomised clinical trial. Lancet 393(10188):2312–2321

Chapter 6
Drugs for the Treatment of COVID-19

Sagheer Ahmed, Halimur Rehman, Rehan Salar, May Nasser Bin-Jumah, M. Tauseef Sultan, and Marius Moga

Introduction

Slowing the COVID-19 pandemic, which has already infected over 34 million people worldwide, and ultimately ending it will depend substantially on the scientists' development of effective treatments. The initial effort to repurpose the existing drugs approved for other ailments to kill the virus and treat the COVID-19 makes sense. Worldwide, many laboratories are engaged in carrying out research on existing drugs to develop remedies for COVID-19. This will considerably shorten the total duration of drug discovery and development, which extends from laboratory investigations to initial toxicity testing to animal studies to safety studies in humans and, finally, large-scale clinical trials.

S. Ahmed (✉)
Shifa College of Pharmaceutical Sciences, Shifa Tameer-e-Millat University, Islamabad, Pakistan
e-mail: sagheer.scps@stmu.edu.pk

H. Rehman · R. Salar
Shifa International Hospital, Islamabad, Pakistan

M. N. Bin-Jumah
Biology Department, College of Science, Princess Nourah bint Abdulrahman University, Riyadh, Saudi Arabia

M. T. Sultan
Institute of Food Science and Technology, Bahauddin Zakariya University, Multan, Pakistan

M. Moga
Faculty of Medicine, Transilvania University of Brasov, Brasov, Romania

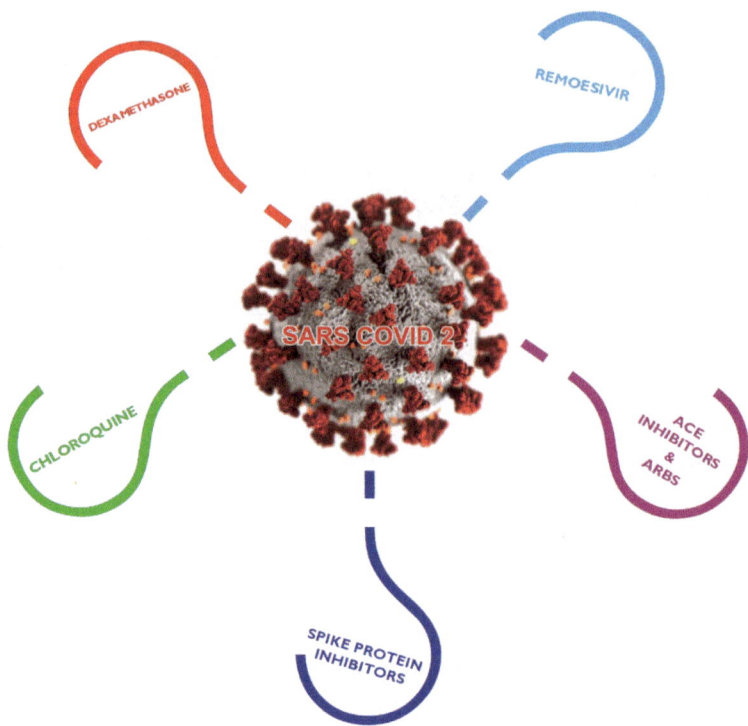

The interruption of COVID-19 spread depends on a combination of pharmacologic and nonpharmacologic interventions. Initial SARS-CoV-2 prevention includes social distancing, face masks, environmental hygiene, and handwashing [1]. Although the most important pharmacologic interventions to prevent SARS-CoV-2 infection are likely to be vaccines, the repurposing of established drugs for short-term prophylaxis is another more immediate option. Here are some of the medicines currently being used in various parts of the world as potential remedies for COVID-19.

Dexamethasone

In June 2020, scientists in the UK announced that an anti-inflammatory agent, dexamethasone, has shown to reduce mortality in a recent clinical trial in which more than 6000 severely sick COVID-19 patients were treated [2]. This represents a significant improvement over the current treatment options available for COVID-19 patients. Immediately, dexamethasone was authorized by the British government to be used in some patients, although its role in treating less severe COVID-19 infection was still not clear. The news of a preexisting, inexpensive medication reducing mortality in COVID-19 patients came as a pleasant surprise after the continuous negative news of what seemed to be an unstoppable spread of novel coronavirus.

Although the complete clinical data were not available immediately, it was quite clear that this could be a breakthrough. The drug is available in large quantities in even developing countries and can be administered orally.

This trial's results were compelling, although they have surprised many people who are actively involved in finding a cure for COVID-19. Many experts in the field suggested that if these results hold up in larger, more powered clinical trials, this drug could be a game-changer, especially for those COVID-19 patients who are critically ill.

Dexamethasone is an important drug, a steroid used to treat general inflammation and a host of other conditions. It is well known in medicine for many years. The drug has an excellent profile with a half-life of up to 54 hours. Since it is a nonspecific treatment because it does not block a specific inflammatory pathway, it has certain adverse effects. This is the drawback of this drug.

The results of the RECOVERY trial are particularly encouraging for severely ill COVID-19 patients. After less than a month of treatment, there was a 35% reduction in the mortality of severely ill patients who required ventilators. In patients who required supplemental oxygen but not ventilators, a 20% reduction in mortality was observed. However, it appears that the drug is not beneficial in less severely ill patients and may exacerbate the disease. Some studies also suggested that patients with acute respiratory distress syndrome can also benefit from the use of dexamethasone.

Two factors are crucial when managing COVID-19 patients with dexamethasone – selectiveness and timing. Initial investigations suggest that dexamethasone should not be used in patients who do not have severe symptoms of the disease and do not need additional oxygen, such as those requiring a ventilator. Even in patients in which dexamethasone is useful, it has certain drawbacks. It decreases an individual's ability to fight the virus because it dampens the overall immunity. Its lack of selectivity in inhibiting inflammatory pathways makes patients susceptible to several untoward effects, which include but are not limited to psychosis, emotional disturbances, and worsening diabetes mellitus. Furthermore, dexamethasone should not be used in the early stages of the disease as it may hinder the immune system in clearing the virus from the body.

Comparing with remdesivir, which is the only other drug showing beneficial effects in COVID-19 patients in a clinical trial, the results produced by dexamethasone are more impressive. Remdesivir is not reported to decrease the mortality in COVID-19 patients but only reduced the number of days patients had to remain hospitalized. On the other hand, dexamethasone reduced the death rate in severely ill COVID-19 patients.

After the trials' results were announced, many clinicians were confident that they were more likely to give dexamethasone to severely ill COVID-19 patients, especially those on ventilators and whose condition is not improving. However, some were not happy with the full results still not available. Scientists need additional information on the trial to help identify further subsets of patients who could benefit the most from the dexamethasone treatment. The trial investigators responded that they acknowledge and understand the scientific community's concern and will release the full data shortly.

The RECOVERY trial was evaluating many other treatment options besides dexamethasone. For example, Kaletra, an HIV drug combination, antimalarial drug hydroxychloroquine, and convalescent plasma were also investigated. The trial's hydroxychloroquine arm was discontinued after 2 weeks when it became apparent that the patient outcome is not improving. The rest of the trial was continued until 6000 patients were enrolled in the trial. Investigators were looking for a drug that could reduce deaths by about 18% with 90% confidence. They found out that dexamethasone was the first one to achieve this target. After that, they started analyzing the data, and very early in the analysis, they found the clear advantage dexamethasone was offering to severely ill COVID-19 patients.

Chloroquine

The U.S. Food and Drug Administration revoked its emergency authorization for hydroxychloroquine, the controversial antimalarial drug promoted by President Donald Trump for treating the coronavirus. The agency said in a letter that the decision is based on new evidence that made it unreasonable to believe hydroxychloroquine and chloroquine "may be effective in diagnosing, treating or preventing" COVID-19, the illness caused by the virus. Citing reports of heart complications, the FDA said the drugs pose a greater risk to patients than any potential benefits.

Chloroquine is a synthetic 4-aminoquinoline that has been the mainstay of antimalarial therapy. It specifically binds to heme, preventing its polymerization to hemozoin. Its use to treat COVID-19 patients started after some studies reported its beneficial effects in treating the disease. Chloroquine and hydroxychloroquine have been previously promoted by certain study groups to prevent and treat SARS-CoV, among other conditions [3]. However, it was recently discovered that in vitro replication of SARS-CoV-2 can be inhibited by hydroxychloroquine [4]. However, other studies did not yield many encouraging results, and their outcome was mixed [5].

A recent clinical trial tested hydroxychloroquine for postexposure prophylaxis, and the results are now reported [6]. In this trial, all the data were reported by the participants who were recruited by the investigators through social media. Placebo or hydroxychloroquine was provided through the mail to participants who reported severe or moderate exposure to COVID-19 patients at the workplace or at home. Only 13% of the enrolled participants developed COVID-19-like symptoms but were confirmed by the PCR in only 3% of the participants. In this setting, the use of hydroxychloroquine did not offer any benefit compared to placebo in protecting participants from developing Covid-like symptoms in the Covid-naïve participants. On the other hand, hydroxychloroquine users reported common adverse effects much more frequently than those who received a placebo (40% vs 17%).

The study design of the trial itself had many limitations. There were no consistent means of verifying whether the participants reporting Covid-like symptoms have the infection. That is why the reported symptoms of Covid patients in the study were less specific [7]. Therefore, the exact number of Covid cases in the study is hard to know. Another problem was the inability to monitor patient compliance and adherence to the dosing schedule. In the study, patients receiving hydroxychloroquine reported less than optimal compliance. The median age of the participants enrolled in the study was 40 years, and most of them were free of coexisting diseases. Therefore, these were a rather low-risk participant group because those at high risk are much older and have comorbid conditions [8].

Another problem with this clinical trial was that hydroxychloroquine was started much later after symptoms appear in the patients (more than 3 days). This indicates that the aim of the trial was to prevent the COVID-19 symptoms rather than the prevention of SARS-CoV-2 infection.

Hydroxychloroquine is a relatively safe drug, which has been emphasized previously [3]. However, careful analysis of such studies indicates that the potential for cardiotoxicity was high, especially in patients with comorbid conditions [9]. Most of the COVID-19 patients are older and with other preexisting diseases. This trial, although did report mild adverse effects of hydroxychloroquine, did not mention any cardiac effect as these could not be assessed.

The fear of SARS-CoV-19 justifies the urgent search for treatments, and therefore advocacy of hydroxychloroquine and its widespread use makes sense. But it should be the scientific evidence that should drive the clinical decisions rather than social media and unfounded political remarks. It appeared for a while that social and political forces were driving the global COVID-19 research agenda [10]. At one time, more than 200 clinical trials were listed on ClinicalTrials.gov website, many of which were intended for the prevention of the disease. What would be the fate of these trials in the face of results by Boulware and colleagues? Since Boulware and colleagues' results are inconclusive and even provocative, the results of these trials, which are much larger and more scientifically sound designs, would be interesting.

On June 15, 2020, FDA has revoked the emergency authorization to use chloroquine and hydroxyl chloroquine in hospitalized COVID-19 patients when a clinical trial is unavailable or participation is not feasible. This decision was based on recent data and a large clinical trial that did not show any benefit of these medications in the speed of recovery or in decreasing the likelihood of death. These outcomes of the clinical trial make sense and are consistent with the recent studies which show that these drugs are unable to inhibit viral mRNA or kill the virus. Hence, FDA determined that the legal criteria for the emergency use authorization are no longer met.

Remdesivir

Viruses from the corona family hijack the cellular machinery after entering the host cell. To replicate inside the host cell, their RNA-dependent RNA polymerase (RDRP) uses the host raw material to replicate [11]. This enzyme is an especially attractive drug target for it is highly conserved because of the evolutionary constraints on its function. This approach of blocking the replicating enzyme in the virus has previously yielded dividends in the treatment of hepatitis C infection, as illustrated by the discovery and development of sofosbuvir [12].

The virus that causes COVID-19-SARS-CoV-2 is a member of coronaviruses. Its treatment targets would be similar to other coronaviruses such as those that caused SARS and MERS, namely, SARS-CoV-1 and MERS-CoV, respectively. A recent study was conducted on the potential of remdesivir to treat COVID-19 [13]. This study has resulted in some confusion regarding the outcome of the trial due to the study design issues.

Remdesivir is a prodrug and must be activated to its active form by several enzymes, including esterases, amidases, and kinases. Once inside, the active drug inhibits viral RDRP [14]. This approach has been historically very effective, although coronaviruses can detect nucleotide sequences errors and correct it [15, 16]. However, remdesivir has the ability to escape from this corrective action of coronaviruses. Similarly, laboratory experiments indicate that a mutation in the viral enzyme would lead to resistance against remdesivir, but such versions of the viruses are less fit and much less pathogenic [15].

Several animal studies show that remdesivir is effective against many viruses. When given postexposure to rhesus monkeys infected with MERS-CoV, it protected the monkeys from severe lung damage, and when given as a pretreatment, it protected the monkeys from infection [17]. Similarly, when given 24 hours postexposure to African green monkeys infected with Nipah virus, it protected monkeys from this infection, which is a cause of fatal encephalitis [18]. It also protected rhesus monkeys from the Ebola virus when given parenterally for 12 days [19].

Remdesivir treatment 12 hours prior to infecting the rhesus monkeys with SARS-CoV-2 protected them from severe lung damage and reduced respiratory symptoms in a randomized control laboratory trial of 12 rhesus monkeys [20]. Whether this efficacy and potency are translated into human clinical trials is a different matter. In another study, Ebola mortality was compared in four groups, one of which was given remdesivir, and the other three were given different antibody treatments. In a randomized control trial in 681 patients, the mortality in the remdesivir group was 53%, which was significantly worse than 35% observed in the most potent antibody group [21]. However, the authors also shared that the patients in the remdesivir group had much severe symptoms.

Many of the initial studies on the effects of remdesivir in COVID-19 stem from small clinical studies or personal experiences [22]. However, the results of some short clinical trials are now being reported. In several studies spanning three contents and over 20 hospitals, the combined results of 53 of 61 patients suggest some protective effects of remdesivir in COVID-19 patients [13]. The mortality rate was 13% in the remdesivir group after 18-day treatment. Among patients who were ventilated, mortality was 5%. The likelihood of patient improvement was 68% after an 18-day treatment period (40–80% with a 95% confidence interval). Common and less severe adverse effects were experienced by 60% of patients, while serious adverse effects were reported by 23% of patients. Rashes, diarrhea, hypotension, and renal impairment were the most frequently encountered adverse effects. This study has some obvious limitations. The trial did not have a randomized control arm; there was no information on the initially treated eight patients, and the follow-up was relatively short. The sample size of the trial was also small precluding generalizability of the study.

A compassionate use program is available for remdesivir as it once existed for penicillin for treating acute infective endocarditis (Dolphin A, Cruickshank R, 1945). However, there are some very important differences between the two. In the case of the compassionate use of penicillin for infective endocarditis, it should be recalled that mortality of endocarditis was high compared to COVID-19. Furthermore, it is hard to establish evidence of treatment benefit in an open, uncontrolled trial. Therefore, the mere existence of a compassionate use program does not guarantee the safety and efficacy of the treatment.

Whether or not remdesivir offers any benefit in COVID-19 could be ascertained by large, well-powered, randomized, double-blind, controlled, well-masked, preferably multicentered trials. Currently, there are more than two dozen different clinical trials; enrolling over 25,000 patients are underway to investigate the effects of remdesivir in COVID-19. However, only about one-fifth of these are double-blind. Many of these are just observational studies. Another thing that could help is preparing standard protocols for these clinical trials for easy comparison of the data. Designing adaptive clinical trials would facilitate interim analyses and decide if a treatment should be stopped or evaluated further. The adaptive design will also help in evaluating multiple treatments at once.

Angiotensin-Converting Enzyme Inhibitors

SARS-CoV-2 infection is not severe in most people, but in some, it can lead to severe inflammation, respiratory depression, and death due to multiorgan failure [23]. The virus enters the cell by endocytosis when the cell-surface protein ACE2 interacts with the viral protein spike in the lung. ADAM metallopeptidase domain 17 (ADAM 17) activity is increased due to this endocytosis, which results in the release of ACE2 from the cell membrane. The loss of ACE2 abolishes its protective role in the cell and causes further release of proinflammatory cytokines in the circulation when the renin-angiotensin-aldosterone system is left uninhibited [24]. In patients with preexisting cardiovascular disease, this inflammation and stress further aggravate the patient's condition [25]. Other factors that play important roles in the observed association between death due to COVID-19 and cardiovascular disease include myocardial depression, viral infection of the heart, endothelial denudation, and cardiomyopathy. These effects may further aggravate preexisting arrhythmias, increase the oxygen demand of the heart, and hasten heart failure [26].

Factors such as old age and patient gender also affect the disease outcome in COVID-19 patients. Older age was found significantly associated with cardiovascular events and death in influenza [25]. Differences in the disease outcome and gender were also observed during the 2003 epidemic of severe acute respiratory syndrome (SARS) [27]. These differences in the disease outcome (death) may be because women can fight infections better than men owing to stronger immunity [28]. Furthermore, it is also known that male mice have a higher susceptibility to SARS-CoV-1 than female mice in the animal models. This could be partly explained by the increased concentrations of inflammatory markers such as neutrophils and macrophages [29]. Animal studies also reveal that female mice's mortality due to SARS-CoV-1 is increased when estrogen receptor antagonists are used or after ovariectomy. This risk due to gender is further enhanced by advanced age [29]. These observations suggest there is a strong protective effect of female sex and younger age as indicated by the increased survival linked to these two variables in COVID-19.

In a recent study, survival among COVID-19 patients is increased when statins or ACE inhibitors were used [26]. Since this study was not a randomized controlled clinical trial, the effects of confounders cannot be overlooked. These could be chance associations. Therefore, it cannot be established that ACE inhibitors or statins are increasing the survival of COVID-19 patients. Hence, in individual patients who do not have any routine indication of these drugs, this study does not provide evidence to initiate those patients on ACE inhibitors or statins. Whether or not these medications can offer any benefit in COVID-19 patients could be established by randomized controlled clinical trials that would evaluate the roles of these medications in detail.

Several studies show that in-hospital deaths, increased risk of severe disease, or increased risk of SARS-CoV-2 infection is not associated with the use of ACE inhibitors or ARBs. These studies alleviate the concerns that ACE inhibitors and ARBs may predispose individuals to severe COVID-19 disease due to the inhibition of the renin-angiotensin-aldosterone system (RAAS). These studies have different study designs and diverse populations. Still, most of these were observational and were published in reputable journals such as the *New England Journal of Medicine* and *JAMA Cardiology*.

Mehra and colleagues [30] showed that various cardiovascular and noncardiovascular adverse effects had been associated with in-hospital deaths. Especially, smoking older age (>65), congestive cardiac failure, coronary disease, cardiac rhythm problems, and chronic obstructive pulmonary disease are independently associated with in-hospital deaths. These adverse effects increased the risk of in-hospital death. On the other hand, the use of ACE inhibitors was not associated with in-hospital deaths. This analysis was described in a study that collected the observational data of 8910 patients from Europe, North America, and Asia, admitted in 169 different hospitals to investigate the link between in-hospital death and drug therapy and cardiovascular events.

The severity of COVID-19 and the risk of SARS-CoV-2 infection were not associated with the use of ACE inhibitors and ARBs in another trial conducted in Italy by Mancia and colleagues [31]. Although the use of these medications was lower in controls compared to cases, no evidence of increased adverse effects or risk of severe infection was found in the control group. This analysis was a result of a population-based case-control study in which 6272 cases and 30,759 matched controls were recruited to investigate the possible link between these drugs (ACE inhibitors and ARBs) and the severity of the disease and the risk of infection.

Previous use of ACE inhibitors and ARBs was not associated with the risk of infection or its severity in another observation controlled trial [32]. Propensity score models and Bayesian analysis revealed that both ACE inhibitors and ARBs are not associated with a high risk of severe COVID-19 infection or a higher risk of disease. This study was conducted in 12,594 patients enrolled in a health network in New York, USA, and were screened for COVID-19 infection.

In another cohort of patients, overlap propensity score weighting did not show an association between the use of cardiovascular medications (ACE inhibitors and ARBs) and the risk of a positive Covoid-19 infection [33]. These were the results of a study conducted by Mehta and colleagues in Florida, USA, in a cohort of 18,472 COVID-19 patients.

Another case-population study conducted in Spain also ruled out any link between ACE inhibitors and ARBs and increased risk to COVID-19 [34]. Conditional logistic regression analyses revealed that the use of ACE inhibitors and ARBs to blunt RAAS was not associated with a higher risk of the disease compared to other cardiovascular drugs. This study, which was conducted in 1139 COVID-19 patients admitted to seven hospitals in Spain, also revealed that ACE inhibitors and ARBs are associated with a lower risk of COVID-19 in diabetic patients. Taken together, these investigations indicate that the use of RAAS inhibitors is not associated with an enhanced risk of COVID-19 severity or other unusual untoward effects.

Spike Protein Inhibitors

Novel coronavirus targets human cells through a highly glycosylated protein called S protein, which binds human angiotensin-converting enzyme 2 (ACE-2) [35]. This initial binding step is followed by the internalization of the virus in human cells. This binding of S protein, through its receptor-binding domain, to ACE-2 is a prerequisite for viral entry and subsequent pathogenesis of the disease [36]. The high affinity of S protein toward ACE-2 is thought to be an important underlying factor in the rapid spread of the disease. The absence of an early high-resolution crystal structure of S protein made the initial efforts to virtually screen or design S protein inhibitors extremely hard [37]. Publishing of a high-resolution (3.5 A), CryoEM structure of the S protein was recently reported [36]. This has made the design and virtual screening of the S protein inhibitors relatively easier.

It was recently shown that Affimer reagents were able to bind S protein from SARS-COV-2, which protected the human cells from COVID-19 infection. This work is a collaboration between the Center for Virus Research, University of Glasgow, and Affimer biotherapeutics and reagents manufacturer Avacta Group plc. These laboratory findings could become a potential therapy for COVID-19.

In a recent statement, the company stated that they had generated more several potent inhibitors of S protein, which could block the first step of viral entry into human cells. They said that their preliminary data is highly encouraging. Their collaborator, Professor David Bhella, at the University of Glasgow, seems to agree. According to him, these spike protein inhibitory compounds could neutralize any attempt of viral entry in the human cell through the ACE-2 receptor.

Compared to antibodies, small-molecule inhibitors, such as those analyzed by the Avacta Group and University of Glasgow team, offer many advantages. Because of their small sizes, high concentrations of these inhibitors could be used without significant solubility issues. Since these compounds are made through combinatorial chemistry techniques, scaling up their production will not be a big challenge. Another critical advantage of Avacta Group compounds is that many of these investigative therapies are bispecific or trispecific, meaning they bind to more than place on the spike protein, ensuring maximum effectiveness of the therapy.

Further work is underway to understand the nature of binding Avacta compounds to spike protein, which could provide insights into the mechanistic aspects of the inhibition. On the other hand, Avacta is actively looking for a large pharmaceutical industry partner for the large-scale production of the compounds once its efficacy and potency are established. The promising aspect of these compounds is that they could be given to the healthcare worker for prophylaxis and the patients for treating the disease.

There is also an increasing interest of the research community in targeting S protein for furin-like cleavage site to generate the so-called furin inhibitors. The presence of a furin-like cleavage site in the genomic sequence of SARS-COV-2 further elevated the interest in designing and developing such inhibitors. This furin-like cleavage site in the S protein has implications for the pathogenicity of the virus and its life cycle.

Conclusions

This chapter discusses several drugs currently in clinical and laboratory studies in addition to several landmark trials and combines the available information with the personal experiences of clinicians with COVID-19 patients. We discussed important clinical trials, including the one that demonstrates a significant decrease in mortality with dexamethasone use. Another important clinical trial that was discussed showed that the use of remdesivir decreases in the number of days a patient with COVID-19 had to spend in a hospital. However, many of the studies show no benefit to the patients, and some are inconclusive. Other studies demonstrate improvement in patients' symptoms, but the evidence was low. One of the major problems in comparing the results from different studies and trials is the lack of uniform study design. However, based on the available evidence, dexamethasone and remdesivir are perhaps the best of the available remedies for COVID-19. ACE inhibitors and ARBs could also be used in COVID-19 patients for cardiovascular indications, and they do not increase the severity or the frequency of known adverse effects. We could not find any worthwhile evince that supports the use of chloroquine or hydroxychloroquine in COVID-19 patients. Many new and exciting molecules are being tested in the laboratories. Some old and new drugs are being investigated in the large, randomized, controlled trial worldwide to treat and prevent COVID-19. The research and medical community's hard work represents a great hope for humanity to overcome this pandemic ultimately.

References

1. Cohen MS, Corey L (2020) Combination prevention for COVID-19. Science 368:551
2. Recovery Trial Group (2020) Dexamethasone in hospitalized patients with COVID-19 — preliminary report. N Engl J Med. https://doi.org/10.1056/NEJMoa2021436
3. Rolain J-M, Colson P, Raoult D (2007) Recycling of chloroquine and its hydroxyl analogue to face bacterial, fungal and viral infections in the 21st century. Int J Antimicrob Agents 30:297–308
4. Yao T-T, Qian J-D, Zhu W-Y, Wang Y, Wang G-Q (2020) A systematic review of lopinavir therapy for SARS coronavirus and MERS coronavirus – a possible reference for coronavirus disease-19 treatment option. J Med Virol 92:556–563
5. Hernandez AV, Roman YM, Pasupuleti V, Barboza JJ, White CM (2020) Hydroxychloroquine or chloroquine for treatment or prophylaxis of COVID-19: a living systematic review. Ann Intern Med 173:287–296
6. Boulware DR, Pullen MF, Bangdiwala AS et al (2020) A randomized trial of hydroxychloroquine as postexposure prophylaxis for COVID-19. N Engl J Med 383:517–525
7. Tostmann A, Bradley J, Bousema T et al (2020) Strong associations and moderate predictive value of early symptoms for SARS-CoV-2 test positivity among healthcare workers, the Netherlands, March 2020. Euro Surveill Bull Eur Sur Mal Transm Eur Commun Dis Bull. https://doi.org/10.2807/1560-7917.ES.2020.25.16.2000508
8. Wu Z, McGoogan JM (2020) Characteristics of and important lessons from the coronavirus disease 2019 (COVID-19) outbreak in China: summary of a report of 72 314 cases from the Chinese Center for Disease Control and Prevention. JAMA 323:1239–1242
9. Magagnoli J, Narendran S, Pereira F, Cummings TH, Hardin JW, Sutton SS, Ambati J (2020) Outcomes of hydroxychloroquine usage in United States Veterans Hospitalized with COVID-19. Med N Y N. https://doi.org/10.1016/j.medj.2020.06.001
10. Sayare S (2020) He was a science star. Then he promoted a questionable cure for COVID-19. N. Y. Times
11. Amirian ES, Levy JK (2020) Current knowledge about the antivirals remdesivir (GS-5734) and GS-441524 as therapeutic options for coronaviruses. One Health Amst Neth 9:100128
12. Xie Y-C, Ogah CA, Jiang X, Li J, Shen J (2016) Nucleoside inhibitors of hepatitis C virus NS5B polymerase: a systematic review. Curr Drug Targets. https://doi.org/10.2174/13894501 17666151209123751
13. Grein J, Ohmagari N, Shin D et al (2020) Compassionate use of remdesivir for patients with severe COVID-19. N Engl J Med 382:2327–2336
14. Gordon CJ, Tchesnokov EP, Feng JY, Porter DP, Gotte M (2020) The antiviral compound remdesivir potently inhibits RNA-dependent RNA polymerase from Middle East respiratory syndrome coronavirus. J Biol Chem AC120:013056
15. Agostini ML, Andres EL, Sims AC et al (2018) Coronavirus Susceptibility to the Antiviral Remdesivir (GS-5734) Is Mediated by the Viral Polymerase and the Proofreading Exoribonuclease. mBio. https://doi.org/10.1128/mBio.00221-18

16. Pruijssers AJ, Denison MR (2019) Nucleoside analogues for the treatment of coronavirus infections. Curr Opin Virol 35:57–62
17. de Wit E, Feldmann F, Cronin J, Jordan R, Okumura A, Thomas T, Scott D, Cihlar T, Feldmann H (2020) Prophylactic and therapeutic remdesivir (GS-5734) treatment in the rhesus macaque model of MERS-CoV infection. Proc Natl Acad Sci U S A 117:6771–6776
18. Lo MK, Feldmann F, Gary JM et al (2019) Remdesivir (GS-5734) protects African green monkeys from Nipah virus challenge. Sci Transl Med. https://doi.org/10.1126/scitranslmed.aau9242
19. Warren TK, Jordan R, Lo MK et al (2016) Therapeutic efficacy of the small molecule GS-5734 against Ebola virus in rhesus monkeys. Nature 531:381–385
20. Williamson BN, Feldmann F, Schwarz B et al (2020) Clinical benefit of remdesivir in rhesus macaques infected with SARS-CoV-2. bioRxiv. https://doi.org/10.1101/2020.04.15.043166
21. Mulangu S, Dodd LE, Davey RT et al (2019) A randomized, controlled trial of ebola virus disease therapeutics. N Engl J Med 381:2293–2303
22. Bhatraju PK, Ghassemieh BJ, Nichols M et al (2020) COVID-19 in critically ill patients in the Seattle Region — case series. N Engl J Med. https://doi.org/10.1056/NEJMoa2004500
23. Siddiqi HK, Mehra MR (2020) COVID-19 illness in native and immunosuppressed states: a clinical–therapeutic staging proposal. J Heart Lung Transplant 39:405–407
24. Wang K, Gheblawi M, Oudit GY (2020) Angiotensin converting enzyme 2: a Double-Edged Sword. Circulation. https://doi.org/10.1161/CIRCULATIONAHA.120.047049
25. Nguyen JL, Yang W, Ito K, Matte TD, Shaman J, Kinney PL (2016) Seasonal influenza infections and cardiovascular disease mortality. JAMA Cardiol 1:274–281
26. Mehra MR, Ruschitzka F (2020) COVID-19 illness and heart failure. JACC Heart Fail 8:512–514
27. Karlberg J, Chong DSY, Lai WYY (2004) Do men have a higher case fatality rate of severe acute respiratory syndrome than women do? Am J Epidemiol 159:229–231
28. Klein SL, Flanagan KL (2016) Sex differences in immune responses. Nat Rev Immunol 16:626–638
29. Channappanavar R, Fett C, Mack M, Ten Eyck PP, Meyerholz DK, Perlman S (2017) Sex-based differences in susceptibility to SARS-CoV infection. J Immunol Baltim Md 1950(198):4046–4053
30. Mehra MR, Desai SS, Kuy S, Henry TD, Patel AN (2020) Cardiovascular disease, drug therapy, and mortality in COVID-19. N Engl J Med 382:e102
31. Mancia G, Rea F, Ludergnani M, Apolone G, Corrao G (2020) Renin–angiotensin–aldosterone system blockers and the risk of COVID-19. N Engl J Med 382:2431–2440
32. Reynolds HR, Adhikari S, Pulgarin C et al (2020) Renin–angiotensin–aldosterone system inhibitors and risk of COVID-19. N Engl J Med 382:2441–2448
33. Mehta N, Kalra A, Nowacki AS et al (2020) Association of use of angiotensin-converting enzyme inhibitors and angiotensin II receptor blockers with testing positive for coronavirus disease 2019 (COVID-19). JAMA Cardiol. https://doi.org/10.1001/jamacardio.2020.1855
34. de Abajo FJ, Rodríguez-Martín S, Lerma V et al (2020) Use of renin–angiotensin–aldosterone system inhibitors and risk of COVID-19 requiring admission to hospital: a case-population study. The Lancet 395:1705–1714
35. Lu R, Zhao X, Li J et al (2020) Genomic characterisation and epidemiology of 2019 novel coronavirus: implications for virus origins and receptor binding. Lancet Lond Engl 395:565–574
36. Wrapp D, Wang N, Corbett KS, Goldsmith JA, Hsieh C-L, Abiona O, Graham BS, McLellan JS (2020) Cryo-EM structure of the 2019-nCoV spike in the prefusion conformation. Science 367:1260–1263
37. Gruber CC, Steinkellner G (2020) Wuhan coronavirus 2019-nCoV – what we can find out on a structural bioinformatics level. https://doi.org/10.6084/m9.figshare.11752749.v3

Chapter 7
COVID-19 Pandemic and Vaccines

Hina Qaiser, Roheena Abdullah, Tehreema Iftikhar, Hammad Majeed, and Imran Imran

Introduction

Viral diseases regularly pose threats to the human health, says the World Health Organization (WHO). The mankind has witnessed some serious epidemics, viz. severe acute respiratory syndrome (SARS), H1N1 flu and MERS within the past 20 years. As the year 2019 was coming to its conclusion, a few cases of pneumonia with intense respiratory problems emerged in Wuhan City, Hubei Province, China. The doctors were unable to identify the aetiology of the disease [1]. As the cases spread, the Centers for Disease Control and Prevention (CDC) intervened and discovered that infection was being caused by a new type of coronavirus that was named as the 2019 novel coronavirus disease (2019-nCoV). It was also referred as SARS-CoV-2 and human coronavirus disease 19 (hCoV-19) at the time [2]. Later on, the official name of the disease was announced as coronavirus disease 2019 (COVID-19) by the WHO. As stated above, coronaviruses are already accountable for causing two recent epidemics, namely, SARS-CoV and

H. Qaiser
Department of Biology, Lahore Garrison University, Lahore, Pakistan

R. Abdullah
Department of Biotechnology, Lahore College for Women University, Lahore, Pakistan

T. Iftikhar (✉)
Applied Botany Lab, Department of Botany, Government College University, Lahore, Pakistan

H. Majeed
Knowledge Unit of Science, University of Management and Technology, Iqbal Campus Sialkot, Sialkot, Punjab, Pakistan

I. Imran
Faculty of Pharmacy, Bahauddin Zakariya University, Multan, Pakistan

© The Author(s), under exclusive license to Springer Nature Switzerland AG 2021
M. Zia-Ul-Haq et al. (eds.), *Alternative Medicine Interventions for COVID-19*, https://doi.org/10.1007/978-3-030-67989-7_7

MERS-CoV. SARS-CoV epidemic emerged from China and spread to 24 nations infecting 8000 individuals with a fatality rate of 9.6% [3]. The MERS-CoV initiated in Saudi Arabia and is accountable for 2500 cases with a much higher fatality rate of 35%. It is also causing periodic cases even in the present [4]. COVID-19 is being transmitted chiefly by respiratory droplets, while person-to-person transmission is also taking place through close contacts since mid-December 2019 [5]. COVID-19 was given the status of pandemic by the WHO on 11 March 2020.

COVID-19 Vaccines: A Need of Time

The CoVs have become the significant microbes of emerging respiratory ailments. They are wrapped non-fragmented positive-sense RNA viruses and have a place with the family Coronaviridae. They are comprehensively circulated in people and other animals [6]. Beforehand, six COVID types have been found to cause human illnesses (Fig. 7.1) [7]. The first four species resulted in minor indications; however, the last two SARS-Co and MERS-CoV created havoc with death rates of 10% and 34% for SARS-Co and MERS-CoV, respectively. These two viruses have resulted in death of more than 10 thousand people in the past 20 years [7–9]. SARS-CoV-2 is number 7 in the list of COVID that can affect people [10]. In spite of the fact that the death rate of SARS-COV-2 is lesser compared to MERS-CoV and SARS-CoV, it is much more contagious than MERS-CoV or SARS-CoV (Table 7.1).

Genome sequencing examination of clinical samples obtained from patients exhibited that 88% of the SARS-CoV-2 genome was homologous to two SARS-like COVIDs present in bats: bat-SL-CoVZC45 and bat-SLCoVZXC21. It also showed 79% and 50% nucleotide sequence homology to SARS-CoV and MERS-CoV, respectively [11].

The genome also consists of 16 nonstructural proteins (nsp1–16) along with 5–8 accessary proteins [12]. Phylogenetic assessment proposed that bats can be the basic source for SARS-CoV-2 [13]; the middle host is as yet under scrutiny. The illness essentially influences the respiratory organs, and sickness seriousness can extend from minor rhinorrhoea to acute respiratory syndrome followed by patient death [14–16]. Other documented symptoms include diarrhoea, rash, anosmia, thromboembolic issues, vasculitis and myocarditis [1, 16–22]. The median incubation time frame is assessed to be 5 days with a greater part creating side effects by 11.5 days [23]. The patients of COVID-19 are believed to shed highest levels of viral load at the beginning of symptoms [24]. This and other epidemiological information proposes the risk of transmission starting from indistinct presymptomatic

Fig. 7.1 Types of coronaviruses infecting humans

Table 7.1 Contagiousness of coronaviruses

Coronavirus type	Contagiousness
SARS-COV-2	Middle R0: 5.7
MERS-CoV	R0: <1
SARS-CoV	R0: 3

period [25]. The clinical deterioration is deferred till the second week from the infection onset and appears to be the manifestation of cytokine-mediated immune response, accountable for severe inflammation and dispersed intravascular coagulation, often with low-intensity viraemia [26, 27]. Death rate is highest among the elderly people (older than 70 years); however, other factors can also account for the death of the patient [1, 16, 28] which include gender (higher fatality rate in males than females), obesity, hypertension and diabetes. The announced case casualty rates (CFR) have been somewhere in the range of 0.82% and 9.64%, with changeability in CFR likely because of the testing recurrence and access along with numerous other health framework limiting factors in various areas of the world [29]. The infection fatality rate (IFT), death count of all individuals infected, asymptomatic and not tested, is a superior gauge of populace mortality and is demonstrated to be somewhere in the range of 0.1% and 0.41% [29]. The countries of the world have worked out strict measure to counter the effects of COVID-19 that includes shutdowns which alone are believed to save an estimated three million lives in 11 European countries [30]. These measures have been taken on the expense of slowing down of economies and recession. A great many jobs in all economic sectors have been lost because of COVID-19 restrictions like social distancing, travel bans and self-isolation. There have been a closure of education institutions and a sharp decline in consumer spending. Two sectors to have thrived in strict shutdowns have been pharmaceuticals and food sector – as there was a high demand of food due to the ongoing panic in shutdowns to stock pile the food products.

Coronavirus Vaccine Development: Challenges of the Past

The 30 + kb genome of RNA coronaviruses is enveloped by a helical nucleocapsid (N) which is surrounded by an outer envelope consisting of matrix protein (M), envelope protein (E) and spike proteins (S) [31]. The naturally occurring trimeric form of S protein holds the receptor-binding domain (RBD) accounting for the association with the angiotensin-converting enzyme 2 (ACE2) and facilitates entry into the cell. This S protein, among other structural proteins of SARS-CoV, has been found to cause the production of neutralizing antibodies, therefore a chief candidate antigen for vaccine development [32, 33]. Coronavirus vaccine development has met with some major challenges in the past. So far, not a single coronavirus vaccine has been licensed for human use to combat respiratory infections. As for the animals, only infectious bronchitis virus (IBV) vaccines are approved for upper

Table 7.2 Immunopathology associated with coronavirus vaccines

Sr No	Animal model	Vaccine	Complications postinoculation with the virus	Reference
I	Mice	Inactivated whole virus vaccines Recombinant DNA spike protein vaccine Virus-like particle vaccine	Lung eosinophilic infiltration	[36]
II	Mice	SARS-CoV N protein	Severe pneumonia Lung eosinophilic infiltration	[37]
III	Mice	Viral replicon particles expressing glycoprotein	No complications	[37]
IV	Mice	Inactivated MERS-CoV vaccine	Severe pneumonia Lung eosinophilic infiltration	[38]

respiratory CoV infections in chickens. Generally, the coronavirus vaccines mimicking human disease have been shown to be immunogenic when injected in animal models but do not effectually avoid disease acquisition [34]. Additionally, there is a worry that immunization, just like normal coronaviral infection, may not instigate enduring insusceptibility and reinfection might be conceivable [35]. Additionally, concerning has been the immunization-related disease enhancement. Past utilization of COVID immunizations (SARS-CoV and MERS-CoV) in some animal models has shown safety issues with respect to Th2 interceded immunopathology (Table 7.2).

Vaccine Development Strategies

Vaccine development is an efficient measure to counteract infectious diseases. With knowledge gained from the previous vaccine models considered for MERS and SARS, several vaccine development strategies, i.e. DNA, mRNA, recombinant protein, inactivated/attenuated virus and adenoviral vector, are being explored (Fig. 7.2). In this regard, S protein and its fragments (S1, S2, RBD and N protein) are being considered as potential candidates for COVID-19 vaccine development (Fig. 7.3) since they have also been used in case of MERS and SARS vaccine [39, 40].

As soon as the genetic information of SARS-CoV-2 was made public on 11 January 2020, around 40 pharmaceutical establishments and academic organizations from several countries are actively involved in vaccine development to win the fight against SARS-CoV-2, and many of them have reached the clinical trial stages. Various vaccine platforms being investigated have specific salient points (Table 7.3). The route to successful vaccine development can be abetted by taking an elaborate look into pathological process of SARS-CoV-2 in humans and the coordinates and duration of immunity. A comprehensive look into the way the virus affects the target organs and its spread within the body will aid in developing vaccines which can prevent virus dissemination to the organs and abate its infection. It is an important

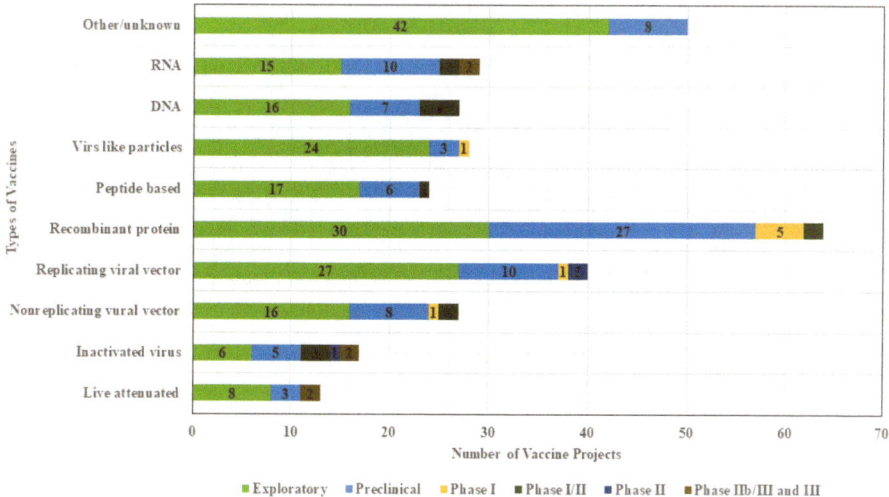

Fig. 7.2 Current standing of COVID-19 vaccine candidates by developmental stage and vaccine type

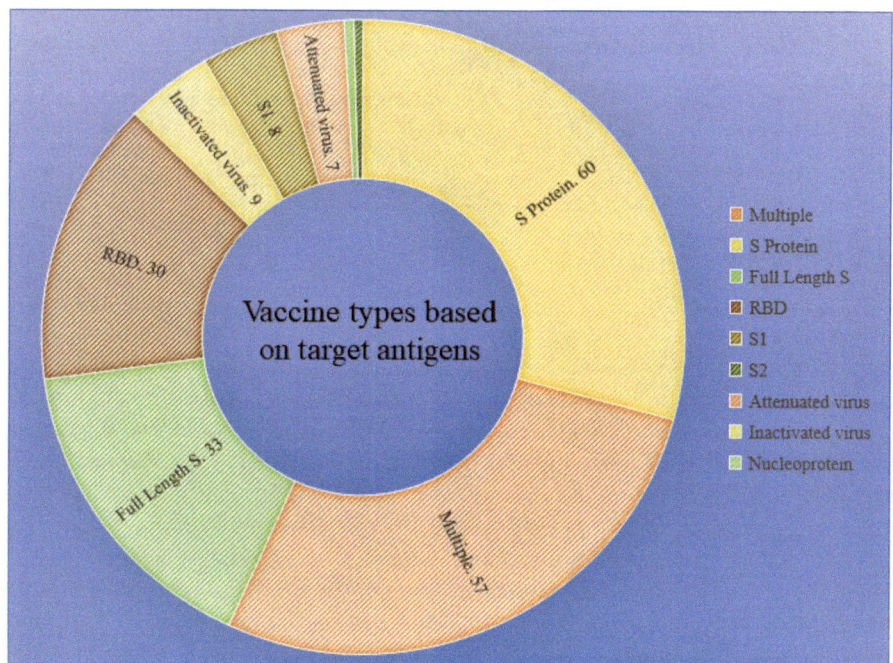

Fig. 7.3 COVID-19 vaccine candidates with respect to target antigen

Table 7.3 A comparison of COVID-19 candidate vaccine platforms

Vaccine types	Nucleic acid	Live-attenuated virus	Inactivated virus	Subunit	Viral vector
Working	DNA/RNA expresses viral proteins which trigger the immune responses	Weakened version of the virus	Virus killed by heat or chemical treatment acts as inactivated vaccine	Viral surface proteins utilized which are targeted by immune system	Nonpathogenic viral vectors used to deliver viral genes to confer immunity
Pros	Simple and quick to engineer	Elicits a strong immune response without causing serious infection	Simple to engineer and safe because of the dead nature of the virus	Directs the immune response towards the significant entity of virus without any infection	Live viruses stimulate robust immune reactions
Cons	No approved DNA/RNA vaccines exist so far	Not a viable option for immunocompromised persons	Not as effectual as a live virus, may escalate the infection	Do not elicit a sturdy immune response, may be used in combination with adjuvants to confer long-lasting immunity	Choice of a safe viral vector is crucial; immune response directed at the vector may weaken the vaccine effectiveness
Existing vaccines	None	Measles, mumps, rubella (MMR) Chicken pox	Polio	Hepatitis B Pertussis Papillomaviruses	Ebola Veterinary medicines

point to consider that as SARS-CoV-2 affects the lungs, it will result in pneumonia by viraemia or after the infection of upper respiratory tract. In case of the latter, vaccine systems based on live replicating vectors or attenuated viruses would be effective as they bring out localized mucosal immunity and shelter the upper and lower respiratory tracts and diminish adenoidal emissions. The use of live-attenuated influenza virus vaccine is a good example since it essentially produces localized IgA antibodies and scarcer systemic antibodies and causes prompt defence [41].

On the contrary, if lungs and other organs are being infected via viraemia, then parenteral (intramuscular) vaccines would serve better as they result in the production of substantial amount of virus neutralizing (VN) antibodies in serum blocking viraemia. Upon reaching the lungs and other target organs, VN antibodies could also efficiently wedge the infection. Furthermore, when COVID-19 recovered patients are exposed to the virus again (like seasonal influenza), an annual booster dose of a parental vaccine alone, such as a subunit S or RBD protein, would suffice. This would result in strong memory B and T cell responses and enable long-term immunity preventing subsequent virus infections. There have been reports of virus shedding in diarrhoea and faeces in few COVID-19 patients, so oronasal vaccines may also be effectual [42]. Therefore, the recipients of COVID-19 vaccines could be divided into three groups, viz. susceptible persons without immunity, healed individuals with varying level of immunity and individuals having immunity (exposed and recovered) against SARS and MERS. Consequently, the immunogenic response, level of protection and lethal effects of candidate vaccines will be different from person to person depending on the above categories. Evaluation of pre-existing levels of immunity with respect to population type and age will be imperative to corroborate the vaccine efficacy and safety.

Vaccine development and its scale-up in the middle of a worldwide pandemic is a demanding and tedious task since it entails the coordination of various activities taking place side by side in comparison to the decennium long, step-by-step vaccine production which includes preclinical trials, clinical trials, scale-up studies and distribution. This situation creates a blend of invested resources and increased financial risk having high stakes [43]. Previous disease outbreaks have witnessed high mortality and morbidity rates due to unorganized and delayed vaccine distribution as in the case of West African Ebola epidemic (2013/2014) which claimed 11,000 casualties [44] causing enormous economic and social losses of an estimated 53 billion dollars [45]. Unfortunately, an effective vaccine developed subsequently might have cut down the social and economic losses at the time controlling the outbreak [46, 47]. Appallingly, the SARS 2003 outbreak finished before an effective vaccine was developed. Disappointingly, funding organizations at that point reallocated reserves that had been focused on vaccine development, leaving producers with monetary misfortune and slowing down other vaccine advancement programmes [43]. In 2017, the Coalition of Epidemic Preparedness Innovation (CEPI) was shaped to answer these previous disappointments with an undertaking to build up a planned reaction to rising irresistible sickness dangers to guarantee robust immunization programmes and early deployment in light of pestilences [48].

Nucleic Acid Vaccines

This strategy involves the use of nucleic acid sequences to express viral antigens. As the cell uptakes the nucleic acid fragments, viral protein expression is initiated which further elicits immune responses (humoral and cellular) similar to the ones taking place in course of natural infections. Such type of vaccines has been tested for animal ailments and exhibited immune protection, for instance, for foot and mouth ailment, deer powassan infection and rabies infection [49, 50]. Nucleic acid vaccines intended for Ebola, influenza and Zika virus have already entered phase I of the clinical trials [51]. The advantage of this platform resides in its ease of antigen control, rapid synthesis and cell-free engineering which evades the requirement of biosafety level 2 laboratory. The major hindrances are the delicate nature of nucleic acids such as mRNA which require continuous cold conditions during handling and capacity [52]. Stage I clinical studies have been carried on SARS-CoV and MERS-CoV DNA vaccine candidates. A recombinant SARS DNA vaccine synthesizing the SARS-CoV N protein, created by the National Institute of Allergy and Infectious Diseases (NIAID), was explored in ten adult individuals [53]. Clinical trial undertaken by GeneOne Life Science/Inovio for a MERS-CoV DNA vaccine (GLS-5300) encoding full-length S protein genome contained a high number of participants ($n = 75$) [52]. Both indicated satisfactory safety profiles and initiated humoral and cellular reactions; the MERS-CoV DNA vaccine has progressed into a stage 2 of clinical trials [29]. The only SARS vaccine to have entered a phase I trial is an inactivated immunization (ISCV) created by Sinovac Biotech [54]. There were no reports of human investigations wherein inoculated subjects were tested by the characteristic infection.

A few significant biotechnology organizations are working on advanced models of nucleic acid vaccines for COVID-19. The Innovation and Value Initiative (IVI), Inovio and the Korea National Institute of Health (KNIH) are working together with the Coalition for Epidemic Preparedness Innovations (CEPI) to safety evaluation and immunogenicity of a DNA vaccine named INO-4800 in phase I/II clinical trial in South Korea. Both Moderna/NIH and CureVac are zeroing in on mRNA vaccine improvement, and a safety evaluation trial of Moderna's mRNA-1273 vaccine was conducted on 45 individuals in March 2020 [55].

Protein Subunit Vaccines

Subunit-based immunizations dependent on recombinant S or S1 protein of SARS-CoV and MERS-CoV have been exhibited to be viable in numerous investigations [56] [57–59]. Clover Biopharmaceuticals is building up a vaccine comprising a trimerized SARS-CoV-2 S protein utilizing their patented Trimer-Tag innovation [59]. The receptor-binding domain (RBD) in SARS-CoV-2 S protein was distinguished, and it was additionally shown that SARS-CoV-2 RBD displayed fundamentally higher binding affinity to ACE2 receptor contrasted with interaction between SARS-CoV RBD and ACE2 [60], recommending that the RBD-based SARS-CoV immunizations can possibly be created for counteraction of SARS-CoV-2 infections. RBD-based vaccines are currently being worked on by a few associations through worldwide coordinated efforts [61]. The pneumonic surfactant-biomimetic nanoparticles used to potentiate heterosubtypic flu resistance can be utilized as adjuvant to improve the immunogenicity of SARS-CoV-2 subunit vaccines [62].

Inactivated or Live-Attenuated Virus Vaccines

Whole inactivated or live-attenuated virus-based immunizations speak to a customary vaccine technique. Scientists at the University of Hong Kong have built up a live influenza vaccine that produces SARS-CoV-2 proteins [63]. Codagenix has built up a "codon deoptimization" innovation to weakened viruses, and the organization is investigating COVID-19 vaccine systems [64].

Virus Vector-Based Vaccines

Immunizations dependent on viral vectors offer a significant level of protein expression and long haul dependability and elicit strong immune reactions [65]. Johnson & Johnson is building up an adenovirus-vectored antibody utilizing AdVac®/PER.C6® immunization technique [55]. The first COVID-19 vaccine candidate dependent on adenovirus-vectored immunization created by Chen Wei bunch entered human clinical testing (NCT04313127) with extraordinary quickness right off the bat on 16 March 2020. Another stage I safety evaluation study of a recombinant adenovirus vaccine candidate (Cansino Biologics Inc., Tianjin, China), Ad5-nCoV, selected 108 adult volunteers in Wuhan, China, in March 2020. *IVI, INOVIO and KNIH cooperate with CEPI in stage 1/2 clinical studies of INOVIO's COVID-19 DNA immunization in South Korea, 2020 [38]. Aside from adenovirus vector-based immunization, two lentivirus vector-based vaccine candidates, COVID-19/aAPC and LVSMENP-DC, have been created by Shenzhen Geno-Immune Medical Institute. The COVID-19/aAPC immunization was created by applying lentivirus alteration including the SARS-CoV-2 minigenes and immune modulatory genes, to the counterfeit antigen presenting cells (aAPCs). The phase I clinical study comprising of 100 members began on 15 February 2020, and the assessed study completion date was 31 December 2024 (NCT04299724). The LVSMENP DC vaccine was created by changing DC with lentivirus vectors synthesizing SARS-CoV-2 minigene SMENP and immune modulatory genes.

We as a whole realize that adjuvants assume a basic part by improving immunogenicity of the immunization candidates and aid in the selection of appropriate dose in some vaccine types. Up until this point, there are around ten developers engineering adjuvanted COVID-19 immunizations. Vaccine developers Dynavax, Seqirus and GlaxoSmithKline have focused on making some licensed adjuvants including MF59, AS03 and CpG 1018 accessible for use [65]. Regardless of which strategy we take to build up the COVID-19 immunizations, analysts need to painstakingly assess the viability and safety of the candidate vaccine at each progression. Under these circumstances, SARS-CoV-2-specific animal models appear to be an important requisite. As of not long ago, some extraordinary animal models are in development stages, including hamsters, ferrets, ACE2-transgenic mice and nonhuman primates [65].

Repurposed Vaccines for COVID-19 and Off-Target Effects of Other Vaccines

Authorized immunizations, for example, BCG and oral polio vaccine, are able to induce vague, immune reactions and have modulatory effect insurance against different infectious ailments [66]. This may hint towards the recommendation that they can be exploited as potential vaccine candidates in the avoidance of COVID-19. Some clinical trials involving BCG vaccines are already underway to determine their capacity to combat COVID-19 in Australia [66], the Netherlands [67] and South Africa [68]. A measles vaccine clinical trial to forestall COVID-19 in medical care labourers in Egypt has been enrolled [58], and oral polio immunizations are being examined in the USA [69].

Plant-Based Vaccines

Presently, a number of bacterial, yeast and mammalian expression systems are available for recombinant protein production including subunit vaccines, but they all have certain shortcomings. They are expensive, have safety concerns and do not guarantee target integrity. In comparison, plant expression systems provide a robust solution to these problems since they are cheap with better yield and offer no risk of pathogenic contamination. Hence, plant-based vaccines can prove to be a convalescent approach towards vaccine development. The expressed protein could adopt the desired conformation with expected post-translational modifications while maintaining its functional integrity in a plant-based system [67].

Since the last three decades, plants have been used as a machinery to manufacture a wide range of biopharmaceuticals such as monoclonal antibodies, edible vaccines, human growth factors and cytokines [68]. Fortunately, the production of a recombinant enzyme named β-glucocerebrosidase in carrot cells for Gaucher's disease treatment has also been licensed by the FDA [69]. The usefulness of plant-based systems resides in their incapacity to allow the growth of human pathogens making the purification process less tedious and contamination-free [70]. So far, various clinical trials are underway for plant-based synthesis of vaccine for swine influenza, rabies and hepatitis B [71]. An eminent example would be that of vector-based influenza vaccine (Medicago Inc.) expressing hemagglutinin (HA) encoded by viral regulatory elements in *Nicotiana benthamiana* [72, 73].

The presently existing plant-based foreign protein expression approaches hold the potential to synthesize many vaccine candidates against COVID-19 in short stretch of time. The chosen target antigen will regulate the choice of expression strategy and targeted organelle. The vaccine antigens may be native viral proteins or fusion proteins premeditated to act as multiepitopic vaccines. This is structured by T cell and B cell epitope selection using the relevant bioinformatics tools and experimental evidence. Presently available genetic engineering tools enable the stable/

transient expression of target antigens which give way to explore diverse immunization approaches. In case of transient expression, the antigens need to be purified and are subsequently formulated as injectable vaccines. Stable expression is useful for edible crops and permits executing oral administration of vaccines shorn of purification. In this regard, already existing vaccine models for MERS and SARS-CoV-1 are crucial leads. The SARS-CoV-1 S1 protein (N-terminal fragment) has been expressed in tomato and tobacco by gene integration into the plant's nuclear genome via *Agrobacterium tumefaciens*-assisted plant transformation. Mice models orally administered with this edible tomato vaccine showed high levels of IgA antibody production, specific to SARS-CoV-1. Mice serum in reaction to transgenic tobacco confirmed the presence of anti-SARS-CoV-1-IgG [74]. In another experiment, a fusion protein consisting of green fluorescent protein and 1-658 amino acid residues of SARS-CoV-1 has been transiently expressed in tobacco leaves. Microscopic studies confirmed the presence of fusion protein in cytosol and nuclear periphery [75].

Another study involving the transient expression of SARS-CoV-1 nucleocapsid (rN) protein (423-aa) *N. benthamiana* documented a substantial yield, i.e. almost 1% of the total soluble protein (TSP). The subsequent inoculation of three doses in mice confirmed the effective immune responses with high levels of IgG1 and IgG2a [75]. The transient expression of nucleocapsid protein (N) and the membrane protein (M) in *N. benthamiana* led to the recovery of 3–4 µg/g fresh leaf weight for the N protein and 0.1–0.15% TSP for M protein. The recovered N protein was able to react with N-specific antibodies in human sera. However, immunization assays were not carried out [76]. These findings can direct the scientist to choose the optimum expression systems for specific SARS-CoV-2 antigens.

Animal Models for SARS-CoV-2 Studies

Appropriate animal models for assessing vaccines for SARS-and MERS-CoV are missing or profoundly restricted, making the vaccine advancement exceptionally challenging [77]. Engineered animal model that imitates the clinical infection can illuminate on pathogenesis just as to create vaccines and therapeutics against these CoVs (Fig. 7.4). A few animal models have been assessed for SARS-and MERS CoVs including mouse, guinea pigs, hamsters, ferrets, bunnies, rhesus macaques, marmosets and felines (Table 7.4) [78, 79].

Early focus was on the development of animal models for SARS-CoV, yet the specificity of the virus to ACE2 (receptor of SARS-CoV) was a significant prevention to such endeavours. Afterwards, a SARS-CoV transgenic mouse model was created by integrating hACE2 gene into the mouse genome. The animal model utilized for building up a MERS-CoV vaccine was rhesus macaques. Challenged animals demonstrated clinical side effects, for example, expanded internal heat level, piloerection, cough, slouched pose and decreased food admission [80]. Another often utilized animal model for MERS-CoV is the regular marmoset, wherein the virus caused deadly pneumonia.

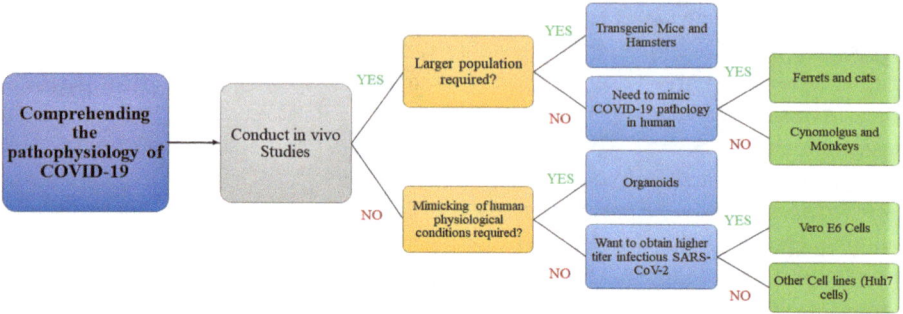

Fig. 7.4 Understanding COVID-19 pathophysiology using animal models, organoids and cell lines

Table 7.4 Animal models being utilized in SARS-CoV-2 research

Sr. No	Animal model	Salient points	Reference
1	Mice (wild-type and human ACE2 transgenic mice)	SARS-CoV-2 could not gain entry into mouse ACE 2 receptors in case of wild-type mice. After challenged with the virus, the mice experienced weight reduction, viral replication in lungs and deteriorated alveoli with widespread apoptosis in case of human ACE2 transgenic mice model	[20, 84]
2	Syrian hamster	After infection, animal exhibited weight loss, viral load in lungs and damaged alveoli with apoptosis	[85]
3	Ferrets	Acute bronchitis developed in the ferrets after infection	[86]
4	Cats	After SARS-CoV-2 infection, following complications were observed: Intra-alveolar oedema Congestion in the inter-alveolar septa Abnormal arrangement of the epithelium with loss of cilia Lymphocytic infiltration into the lamina propria	[87]
5	Cynomolgus macaques	Both type I and II pneumocytes were damaged by infection. The animal gathered oedema fluid in alveolar lumina along with pneumonia and pulmonary consolidation postinfection	[88]
6	Rhesus macaques	High viral titre in respiratory tract, humoral and cellular immune reactions and pneumonia were observed in the animal model. The findings suggest that therapeutic responses of adenovirus-vectored vaccine and DNA vaccine candidates expressing S protein could be evaluated	[89–91]

Humoral and cell-mediated immune response could be identified in both rhesus macaques and basic marmoset when challenged with MERS-CoV virus [81]. Roberts et al. engineered golden Syrian hamsters (strain LVG) to examine immunization protection against various SARS-CoV strains [82]. These hamsters are useful model for contemplating CoV pathology and pathogenesis and vaccine viability. The weakened NSP16 CoV vaccine was inoculated in mice [83].

Endeavours to create animal models for MERS-CoV, for example, mice, hamsters and ferrets, face constraints since MERS-CoV cannot multiply inside the respiratory organs of these species. Animals such as mice or hamsters naturally insusceptible to MERS-CoVs but prone to SARS-CoV have been genetically altered to become more humanized, e.g. hDPP4 human, hDPP4-transduced and hDPP4-Tg mice (transgenic for expressing hDPP4), and liable to MERS-CoV [92]. Modification in the mouse genome utilizing the CRISPR-Cas9 system could make the animals vulnerable to CoV viruses and infection [93]. Genetically engineered 288–330+/+ MERS-CoV mouse model has been designed for the assessment of novel MERS-CoV vaccines and medications [94]. Contrasted with the larger animals, small animal models, for example, mice and bunnies, are favoured because of lower cost, simplicity of alteration and promptly accessible adequacy strategies [95]. Further investigations are expected to perceive reasonable models for rising SARS-CoV-2 by distinguishing receptor specificity of SARS-CoV-2 and examining disease indications and pathologies/viral pathogenesis related with exploratory immunization of the virus in mice, rodents and different models, along with inspection of virus specific immune responses and immunity. This would encourage preclinical assessments of comer COVID-19 vaccine runners and medications.

Cell Culture Systems for SARS-CoV-2 Studies

A few permissive cell lines to hCoVs such as monkey epithelial cell lines (LLC-MK2 and Vero-B4) have been utilized in neutralization tests for evaluating the required titres of neutralization antibodies (Table 7.5). Goat lung cells, alpaca kidney cells and dromedary umbilical line cells have been discovered to allow MERS-CoV replication [96]. Progressed ex vivo 3D tracheobronchial tissue (mirroring epithelium of conductive aviation route) has been utilized for human CoVs [97]. In addition, VLPs showing SARS-CoV S protein were discovered skilful for passage to permissive cells or transfected cells that overexpress infection receptors [98]. SARS-CoV-2 cultures have been purified in Vero and the Huh-7 cells (human liver malignancy cells) [99]. Pseudotyped virions/VLPs expressing reporter genes, for example, GFP or luciferase, can be utilized for measurement and assessment of the viability of mAbs and medications in hindering the entry of CoVs into the cells [100].

Clinical and Immunological Endpoints

The essential endpoints for characterizing the adequacy of a COVID vaccine likewise require conversation (Fig. 7.5). The two most regularly referenced are:

Table 7.5 Cell lines being utilized in SARS-CoV-2 research

Sr. No	Cell lines	Source	Salient points	Reference
1	Human airway epithelial cells	Commercially supplied by Lonza, Pomocell	Allows culturing of SARS-CoV-2 Mimicking of infected human cell lungs Cytopathic effects observed after infection	[10]
2	Vero E6 cells (wild-type and TMPRSS-2 overexpressing cells)	Epithelial cells isolated from the kidneys of African green monkey	Most popular cell line for culturing SARS-CoV-2 (E6 cells) Allows 100 times higher replicating RNA copies than the wild type (TMPRSS-2 overexpressing cells)	[20, 101]
3	Caco-2 cells	Human colon adenocarcinoma cells	Replication of SARS-CoV-2 possible	[102]
4	Calu-3 cells	Non-small cells from lung cancer	SARS-CoV-2 S pseudovirions exhibited 500 times greater luciferase activity when cultured in Calu-3 cells in comparison to the controls	[103]
5	HEK293T cells	In vitro cultures of human embryonic kidney cells (HEK)	Moderate SARS-CoV-2 replication	[104]
6	Huh7 cells	Hepatocyte originated cellular carcinoma cells	Allows tenfold increase in luciferase activity when transduced by SARS-CoV-2 S pseudovirions	[103]

(i) Safekeeping from infection as manifested in case of seroconversion
(ii) Avoidance of clinically symptomatic illness, particularly enhancement of disease seriousness, including the recurrence of malady requiring high-force clinical consideration and hospitalization.

This requires the deep assessment of the impact of vaccination on the seriousness of COVID-19 illness in a wide assortment of epidemiological and clinical settings among both more youthful and old people and also underserved minorities. Initial efficacy trials are to be evaluated based on these parameters. Accomplishing these endpoints will also lead to a diminished transmission rate of COVID-19 among the population.

Primary endpoints attributed to reduction of the disease require a higher number of people enrolled for clinical trials, given that asymptomatic infection accounts for 20% to 40% of the total instances of COVID-19 [105]. Initial efficacy trials will hence require a large number of persons with monitoring of both clinical and serologic endpoints. Lack of data of incidence rates will pose a major challenge that will add to complications in formulating clinical trial procedures for serological

Clinical trials	Phase I	Phase II	Phase III	Phase IV
OBJECTIVES	Safety Evaluation	Randomized trials to understand the underlying immune reactions to vaccines in comparison to placebo treament	Efficacy evaluation with larger sample size and statistical optimization	Risk assessment, harmful effects and after vaccine approval and marketing
ENDPOINTS	Determine lethal effects and reactions such fever, pain at site of injection Identify immune response	Pre-defined immune reactions such as antibodies and immune cell activation	Overall safety, response to infection or disease manisfestation	Ongoing

Classified as Investigational New Drug (IND) ▼ ▼ Receives FDA Approval

CONTINUOUS MONITORING: SAFETY, HOST RESPONSES, LETHAL EFFECTS

Fig. 7.5 Endpoints of clinical vaccine trials

endpoints [106]. An important requirement for this multi-trial strategy that provides a binding relationship between various vaccinations and vaccine efficacy studies is to establish laboratories with identical validated serologic assays. These laboratories should provide clinical trials and dispersion of crucial samples from a trial. Variables defining the immune reaction coming about because of immunization and from the actual infection are under substantial examination, and it's important to figure measures to manage this issue.

Efficacy studies should be assessed for both positive and negative effects. The probability of SARS-CoV-2 re-exposure is a lot higher than that of SARS-CoV-1, which has vanished from network course, and consequently longer-term assessment of likely upgrade with re-exposure is required. This necessity doesn't block licensure dependent on the endpoints mentioned here; nonetheless, it demonstrates that a more detailed follow-through of the initial immunization entities ought to be embraced. The efficacy of clinical and serologic endpoints will likewise be investigated, as fading of immunity is normal with human COVID diseases [107].

COVIDs have a high transformation rate in their RNA genome. Despite the fact that there has been some hereditary float during the development of the SARS-CoV-2 epidemic, there are no major changes in the spike protein up till now, especially in the parts believed to be significant for balance; this creates a hope that the vaccine to be developed currently will be effective for the coming 6–12 months against the strains [108]. The chance of performing controlled human challenge trials, in which few volunteers are vaccinated and thusly tested with SARS-CoV-2, has been proposed. Such experiments may prove to be useful when performed to define potential immune correlates or filtering out weaker vaccine strategies. In any case, this methodology has deficiencies regarding pathophysiology and safety [109]. The possibility of performing controlled human test studies, in which not many participants are inoculated and subsequently tried with SARS-CoV-2, is being proposed. These tests may be useful when performed to characterize possible immune correlates or filter out weaker immunization approaches. In spite of the fact that the danger of serious sickness or demise in youthful healthy people from COVID-19 is extremely low, it isn't nil, and yet we have not demonstrated successful treatments of the disease to safeguard volunteers having problems from such a test. Almost certainly, a 2019-nCoV test strain will, by configuration, result in gentle sickness in many volunteers and in this way may not sum up the aspiratory pathophysiology found in some patients. Additionally, limited viability in youthful people doesn't foresee comparable viability among older adults with significant cofactors related with COVID-19 malady, nor would it reveal decrease of contagiousness to significant susceptible groups. Regardless of whether such examinations might be deserving of interest or would beneficially affect timetables for immunization improvement needs cautious assessment by an autonomous board of professionals on vaccine progress.

Vaccine Development Landscape

Vaccine developments usually take 5–10 years, and the procedure includes exceptionally controlled preclinical and clinical studies before the vaccine is approved for the masses. Amazingly within just a couple of weeks of the publication of viral genetic information, COVID-19 vaccines had been prepared for trials on patients. Global COVID-19 vaccine R&D landscape as of 28 August includes 176 vaccine candidates, 35 of which have entered clinical evaluation including phase II and III studies (Table 7.6), whereas 143 candidates are in preclinical stages. Existing and new vaccine production technologies are being deployed for the creation of COVID-19 vaccine candidates and the pace of development.

Cutting-edge immunization advancement can be assisted by using sequence data alone instead of depending on in vitro cultures of viruses. Nucleic acid-based vaccines that utilize this cutting-edge approach have been the leaders for vaccine hunt. One such model is an mRNA vaccine (mRNA-1273, encoding for the viral spike protein which locks onto human host cells) created by the National Institutes of Health and Moderna Therapeutics and has just given encouraging indications and has proceeded to phase III of the clinical studies. A new time record has been set by this vaccine by reaching the preliminaries (NCT04283461) in such a brief period of time after the verification of the SARS-CoV-2 as the agent causing the present pandemic [110]. Another leader in stage IIB/III clinical preparation, ChAdOX1 nCoV19, created by Oxford University and AstraZeneca, is a vaccine programme dependent on an adenovirus vector likewise encoding for SARS-CoV-2 spike protein.

Table 7.6 Candidate vaccines under clinical evaluation

	Vaccine platform	Type of candidate vaccine	COVID-19 vaccine developer	Number of doses	Timing of doses	Clinical stage
1.	Non-replicating viral vector	ChAdOx1-S	University of Oxford/AstraZeneca	1		Phase 3
2.	Non-replicating viral vector	Adenovirus Type 5 vector	CanSino Biological Inc./Beijing Institute of Biotechnology	1		Phase 3
3.	Non-replicating viral vector	Adeno-based (rAd26-S+rAd5-S)	Gamaleya Research Institute	2	0, 21 days	Phase 3
4.	Non-replicating viral vector	Ad26COVS1	Janssen Pharmaceutical Companies	2	0, 56 days	Phase 3
5.	Inactivated	Inactivated	Sinovac	2	0, 14 days	Phase 3
6.	Inactivated	Inactivated	Wuhan Institute of Biological Products/Sinopharm	2	0, 14 or 0, 21 days	Phase 3
7.	Inactivated	Inactivated	Beijing Institute of Biological Products/Sinopharm	2	0, 14 or 0, 21 days	Phase 3
8.	RNA	LNP-encapsulated mRNA	Moderna/NIAID	2	0, 28 days	Phase 3
9.	RNA	3 LNP-mRNAs	BioNTech/Fosun Pharma/Pfizer	2	0, 28 days	Phase 3
10.	Protein subunit	Full-length recombinant SARS CoV-2 glycoprotein nanoparticle vaccine adjuvanted with Matrix M	Novavax	2	0, 21 days	Phase 1/2
11.	Protein subunit	Adjuvanted recombinant protein (RBD-Dimer)	Anhui Zhifei Longcom Biopharmaceutical/Institute of Microbiology, Chinese Academy of Sciences	2 or 3	0, 28 or 0, 28, 56 days	Phase 2
12.	RNA	mRNA	Curevac	2	0, 28 days	Phase 2
13.	Inactivated	Inactivated	Institute of Medical Biology, Chinese Academy of Medical Sciences	2	0, 28 days	Phase 1/2
14.	Inactivated	Inactivated	Research Institute for Biological Safety Problems, Rep of Kazakhstan	2	0, 21 days	Phase 1/2

(continued)

Table 7.6 (continued)

	Vaccine platform	Type of candidate vaccine	COVID-19 vaccine developer	Number of doses	Timing of doses	Clinical stage
15.	DNA^	DNA plasmid vaccine with electroporation	Inovio Pharmaceuticals/ International Vaccine Institute	2	0, 28 days	Phase 1/2
16.	DNA	DNA plasmid vaccine + Adjuvant	Osaka University/ AnGes/ Takara Bio	2	0, 14 days	Phase 1/2
17.	DNA^	DNA plasmid vaccine	Cadila Healthcare Limited	3	0, 28, 56 days	Phase 1/2
18.	DNA	DNA vaccine (GX-19)	Genexine Consortium	2	0, 28 days	Phase 1/2
19.	Inactivated	Whole-Virion Inactivated	Bharat Biotech	2	0, 14 days	Phase 1/2
20.	Protein subunit	RBD-based	Kentucky Bioprocessing, Inc	2	0, 21 days	Phase 1/2
21.	Protein subunit	S protein (baculovirus production)	Sanofi Pasteur/GSK	2	0, 21 days	Phase 1/2
22.	RNA	mRNA	Arcturus/Duke-NUS			Phase 1/2
23.	Non-replicating viral vector	Replication defective Simian Adenovirus (GRAd) encoding S	ReiThera/ LEUKOCARE/ Univercells	1		Phase 1
24.	Protein subunit	Native like Trimeric subunit Spike Protein vaccine	Clover Biopharmaceuticals Inc./ GSK/Dynavax	2	0, 21 days	Phase 1
25.	Protein subunit	Recombinant spike protein with Advax™ adjuvant	Vaxine Pty Ltd/Medytox	1		Phase 1
26.	Protein subunit	Molecular clamp stabilized spike protein with MF59 adjuvant	University of Queensland/CSL/ Seqirus	2	0, 28 days	Phase 1
27.	Protein subunit	S-2P protein + CpG 1018	Medigen Vaccine Biologics Corporation/ NIAID/Dynavax	2	0, 28 days	Phase 1
28.	Protein subunit	RBD + adjuvant	Instituto Finlay de Vacunas, Cuba	2	0, 28 days	Phase 1
29.	Protein subunit	Peptide	FBRI SRC VB VECTOR, Rospotrebnadzor, Koltsovo	2	0, 21 days	Phase 1
30.	Protein subunit	RBD (baculovirus production expressed in Sf9 cells)	West China Hospital, Sichuan University	2	0, 28 days	Phase 1

(continued)

7 COVID-19 Pandemic and Vaccines

Table 7.6 (continued)

	Vaccine platform	Type of candidate vaccine	COVID-19 vaccine developer	Number of doses	Timing of doses	Clinical stage
31.	Replicating viral vector	Measles-vector based	Institute Pasteur/Themis/ Univ. of Pittsburgh CVR/Merck Sharp & Dohme	1 or 2	0, 28 days	Phase 1
32.	Replicating viral vector	Intranasal flu-based-RBD	Beijing Wantai Biological Pharmacy/ Xiamen University	1		Phase 1
33.	RNA	LNP-nCoVsaRNA	Imperial College London	2		Phase 1
34.	RNA	mRNA	People's Liberation Army (PLA) Academy of Military Sciences/ Walvax Biotech.	2	0, 14 or 0, 28 Days	Phase 1
35.	VLP	Plant-derived VLP adjuvanted with GSK or Dynavax adjs.	Medicago Inc.	2	0, 21 days	Phase 1

The other progressed competitors moved into clinical development incorporates:

(i) CoronaVac from Sinovac
(ii) Ad5-nCoV from CanSino Biologics
(iii) BNT162 from Pfizer
(iv) BioNtech
(v) BCG live-attenuated vaccine from Radboud University Medical Center, University of Melbourne and Murdoch Children's Research Institute and Faustman Lab at Massachusetts General Hospital.

The movement of advancement is dramatic when contrasted with new ailments causing serious epidemic pronounced by the WHO (Fig. 7.6). The extraordinary larger part of authorized immunizations is situated in inactivation pathogens which stretch the development, finances and creation of the vaccine. Recombinant viral vectored, DNA/RNA and protein innovations are establishing the quickest precedents in vaccine advancement. Only a few have been authorized so far for veterinary utilization, since, for people, a few immunizations have not met some administrative necessities for endorsement and commercialization. The global crises like the current COVID-19 could give a last push towards acquiring licensure.

This features the capability of vaccinology to gain quick ground when fitting worldwide help exists, demonstrating that when there is a will, there is a way. A comparison has been made between the proposed COVID-19 vaccine development timeline and that of the classical one in Fig. 7.7.

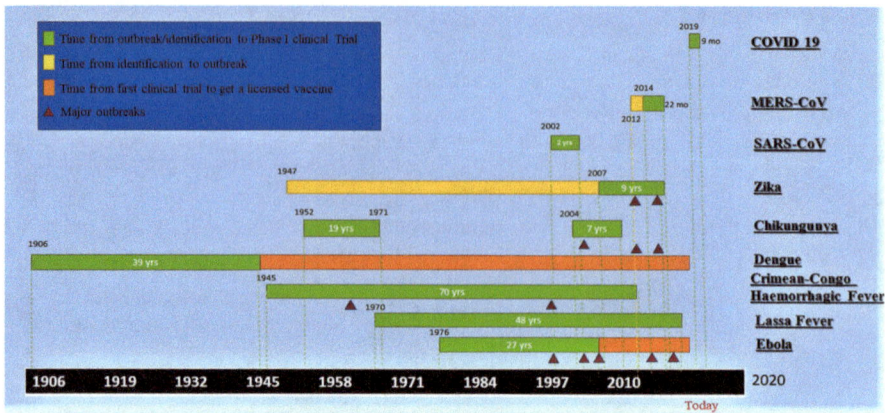

Fig. 7.6 Vaccine development timeline of recently emerged outbreaks

Global and Equitable Distribution of Vaccines

The pandemic has just caused the loss of a huge number of lives and affected the lives of billions of people worldwide. Just as lessening the heartbreaking death toll and assisting with getting the pandemic levelled out, the presentation of a vaccine will forestall the loss of US$ 375 billion to the worldwide economy per month. Worldwide evenhanded admittance to a vaccine, especially securing medical services labourers and those most in danger, is the best way to relieve the general wellbeing and monetary effect of the pandemic. The events and broad supply of COVID-19 clinical medicines are a typical worldwide intrigue. An effective remedy for COVID-19 will be there in a few months' time, and it will take at least a year for vaccine to be developed. To meet the outstanding demand worldwide, once COVID-19 therapeutic and vaccines are approved, they need to be manufactured on a large scale.

This is the ideal time to calculate the production limits, funding and circulation system important to create adequate amounts to address worldwide demand in a reasonable way beneficial for the public health. Countries of the world are currently collaborating with each other, but there is curiosity that which nation will succeed first in the development of therapeutics and vaccines. Governments as of now have the best impetus to team up while vulnerability stays with regard to which countries' immunizations and therapeutics will succeed. In the course of the most recent decade, technological research and production capabilities are widely spread globally. Technological advancements mean that the best cures and vaccines for COVID-19 will be developed in new ways outside the customary drug development centres.

Rich nations cannot depend on outbidding contesters if vaccine and therapeutic supplies are not shared by the countries manufacturing them. For there to be a uniform distribution of therapies for COVID-19, the countries of the world must

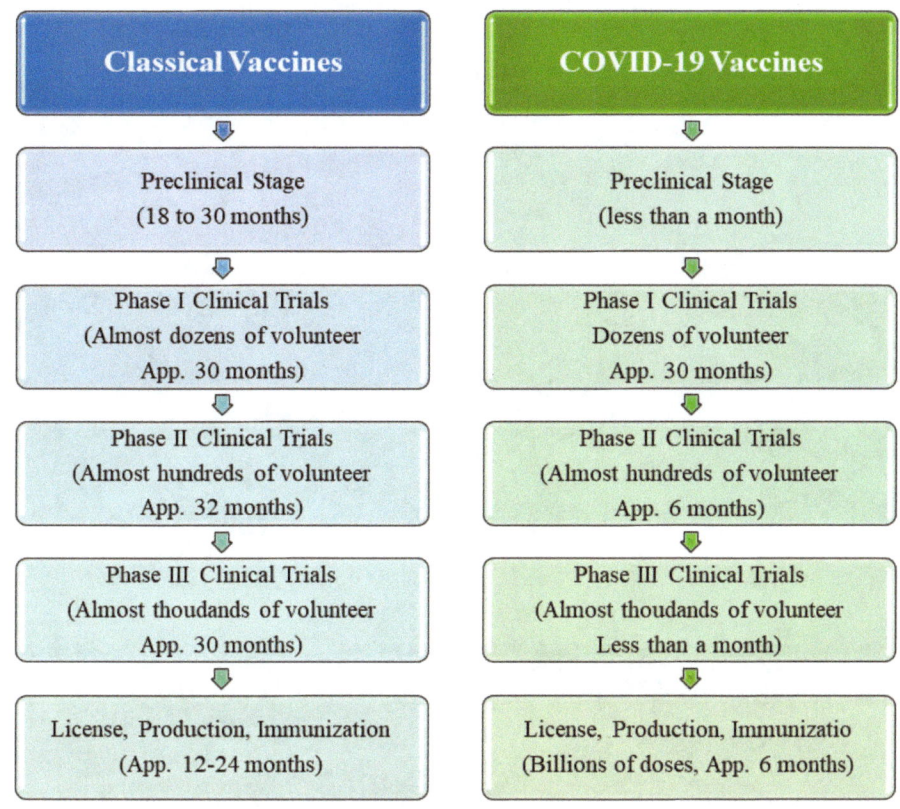

Fig. 7.7 Comparison between the classical vaccine and COVID-19 vaccine development

Challenge studies are likely to expedite the development of vaccines and their availability in the presence of well-coordinated collaboration among the scientists, developers and regulators (Fig. 7.8). Whatsoever, these kinds of studies must be included into extensive research plans which involves large-scale studies to establish accurate safety and efficiency. SARS-CoV-2 challenge studies may also be helpful to other types of vaccine experimentation by providing precise evaluation of infections with no apparent symptoms and swift and standardized testing of various vaccine candidates. WHO has issued eight ethical criteria which should be adhered to for conduction challenge trials (Fig. 7.9)

cooperate with each other. However, this seems to be a difficult task. In the midst of rising populism, governments have opposed multilateral organizations and international accords. Numerous nations have reacted to this pandemic by turning internal: shutting borders, accumulating clinical assets and scapegoating outsiders. Building up a structure for reasonable and evenhanded distribution of COVID-19 vaccines and therapeutics is unquestionably more unpredictable and will require coordination of numerous foundations, funders, governments and drug organizations (Fig. 7.10).

European Commission helped start a programme, Access to COVID-19 Tool (ACT) Accelerator, which will focus on fast development and fair supply of

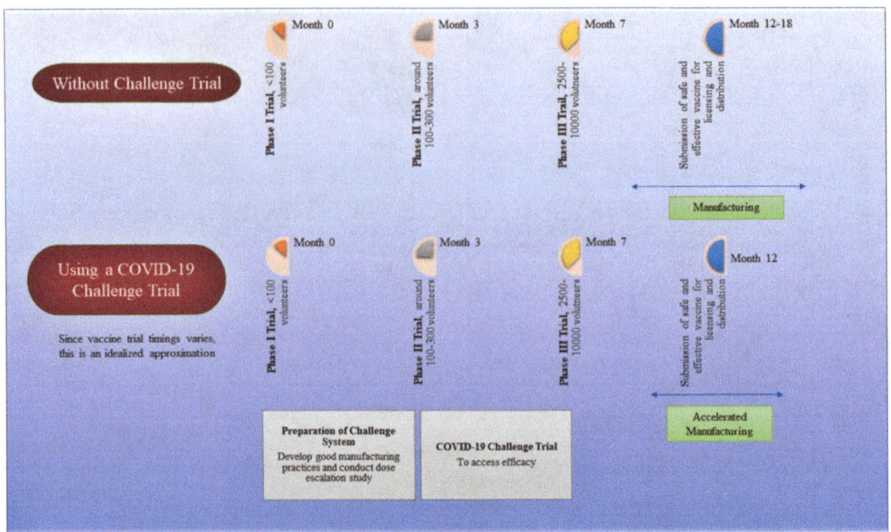

Fig. 7.8 Acceleration of COVID-19 vaccine development using challenge trials

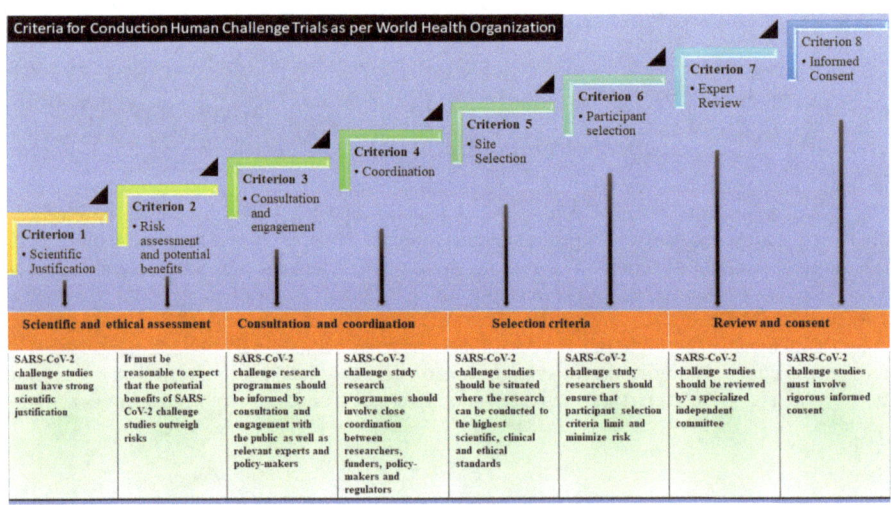

Fig. 7.9 Criteria listed by the WHO for conduction of challenge trials

vaccines, therapeutics and diagnostics. COVAX, co-driven by Gavi, the Coalition for Epidemic Preparedness Innovations (CEPI) and WHO, is the vaccine backbone of the Access to COVID-19 Tools (ACT) Accelerator. The purpose of COVAX is to speed up the process for the development and production of the vaccines for COVID-19 and to ensure their just, fair and equal distribution in all parts of the world (Fig. 7.11).

Currently, a total of 172 countries are working to participate in the COVAX initiative, which has the biggest and most spread COVID-19 vaccine portfolio. This

Fig. 7.10 A general framework for the equitable distribution of vaccines

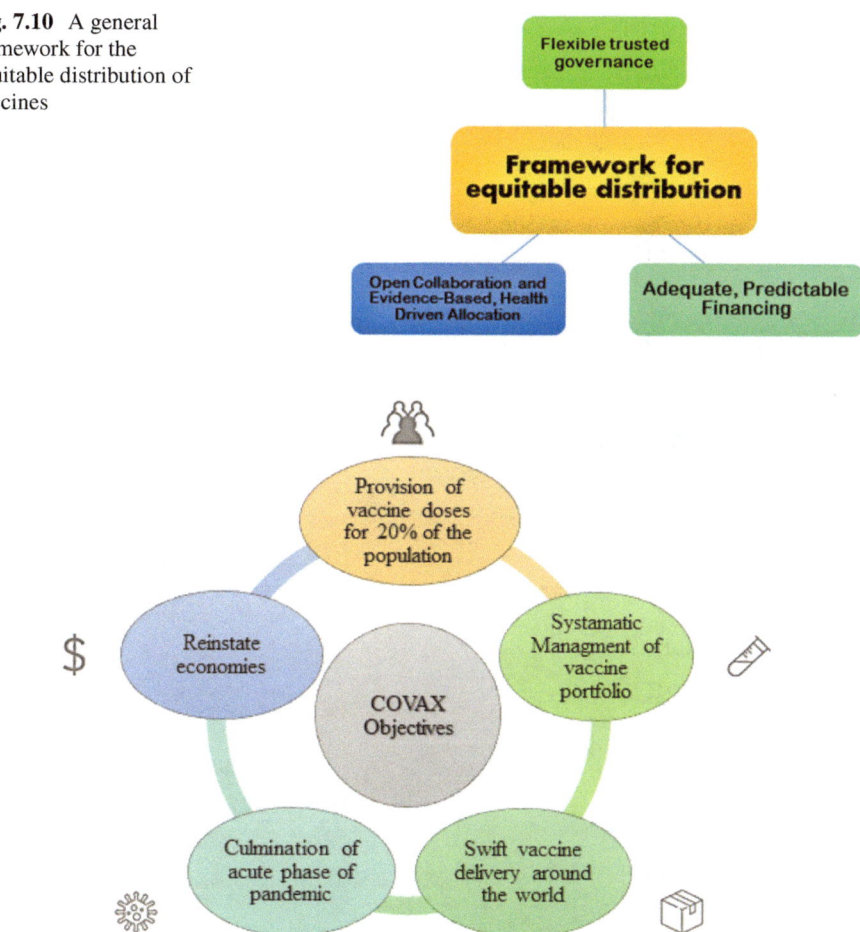

Fig. 7.11 Objectives of COVAX ensuring equitable distribution of vaccines

incorporates nine candidate vaccines, with a further nine under assessment, and discussions are in progress for additional vaccines (Table 7.7). The plan entails the cooperation at worldwide scale, working with governments and manufacturers to guarantee that COVID-19 vaccines are accessible worldwide to both higher-salary and lower-pay nations. So as to tie down equivalent access to COVID-19 vaccines for nations across the globe, the subsequent stage is to affirm self-financing members by 18 September 2020, with the main forthright instalments to be made no later than 9 October 2020. 80 higher-salary economies, which would back the vaccines from their own public money financial plans, have submitted Expressions of Interest in front of the 31 August 2020 cutoff time for affirmation of aim to partake. They will cooperate with 92 low- and middle-salary nations that will be upheld by the AMC on the basis if its funding targets are met. Together, this gathering of 172 nations represents to over 70% of the total population of the world. CEPI has allocated a budget of $ 2 billion for the rapid COVID-19 vaccine development

Table 7.7 Vaccine developers supported by CEPI

Sr. No	Organization	Country	Vaccine clinical stage
1.	Inovio	USA	Phase I/II
2.	Moderna	USA	Phase III
3.	CureVac	Germany	Phase I
4.	Astra Zeneca/University of Oxford	UK and Northern Ireland	Phase III
5.	Institute of Pasteur/Merck/Themis	France/USA/Austria	Preclinical
6.	University of Hong Kong	China	Preclinical
7.	Novavax	USA	Phase I/II
8.	Clover Biopharmaceuticals	China	Phase I
9.	University of Queensland/CSL	Australia	Phase I

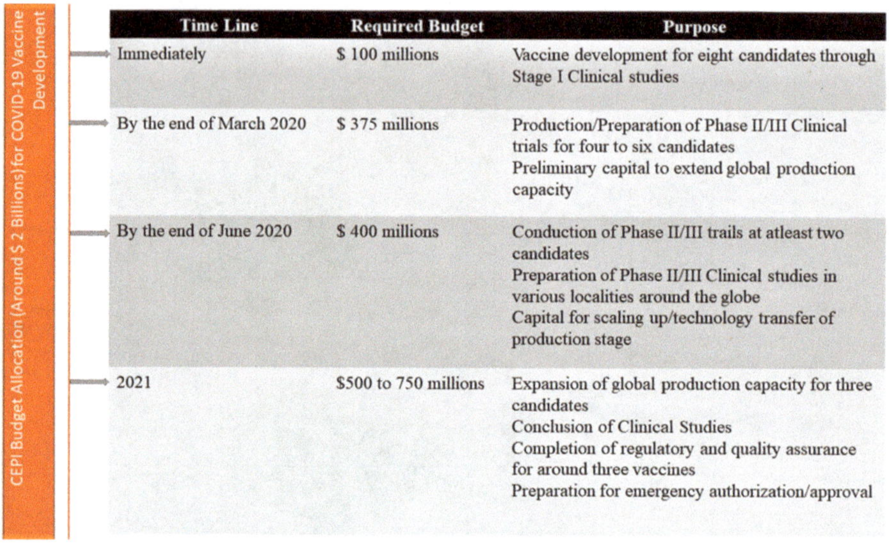

	Time Line	Required Budget	Purpose
CEPI Budget Allocation (Around $ 2 Billions) for COVID-19 Vaccine Development	Immediately	$ 100 millions	Vaccine development for eight candidates through Stage I Clinical studies
	By the end of March 2020	$ 375 millions	Production/Preparation of Phase II/III Clinical trials for four to six candidates Preliminary capital to extend global production capacity
	By the end of June 2020	$ 400 millions	Conduction of Phase II/III trails at atleast two candidates Preparation of Phase II/III Clinical studies in various localities around the globe Capital for scaling up/technology transfer of production stage
	2021	$500 to 750 millions	Expansion of global production capacity for three candidates Conclusion of Clinical Studies Completion of regulatory and quality assurance for around three vaccines Preparation for emergency authorization/approval

Fig. 7.12 CEPI budget allocation for COVID-19 vaccine development

(Fig. 7.12). The world, in any case, is more ready to react than at any other time. The accompanying proposed structure uses existing global discussions to encourage evenhanded circulation of COVID-19 vaccines and therapeutics.

Vaccine Developers and Geographical Distribution

Greater commitment by large multinational companies has brought many great changes in the general outlook of COVID-19 vaccine development since April. Of the vaccine candidates in the clinic, 11 are being created by Chinese associations, and seven are being upheld by the US Operation Warp Speed programme, which plans to convey 300 million vaccine portions for COVID-19 by January 2021 and has so far reported subsidizing of more than US$10 billion to propel vaccine advancement. Eight of the clinical applicants have gotten subsidizing from the Coalition for Epidemic Preparedness Innovations (CEPI) and are presently a part of the COVAX, a joint effort drove by CEPI, Gavi and the WHO that plans to provide 2 billion vaccine dosages for worldwide distribution before the end of 2021.

Conclusions

In this time of a pandemic, fast research, manufacture and organization of first-generation vaccines are crucial. For this, the main applicants are the nucleic acid (DNA, mRNA) and subunit (S and N protein) vaccines for artificial immunity induction to confer strong resistance against the actual infection. A methodology used to speed up immunizations during epidemics is to give restrictive licenses; for COVID-19, they could be based on human clinical trial data affirming wellbeing and satisfactory degrees of assurance to lessen death toll in the most vulnerable people (old people and patients having comorbidities, medical services staff). Meanwhile, second-generation, more powerful or effective immunizations to forestall infection and deaths and decrease shedding, as examined, ought to be created in parallel for future organization.

The presence of virus neutralizing antibodies and cell-mediated immune reactions in response to vaccine inoculations in animal models can serve as the pointers of safeguard, yet the coordinates of immunity to COVID-19 in people are obscure. A point of concern is the effectiveness of the vaccine to combat extreme infections and death rates but its inadequacy to limit the nasal shedding, permitting continuity in the spread of the disease. Immunity at the mucosal level may also lead to diminished viral shedding through nasal secretions. Achieving the mucosal immunity is a key argument to break off/recede the transmission chain; however, this may require booster vaccine doses. Aged individuals having chronic illnesses are prone to acquiring severe infections; however, the symptoms can be relieved by the administration of other vaccines such as those existing for flu, etc. Other ways may include the use of adjuvants and varying vaccine dosages in weak groups to uplift their guard. Animal models additionally need to emulate these criteria.

The way the COVID-19 causes infection is not fully understood yet, and immunization systems might require modifications if the infection taints both the respiratory and intestinal tracts. Oral administration through mouth and nose and the subsequent dose of parenteral vaccine might be ideal to forestall both intestinal and respiratory complications as well as viral shedding. Future overflow of coronavirus transmission from animal repositories is likely. New insights to develop vaccines against beta-CoV lineage that are effective equally in both animal models and humans are the dire need of time. Presently, in the absence of an efficient vaccine, the course of battling the COVID-19 shifts towards passive immunization as means of both the prophylactic measures and clinical therapy. The best approach so far is the convalescent plasma therapy which entails the use of plasma recovered from the recuperated patients. This strategy helps in speedy recovery and lessens the hospitalization ultimately reducing mortality rates. The clinical trials also corroborate its beneficial outcomes when compared to placebo treatment. The establishment of plasma banks having plasma donated by healed patients of COVID-19 would also be effective in suppressing the annihilation caused by this pandemic. Monoclonal antibodies designed against the surface viral proteins which are involved in eliciting immune response can also be engineered and tested in animal models challenged with SARS-CoV-2. These MAB can programme the immune responses and serve as prophylactic therapy. The development of respiratory syncytial virus monoclonal antibody by the name palivizumab is worth mentioning in this regard. This can be inhaled as aerosols through the nose.

References

1. Huang C, Wang Y, Li X, Ren L, Zhao J, Hu Y et al (2020) Clinical features of patients infected with 2019 novel coronavirus in Wuhan, China. The Lancet 395(10223):497–506
2. Jiang S, Shi Z, Shu Y, Song J, Gao GF, Tan W et al (2020) A distinct name is needed for the new coronavirus. Lancet (London, England) 395(10228):949
3. Sanche S, Lin YT, Xu C, Romero-Severson E, Hengartner N, Ke R (2020) Early release-high contagiousness and rapid spread of severe acute respiratory syndrome coronavirus. Emerg. Infect. Dis 26(7):1407–1407
4. Hui DS, Azhar EI, Madani TA, Ntoumi F, Kock R, Dar O et al (2020) The continuing 2019-nCoV epidemic threat of novel coronaviruses to global health—The latest 2019 novel coronavirus outbreak in Wuhan, China. Int J Infect Dis 91:264–266
5. Li Q, Guan X, Wu P, Wang X, Zhou L, Tong Y et al (2020) Early transmission dynamics in Wuhan, China, of novel coronavirus–infected pneumonia. N Engl J Med 382:1199–1207
6. Zhang L, Liu Y (2020) Potential interventions for novel coronavirus in China: a systematic review. J Med Virol 92(5):479–490
7. Su S, Wong G, Shi W, Liu J, Lai AC, Zhou J et al (2016) Epidemiology, genetic recombination, and pathogenesis of coronaviruses. Trends Microbiol 24(6):490–502
8. Raj VS, Mou H, Smits SL, Dekkers DH, Müller MA, Dijkman R et al (2013) Dipeptidyl peptidase 4 is a functional receptor for the emerging human coronavirus-EMC. Nature 495(7440):251–254
9. Lu L, Liu Q, Zhu Y, Chan K-H, Qin L, Li Y et al (2014) Structure-based discovery of Middle East respiratory syndrome coronavirus fusion inhibitor. Nat Commun 5(1):1–12
10. Zhu N, Zhang D, Wang W, Li X, Yang B, Song J et al (2020) A novel coronavirus from patients with pneumonia in China, 2019. N Engl J Med 382:727–733
11. Lu R, Zhao X, Li J, Niu P, Yang B, Wu H et al (2020) Genomic characterisation and epidemiology of 2019 novel coronavirus: implications for virus origins and receptor binding. The Lancet 395(10224):565–574
12. Jiang S, Hillyer C, Du L (2020) Neutralizing antibodies against SARS-CoV-2 and other human coronaviruses. Trends Immunol 41:355–359
13. Perlman S (2020) Another decade, another coronavirus. N Engl J Med 382:760–762
14. Guan W-j, Ni Z-y, Hu Y, Liang W-h, Ou C-q, He J-x et al (2020) Clinical characteristics of coronavirus disease 2019 in China. N Engl J Med 382(18):1708–1720
15. Chen N, Zhou M, Dong X, Qu J, Gong F, Han Y et al (2020) Epidemiological and clinical characteristics of 99 cases of 2019 novel coronavirus pneumonia in Wuhan, China: a descriptive study. The Lancet 395(10223):507–513

16. Bhatraju PK, Ghassemieh BJ, Nichols M, Kim R, Jerome KR, Nalla AK et al (2020) Covid-19 in critically ill patients in the Seattle region—case series. N Engl J Med 382(21):2012–2022
17. Wang D, Hu B, Hu C, Zhu F, Liu X, Zhang J et al (2020) Clinical characteristics of 138 hospitalized patients with 2019 novel coronavirus–infected pneumonia in Wuhan, China. JAMA 323(11):1061–1069
18. Pan L, Mu M, Yang P, Sun Y, Wang R, Yan J et al (2020) Clinical characteristics of COVID-19 patients with digestive symptoms in Hubei, China: a descriptive, cross-sectional, multicenter study. Am J Gastroenterol 115:766–773
19. Giacomelli A, Pezzati L, Conti F, Bernacchia D, Siano M, Oreni L et al (2020) Self-reported olfactory and taste disorders in patients with severe acute respiratory coronavirus 2 infection: a cross-sectional study. Clin Infect Dis 71: 889–890
20. Zhou P, Yang X-L, Wang X-G, Hu B, Zhang L, Zhang W et al (2020) A pneumonia outbreak associated with a new coronavirus of probable bat origin. Nature 579(7798):270–273
21. Song Y, Liu P, Shi X, Chu Y, Zhang J, Xia J et al (2020) SARS-CoV-2 induced diarrhoea as onset symptom in patient with COVID-19. Gut 69(6):1143–1144
22. Recalcati S (2020) Cutaneous manifestations in COVID-19: a first perspective. J Eur Acad Dermatol Venereol 34:e212–e213
23. Lauer SA, Grantz KH, Bi Q, Jones FK, Zheng Q, Meredith HR et al (2020) The incubation period of coronavirus disease 2019 (COVID-19) from publicly reported confirmed cases: estimation and application. Ann Intern Med 172(9):577–582
24. Zou L, Ruan F, Huang M, Liang L, Huang H, Hong Z et al (2020) SARS-CoV-2 viral load in upper respiratory specimens of infected patients. N Engl J Med 382(12):1177–1179
25. He X, Lau EH, Wu P, Deng X, Wang J, Hao X et al (2020) Temporal dynamics in viral shedding and transmissibility of COVID-19. Nat Med 26(5):672–675
26. Zhou F, Yu T, Du R, Fan G, Liu Y, Liu Z et al (2020) Clinical course and risk factors for mortality of adult inpatients with COVID-19 in Wuhan, China: a retrospective cohort study. The Lancet 395:1059–1062
27. Tang N, Li D, Wang X, Sun Z (2020) Abnormal coagulation parameters are associated with poor prognosis in patients with novel coronavirus pneumonia. J Thromb Haemost 18(4):844–847
28. Guan W-j, Liang W-h, Zhao Y, Liang H-r, Chen Z-s, Li Y-m et al (2020) Comorbidity and its impact on 1590 patients with Covid-19 in China: a nationwide analysis. Eur Respir J 55(5):2000547
29. Koirala A, Joo YJ, Khatami A, Chiu C, Britton PN (2020) Vaccines for COVID-19: the current state of play. Paediatr Respir Rev 35:43–49
30. Flaxman S, Mishra S, Gandy A, Unwin HJT, Mellan TA, Coupland H et al (2020) Estimating the effects of non-pharmaceutical interventions on COVID-19 in Europe. Nature 584:257–261
31. Boopathi S, Poma AB, Kolandaivel P (2020) Novel 2019 coronavirus structure, mechanism of action, antiviral drug promises and rule out against its treatment. J Biomol Struct Dyn:1–10
32. Buchholz UJ, Bukreyev A, Yang L, Lamirande EW, Murphy BR, Subbarao K et al (2004) Contributions of the structural proteins of severe acute respiratory syndrome coronavirus to protective immunity. Proc Natl Acad Sci 101(26):9804–9809
33. Walls AC, Park Y-J, Tortorici MA, Wall A, McGuire AT, Veesler D (2020) Structure, function, and antigenicity of the SARS-CoV-2 spike glycoprotein. Cell 181:281–292
34. Roper RL, Rehm KE (2009) SARS vaccines: where are we? Exp Rev Vaccines 8(7):887–898
35. Edridge AW, Kaczorowska JM, Hoste AC, Bakker M, Klein M, Jebbink MF et al (2020) Human coronavirus reinfection dynamics: lessons for SARS-CoV-2. MedRxiv:1–10
36. Tseng C-T, Sbrana E, Iwata-Yoshikawa N, Newman PC, Garron T, Atmar RL et al (2012) Immunization with SARS coronavirus vaccines leads to pulmonary immunopathology on challenge with the SARS virus. PloS One 7(4):e35421
37. Agrawal AS, Tao X, Algaissi A, Garron T, Narayanan K, Peng B-H et al (2016) Immunization with inactivated Middle East Respiratory Syndrome coronavirus vaccine leads to lung immunopathology on challenge with live virus. Human Vaccines Immunother 12(9):2351–2356

38. Li E, Yan F, Huang P, Chi H, Xu S, Li G et al (2020) Characterization of the immune response of MERS-CoV vaccine candidates derived from two different vectors in mice. Viruses 12(1):125
39. Wang N, Shang J, Jiang S, Du L (2020) Subunit vaccines against emerging pathogenic human coronaviruses. Front Microbiol 11:298
40. Jiang S, Du L, Shi Z (2020) An emerging coronavirus causing pneumonia outbreak in Wuhan, China: calling for developing therapeutic and prophylactic strategies. Emerg Microbes Infect 9(1):275–277
41. Rudraraju R, Mordant F, Subbarao K (2019) How live attenuated vaccines can inform the development of broadly cross-protective influenza vaccines. J Infect Dis 219(Supplement_1):S81–S87
42. Woelfel R, Corman V, Guggemos W, Seilmaier M, Zange S, Müller M. et al (2020) Virological assessment of hospitalized cases of coronavirus disease 2019. Nature 581:465–469
43. Lurie N, Saville M, Hatchett R, Halton J (2020) Developing Covid-19 vaccines at pandemic speed. N Engl J Med 382(21):1969–1973
44. Team WER (2016) After Ebola in West Africa—unpredictable risks, preventable epidemics. N Engl J Med 375(6):587–596
45. Huber C, Finelli L, Stevens W (2018) The economic and social burden of the 2014 Ebola outbreak in West Africa. J Infect Dis 218(Suppl_5):S698–S704
46. Jones SM, Feldmann H, Ströher U, Geisbert JB, Fernando L, Grolla A et al (2005) Live attenuated recombinant vaccine protects nonhuman primates against Ebola and Marburg viruses. Nat Med 11(7):786–790
47. Henao-Restrepo AM, Camacho A, Longini IM, Watson CH, Edmunds WJ, Egger M et al (2017) Efficacy and effectiveness of an rVSV-vectored vaccine in preventing Ebola virus disease: final results from the Guinea ring vaccination, open-label, cluster-randomised trial (Ebola Ça Suffit!). The Lancet 389(10068):505–518
48. Brende B, Farrar J, Gashumba D, Moedas C, Mundel T, Shiozaki Y et al (2017) CEPI—a new global R&D organisation for epidemic preparedness and response. The Lancet 389(10066):233–235
49. Saxena S, Sonwane AA, Dahiya SS, Patel CL, Saini M, Rai A et al (2009) Induction of immune responses and protection in mice against rabies using a self-replicating RNA vaccine encoding rabies virus glycoprotein. Vet Microbiol 136(1–2):36–44
50. Pulido MR, Sobrino F, Borrego B, Sáiz M (2010) RNA immunization can protect mice against foot-and-mouth disease virus. Antiviral Res 85(3):556–558
51. Rauch S, Jasny E, Schmidt KE, Petsch B (2018) New vaccine technologies to combat outbreak situations. Front Immunol 9:1963
52. Zhang C, Maruggi G, Shan H, Li J (2019) Advances in mRNA vaccines for infectious diseases. Front Immunol 10:594
53. Martin JE, Louder MK, Holman LA, Gordon IJ, Enama ME, Larkin BD et al (2008) A SARS DNA vaccine induces neutralizing antibody and cellular immune responses in healthy adults in a Phase I clinical trial. Vaccine 26(50):6338–6343
54. Lin J, Zhang J-S, Su N, Xu J-G, Wang N, Chen J-T et al (2007) Safety and immunogenicity from a phase I trial of inactivated severe acute respiratory syndrome coronavirus vaccine. Antiviral Ther 12(7):1107
55. Zhang N, Li C, Hu Y, Li K, Liang J, Wang L et al (2020) Current development of COVID-19 diagnostics, vaccines and therapeutics. Microbes Infect 22:231–235
56. Du L, Yang Y, Zhou Y, Lu L, Li F, Jiang S (2017) MERS-CoV spike protein: a key target for antivirals. Exp Opin Ther Targets 21(2):131–143
57. Wang Q, Wong G, Lu G, Yan J, Gao GF (2016) MERS-CoV spike protein: targets for vaccines and therapeutics. Antiviral Res 133:165–177
58. Du L, Tai W, Yang Y, Zhao G, Zhu Q, Sun S et al (2016) Introduction of neutralizing immunogenicity index to the rational design of MERS coronavirus subunit vaccines. Nature Commun 7(1):1–9

59. Biopharmaceuticals C (2020) Clover initiates development of recombinant subunit-trimer vaccine for wuhan coronavirus (2019-ncov)
60. Tai W, He L, Zhang X, Pu J, Voronin D, Jiang S et al (2020) Characterization of the receptor-binding domain (RBD) of 2019 novel coronavirus: implication for development of RBD protein as a viral attachment inhibitor and vaccine. Cell Mol Immunol 17(6):613–620
61. Chen W-H, Strych U, Hotez PJ, Bottazzi ME (2020) The SARS-CoV-2 vaccine pipeline: an overview. Curr Trop Med Rep 7:61–64
62. Wang J, Li P, Yu Y, Fu Y, Jiang H, Lu M et al (2020) Pulmonary surfactant–biomimetic nanoparticles potentiate heterosubtypic influenza immunity. Science 367(6480):eaau0810
63. Cheung E (2020) China coronavirus: Hong Kong researchers have already developed vaccine but need time to test it, expert reveals. South China Morning Post
64. Shieber J (2020) Codagenix raises $20 million for a new flu vaccine and other therapies. Tech Crunch
65. Le TT, Andreadakis Z, Kumar A, Roman RG, Tollefsen S, Saville M et al (2020) The COVID-19 vaccine development landscape. Nat Rev Drug Discov 19(5):305–306
66. Goodridge HS, Ahmed SS, Curtis N, Kollmann TR, Levy O, Netea MG et al (2016) Harnessing the beneficial heterologous effects of vaccination. Nat Rev Immunol 16(6):392–400
67. Yusibov V, Rabindran S (2008) Recent progress in the development of plant derived vaccines. Exp Rev Vaccines 7(8):1173–1183
68. Fischer R, Buyel JF (2020) Molecular farming–the slope of enlightenment. Biotechnol Adv 40:107519
69. Tekoah Y, Shulman A, Kizhner T, Ruderfer I, Fux L, Nataf Y et al (2015) Large-scale production of pharmaceutical proteins in plant cell culture—the protalix experience. Plant Biotechnol J 13(8):1199–1208
70. Salazar-González JA, Bañuelos-Hernández B, Rosales-Mendoza S (2015) Current status of viral expression systems in plants and perspectives for oral vaccines development. Plant Mol Biol 87(3):203–217
71. Rosales-Mendoza S (2014) Genetically engineered plants as a source of vaccines against wide spread diseases. Springer, New York
72. McNulty MJ, Gleba Y, Tusé D, Hahn-Löbmann S, Giritch A, Nandi S et al (2020) Techno-economic analysis of a plant-based platform for manufacturing antimicrobial proteins for food safety. Biotechnol Prog 36(1):e2896
73. Peyret H, Brown JK, Lomonossoff GP (2019) Improving plant transient expression through the rational design of synthetic 5′ and 3′ untranslated regions. Plant Methods 15(1):108
74. Pogrebnyak N, Golovkin M, Andrianov V, Spitsin S, Smirnov Y, Egolf R et al (2005) Severe acute respiratory syndrome (SARS) S protein production in plants: development of recombinant vaccine. Proc Natl Acad Sci 102(25):9062–9067
75. Li H-Y, Ramalingam S, Chye M-L (2006) Accumulation of recombinant SARS-CoV spike protein in plant cytosol and chloroplasts indicate potential for development of plant-derived oral vaccines. Exp Biol Med 231(8):1346–1352
76. Demurtas OC, Massa S, Illiano E, De Martinis D, Chan PK, Di Bonito P et al (2016) Antigen production in plant to tackle infectious diseases flare up: the case of SARS. Front Plant Sci 7:54
77. Gretebeck LM, Subbarao K (2015) Animal models for SARS and MERS coronaviruses. Curr Opin Virol 13:123–129
78. Martina BE, Haagmans BL, Kuiken T, Fouchier RA, Rimmelzwaan GF, Van Amerongen G et al (2003) SARS virus infection of cats and ferrets. Nature 425(6961):915–915
79. Dhama K, Sharun K, Tiwari R, Dadar M, Malik YS, Singh KP et al (2020) COVID-19, an emerging coronavirus infection: advances and prospects in designing and developing vaccines, immunotherapeutics, and therapeutics. Human Vaccines Immunother 16:1232–1238
80. Munster VJ, De Wit E, Feldmann H (2013) Pneumonia from human coronavirus in a macaque model. N Engl J Med 368(16):1560

81. Falzarano D, de Wit E, Feldmann F, Rasmussen AL, Okumura A, Peng X et al (2014) Infection with MERS-CoV causes lethal pneumonia in the common marmoset. PLoS Pathog 10(8):e1004250
82. Roberts A, Lamirande EW, Vogel L, Jackson JP, Paddock CD, Guarner J et al (2008) Animal models and vaccines for SARS-CoV infection. Virus Res 133(1):20–32
83. Menachery VD, Gralinski LE, Mitchell HD, Dinnon KH, Leist SR, Yount BL et al (2018) Combination attenuation offers strategy for live attenuated coronavirus vaccines. J Virol 92(17):e00710
84. Bao L, Deng W, Huang B, Gao H, Liu J, Ren L et al (2020) The pathogenicity of SARS-CoV-2 in hACE2 transgenic mice. Nature 583:830–833
85. Chan JF-W, Zhang AJ, Yuan S, Poon VK-M, Chan CC-S, Lee AC-Y et al (2020) Simulation of the clinical and pathological manifestations of Coronavirus Disease 2019 (COVID-19) in golden Syrian hamster model: implications for disease pathogenesis and transmissibility. Clin Infect Dis
86. Kim Y-I, Kim S-G, Kim S-M, Kim E-H, Park S-J, Yu K-M et al (2020) Infection and rapid transmission of SARS-CoV-2 in ferrets. Cell Host Microbe 27:7014–709
87. Chan JF-W, Zhang AJ, Yuan S, Poon VK-M, Chan CC-S, Lee AC-Y et al (2020) Simulation of the clinical and pathological manifestations of Coronavirus Disease 2019 (COVID-19) in golden Syrian hamster model: implications for disease pathogenesis and transmissibility. Clin Infect Dis 71:2428–2446
88. Rockx B, Kuiken T, Herfst S, Bestebroer T, Lamers MM, Munnink BBO et al (2020) Comparative pathogenesis of COVID-19, MERS, and SARS in a nonhuman primate model. Science 368(6494):1012–1015
89. Yu J, Tostanoski LH, Peter L, Mercado NB, McMahan K, Mahrokhian SH et al (2020) DNA vaccine protection against SARS-CoV-2 in rhesus macaques. Science 369:806–811
90. van Doremalen N, Lambe T, Spencer A, Belij-Rammerstorfer S, Purushotham JN, Port JR et al (2020) ChAdOx1 nCoV-19 vaccination prevents SARS-CoV-2 pneumonia in rhesus macaques. Nature 586:578–582
91. Chandrashekar A, Liu J, Martinot AJ, McMahan K, Mercado NB, Peter L et al (2020) SARS-CoV-2 infection protects against rechallenge in rhesus macaques. Science 369:812
92. Zhou Y, Jiang S, Du L (2018) Prospects for a MERS-CoV spike vaccine. Expert Rev Vaccines 17(8):677–686
93. Cockrell AS, Yount BL, Scobey T, Jensen K, Douglas M, Beall A et al (2016) A mouse model for MERS coronavirus-induced acute respiratory distress syndrome. Nat Microbiol 2(2):1–11
94. Leist SR, Cockrell AS (2020) Genetically engineering a susceptible mouse model for MERS-CoV-induced acute respiratory distress syndrome. In: MERS coronavirus. Springer, Cham, pp 137–159
95. Yong CY, Ong HK, Yeap SK, Ho KL, Tan WS (2019) Recent advances in the vaccine development against Middle East respiratory syndrome-coronavirus. Front Microbiol 10:1781
96. Eckerle I, Corman VM, Müller MA, Lenk M, Ulrich RG, Drosten C (2014) Replicative capacity of MERS coronavirus in livestock cell lines. Emerg Infect Dis 20(2):276
97. Milewska A, Nowak P, Owczarek K, Szczepanski A, Zarebski M, Hoang A et al (2018) Entry of human coronavirus NL63 into the cell. J Virol 92(3):e01933
98. Fukushi S, Mizutani T, Saijo M, Kurane I, Taguchi F, Tashiro M et al (2006) Evaluation of a novel vesicular stomatitis virus pseudotype-based assay for detection of neutralizing antibody responses to SARS-CoV. J Med Virol 78(12):1509–1512
99. Wang M, Cao R, Zhang L, Yang X, Liu J, Xu M et al (2020) Remdesivir and chloroquine effectively inhibit the recently emerged novel coronavirus (2019-nCoV) in vitro. Cell Res 30(3):269–271
100. Kilianski A, Baker SC (2014) Cell-based antiviral screening against coronaviruses: developing virus-specific and broad-spectrum inhibitors. Antiviral Res 101:105–112

101. Matsuyama S, Nao N, Shirato K, Kawase M, Saito S, Takayama I et al (2020) Enhanced isolation of SARS-CoV-2 by TMPRSS2-expressing cells. Proc Natl Acad Sci 117(13):7001–7003
102. Kim J-M, Chung Y-S, Jo HJ, Lee N-J, Kim MS, Woo SH et al (2020) Identification of coronavirus isolated from a patient in Korea with COVID-19. Osong Public Health Res Perspect 11(1):3
103. Ou X, Liu Y, Lei X, Li P, Mi D, Ren L et al (2020) Characterization of spike glycoprotein of SARS-CoV-2 on virus entry and its immune cross-reactivity with SARS-CoV. Nat Commun 11(1):1–12
104. Harcourt J, Tamin A, Lu X, Kamili S, Sakthivel SK, Murray J et al (2020) Isolation and characterization of SARS-CoV-2 from the first US COVID-19 patient. BioRxiv: the preprint server for biology 2020.03.02.972935. 7 Mar. 2020
105. McMichael T (2020) Epidemiology of Covid-19 in long-term facility in King County, Washington New England. J Med 382:2005–2011
106. Kissler SM, Tedijanto C, Goldstein E, Grad YH, Lipsitch M (2020) Projecting the transmission dynamics of SARS-CoV-2 through the postpandemic period. Science 368(6493):860–868
107. Liu W, Fontanet A, Zhang P-H, Zhan L, Xin Z-T, Baril L et al (2006) Two-year prospective study of the humoral immune response of patients with severe acute respiratory syndrome. J Infect Dis 193(6):792–795
108. Corey L, Mascola JR, Fauci AS, Collins FS (2020) A strategic approach to COVID-19 vaccine R&D. Science 368(6494):948–950
109. Shah SK, Miller FG, Darton TC, Duenas D, Emerson C, Lynch HF et al (2020) Ethics of controlled human infection to address COVID-19. Science 368(6493):832–834
110. Wang C, Horby PW, Hayden FG, Gao GF (2020) A novel coronavirus outbreak of global health concern. The Lancet 395(10223):470–473

Chapter 8
Updates in Vaccine Development Against COVID-19

Sagheer Ahmed, Rehan Salar, Halimur Rehman, Sarah I. Alothman, M. Riaz, and Muhammad Zia-Ul-Haq

Introduction

As the world is passing through COVID-19 pandemic and many international organizations are joining hands to lead the response against the virus, the efforts to find a safe and effective vaccine are intensifying. Currently, no approved vaccines exist to prevent infection with SARS-CoV-2. Therefore, finding a safe and effective vaccine to prevent infection with SARS-CoV-2 is an urgent public health priority. Vaccines are very effective in developing immunity in the body against bacteria or viruses or any other foreign proteins against which they are directed. Therefore, when these foreign organisms later infect human body, it recognizes it and can efficiently neutralize it. Every year, vaccines save many million lives. Vaccines are available to protect against more than 20 deadly diseases, and the ones available against influenza, diphtheria, pertussis, tetanus, and measles are saving millions of

---

S. Ahmed (✉)
Shifa College of Pharmaceutical Sciences, Shifa Tameer-e-Millat University, Islamabad, Pakistan
e-mail: sagheer.scps@stmu.edu.pk

R. Salar · H. Rehman
Shifa International Hospital, Islamabad, Pakistan

S. I. Alothman
Faculty of Science, Biology Department, Princess Nourah bint Abdulrahman University, Riyadh, Saudi Arabia

M. Riaz
Department of Pharmacy, Shaheed Benazir Bhutto University, Sheringal, Dir Upper, Khyber Pakhtunkhwa, Pakistan

M. Zia-Ul-Haq
Office of Research, Innovation and Commercialization, Lahore College for Women University, Lahore, Pakistan

© The Author(s), under exclusive license to Springer Nature Switzerland AG 2021
M. Zia-Ul-Haq et al. (eds.), *Alternative Medicine Interventions for COVID-19*, https://doi.org/10.1007/978-3-030-67989-7_8

life every year. The efforts are being made at unprecedented pace to make a vaccine that could prevent the infection of SARS-COV2.

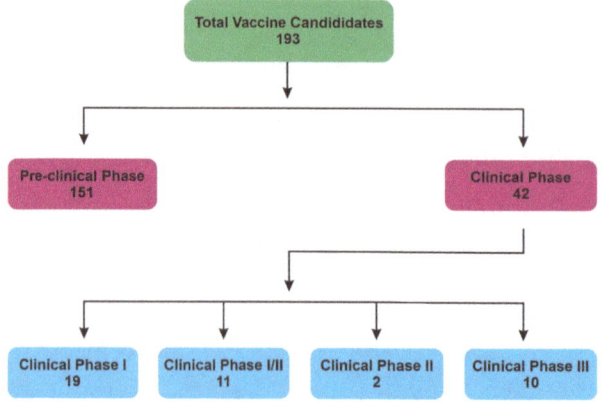

Sinovac Biotech, China

An inactivated virus vaccine manufactured by Sinovac Biotech, a Chinese biotechnology and pharmaceutical company, is being investigated in humans. The vaccine is called CoronaVac. A randomized clinical trial sponsored by the Butantan Institute is currently underway. The official title of the trial is, "Double-Blind, Randomized, Placebo-Controlled Phase III Clinical Trial to Evaluate Efficacy and Safety in Healthcare Professionals of the Adsorbed COVID-19 (Inactivated) Vaccine Manufactured by Sinovac." The study was started on July 21, 2020, and was expected to be complete by September 21, 2020 [1]. However, the trial is still recruiting and now expected to be complete by October or November 2020. CoronaVac is one of four vaccines, developed by Chinese companies, in various clinical trials.

In this clinical trial, which is a Phase III trial, CoronaVac is being assessed for its efficacy and safety. CoronaVac is an adsorbed COVID-19 (inactivated) vaccine. The primary purpose of this trial is the prevention of recruited population from COVIC-19 infection. This trial is estimated to recruit about 8870 participants. In this trial, equal number of patients will be recruited to test and control arms. Patients in the test group will receive the CoronaVac vaccine, while patients recruited in the control arm will be given placebo. The dosing protocol involves administering two doses of CoronaVac through intramuscular route, 14 days apart. Efficacy of the vaccine will be measured by detecting symptomatic COVID-19 cases in the second week after administration of the vaccine. Patients will be divided into two categories for evaluation safety of the vaccine. Adverse effects will be measured separately in the two groups, adults (18–59 years) and elderly (60 years and above). Follow-up period will be 1 year. The trial intends to detect adverse effects whose frequency is

1 in 1000 or higher for adult group and 1 in 500 in elderly patients' group. After reaching a target number of 150 patients, an interim preliminary analysis will be carried out to find out the efficacy of the vaccine.

The Chinese company which has manufactured this vaccine announces that CoronaVac will be available to be distributed throughout the world by early 2021. The company is also interested in distributing CoronaVac in the USA, for which they will require to apply to US Food and Drug Administration (FDA). The company has recently shown its intention to apply to the US FDA, once Phase III clinical trials of the vaccine are complete. The CEO of the company, Mr. Yin Weidong, is confident that they will succeed in getting the approval from the US FDA. According to him, the company was China-centric initially, but then it modified its strategy significantly in June and July 2020 and started focusing on other parts of the world as well. Now the company intends to distribute its vaccine to EU, Australia, South America, and the USA, among others. The CEO of Sinovac is confident that the quality of their product and clinical data would be sufficient to clear regulatory and safety hurdles in these countries which have historically blocked Chinese vaccines. The company is convinced that all this is to change.

1. *Wuhan Institute of Biological Products/Sinopharm, China*

    Wuhan Institute of Biological Products Co., LTD, and Sinopharm are the manufacture of this vaccine, called Vero cells. This vaccine is being tested in a Phase III trial. It is a vaccine made from inactivated novel coronavirus (nCoV). The official title of the clinical trial is, "Randomized, Double Blind, Parallel Placebo Controlled, Phase III Clinical Trial to Evaluate the Safety and Protective Efficacy of Inactivated SARS-CoV-2 Vaccine in Healthy Population Aged 18 Years and Above." This trial is being conducted in the United Arab Emirates and is approved by the Abu Dhabi Health COVID-19 Research Ethics Committee [2].

    The trial intends to establish safety and efficacy of the vaccine under investigation by evaluating it in healthy participants. This is placebo-controlled clinical trial in which one group of participants will receive investigational vaccine and the other group will receive placebo, serving as a control arm. Two doses of the vaccine or placebo will be administered. There will be 21 days interval between the first and the second dose. The estimated number of participants in each group is 5000. The primary outcome of the trial is to test the protective ability of this vaccine against COVID-19 infection at least 14 days after vaccination.

    Secondary outcomes of the trial include evaluation of preventive potential of two doses of vaccine against pneumonia and death caused by SARS-COV-2 after 14 days of the last dose; incidence of adverse effects inside 30 minutes of the vaccination and then between 0–7, 8–21, and 28 days after vaccination; incidence of severe adverse effects after 12 months of administration of the last vaccine dose; and measuring COVID-19 antibody levels in vaccinated individuals after complete vaccination.

    China is moving fast in the global completion to develop an effective vaccine to combat SARS-COV-2. It was recently announced that several Chinese citizens have already received this vaccine although the phase III clinical trial is still underway. These individuals were vaccinated under the government-authorized, emergency use program. The provision of such an emergency use of vaccine is available in Chinese government regulations, although their scope and timing are restricted to a health emergency. It was also announced by the Chinese government officials that high-risk individuals will be given priority, implying frontline healthcare workers and elderly would be the first groups to receive the experimental vaccine.

2. *CanSino Biological Inc./Beijing Institute of Biotechnology, China*

CanSino Biological Inc. and its collaborator Beijing Institute of Biotechnology are sponsoring a vaccine candidate, Ad5-nCoV, which is in Phase III clinical trial to assess its safety, efficacy, and immunogenicity. Ad5-nCoV vaccine is manufactured by CanSino Biological Inc. and its collaborator Beijing Institute of Biotechnology to be tested in health adults aged 18 years old and above. The official title of the trial is, "A Global Multicenter, Randomized, Double-blind, Placebo-Controlled, Adaptive Designed Phase III Clinical Trial to Evaluate the Efficacy, Safety and Immunogenicity of Ad5-nCoV in Adults 18 Years of Age and Older." The actual starting date of the trial is September 15, 2020, and estimated completion date is January 30, 2022. The study is aimed at enrolling an estimated 40,000 participants, of which 20,000 will be in experimental group and remaining 20,000 in the placebo arm of the study. Ad5-nCoV is a single dose vaccine and given through intramuscular route of administration [3, 4]. The Phase III trial recruiting locations include Shaukat Khanum Memorial Cancer Hospital and Research Center, Lahore, and Shifa International Hospital, Islamabad.

The efficacy of Ad5-nCoV is determined by measuring its ability to prevent COVID-19 disease. Therefore, the primary outcome measures of the phase III trial include prevention of COVID-19 cases from day 28 to 12 months postvaccination. Timeframe for evaluating severe adverse event is 12 months. Secondary outcome measures include incidence of severe COVID-19 cases from day 14 to 12 months postvaccination. The trial will also evaluate the efficacy of Ad5-nCoV in preventing severe COVID-19 caused by SARS-CoV-2 infection. Incidence of solicited adverse reaction within day 0 to day 7 postvaccination will also be measured as well as the incidence of unsolicited adverse reaction with in day 0 to day 28 postvaccination. After 28 days of postvaccination, immunogenicity of S-RBD IgG antibody through ELISA method will be determined. The seroconversion rate of S-RBD IgG antibody after vaccination and cell-mediated immune response profile from day 28 postvaccination will be measured as secondary outcomes.

3. *University of Oxford/AstraZeneca, UK*

AZD1222 candidate vaccine was co-invented by the University of Oxford and AstraZeneca. It is being manufactured by AstraZeneca for using in clinical trials. AZD1222 is a nonreplicating viral vector vaccine. The official title of the currently undergoing advanced clinical trial is, "A Phase III Randomized, Double-Blind, and Placebo-Controlled Multicenter Study in Adults to Determine the

Safety, Efficacy, and Immunogenicity of AZD1222, a Nonreplicating ChAdOx1 Vector Vaccine, for the Prevention of COVID-19" [5–8].

This clinical trial started in August 2020, and its estimated primary completion date is December 2, 2020. The trail is estimated to enroll approximately 30,000 participants. The eligibility criteria for inclusion in the trial demand for individuals older than 18 year of age, both male and female. The study objective is to assess the protective efficacy of inactivated SARS-CoV-2 vaccine (Vero cell) after complete course of immunization in preventing disease caused by SARS-CoV-2 in healthy subjects of 18 years and above. The vaccine consists of a single dose given intramuscular route. The primary outcome of the trial is to test the protecting ability of this vaccine against COVID-19 infection at least 14 days after vaccination. Secondary outcomes include measuring growth rate and antibody level (GMT, GMI) of serum antibody against COVID-19 during 12 months from first to last vaccination schedule.

4. *Janssen Pharmaceutica, USA*

Janssen Vaccines & Prevention BV is sponsoring the Phase III clinical trial of adenovirus serotype 26 vector-based vaccine (Ad26.COV2.S). It is a nonreplicating viral vector vaccine (which uses live viruses to carry DNA into human cells), also known as JNJ-78436735. This viral vector vaccine is being assessed for its efficacy in the prevention of molecularly confirmed moderate to severe/critical coronavirus infection. The study is officially named as "A Randomized, Double-blind, Placebo-Controlled Phase 3 Study to Assess the Efficacy and Safety of Ad26.COV2.S for the Prevention of SARS-CoV-2-Mediated COVID-19 in Adults Aged 18 Years and Older." The trial was initiated on September 7, 2020, and is expected to complete by March 10, 2023, with an approximated enrollment of 60,000 participants [9].

This interventional study will compare the efficacy of *Ad26.COV2.S* to a placebo, in preventing molecularly confirmed moderate to severe/critical COVID-19, among adult participants, with or without stable comorbidities which are related to the progression of coronavirus infection. The participants in the experimental arm will receive a single intramuscular (IM) injection of *Ad26.COV2.S* at a dose level of $5*10^{10}$ virus particles (vp) on day 1, whereas participants in the control arm will receive an IM injection of placebo on day 1.

Primary endpoint measures include the number of participants with the first incidence of molecularly confirmed moderate to severe coronavirus infection having seronegative status. The secondary endpoint measures will be including the number of participants with molecularly confirmed coronavirus infection regardless of serostatus, participants requiring medical intervention, and assessment of SARS-CoV-2 viral load by quantitative reverse-transcriptase PCR. Other secondary outcomes consist of finding the number of participants with molecularly confirmed mild infection as per FDA harmonized case definition, burden of disease (BOD) based on symptomatic coronavirus infection, seroconversion between pre- and postvaccination using ELISA, and the number of participants with serious local and systemic adverse events. Secondary endpoint measures will also include assessment of SARS-CoV-2 neutralizing antibody titers by virus neutralization assay (VNA) and SARS-CoV-2 binding antibodies by ELISA.

Earlier this year, Johnson & Johnson (J&J) and Janssen Pharmaceutical research panel, in collaboration with Beth Israel Deaconess Medical Center, produced and analyzed multiple vaccine candidates using the Janssen AdVac® technology and selected its lead vaccine candidate *Ad26.COV2-S* for the preclusion of coronavirus infection. On June 10, 2020, J&J announced that in Phase I/IIa of clinical trial, the safety, reactogenicity, and immunogenicity of adenovirus serotype 26 (Ad26) vector-based vaccine will be assessed. Data published in Nature on July 30, 2020, reported that the investigational vaccine elicited a robust immune response and successfully averted the subsequent infection and in the preclinical study provided protection in the lungs from the virus in nonhuman primates (NHPs). The study published in Nature on September 3, 2020, reported high-dose intranasal SARS-CoV-2 infection in hamsters caused severe clinical disease, high levels of virus replication, extensive pneumonia, weight loss, and mortality. A single immunization with Ad26.COV2.S protected against SARS-CoV-2-induced pneumonia, weight loss, and mortality.

5. *Moderna/NIAID, USA*

Moderna, Inc., a clinical-stage biotechnology company in the USA, is sponsoring Phase III clinical trial of its mRNA vaccine candidate *mRNA-1273* against COVID-19. The trial will be primarily evaluating the efficacy, safety, and immunogenicity of *mRNA-1273* to prevent coronavirus infection up to 2 years after the second dose of the investigational vaccine. With an estimated enrollment of 30,000 participants, this interventional study is officially named as "A Phase 3, Randomized, Stratified, Observer-Blind, Placebo-Controlled Study to Evaluate the Efficacy, Safety, and Immunogenicity of mRNA-1273 SARS-CoV-2 Vaccine in Adults Aged 18 Years and Older." It was started on July 27, 2020, and is expected to be completed by October 27, 2022. Moderna's vaccine candidate (mRNA-1273) comprises nucleoside-modified messenger RNA (mRNA) which is lipid nanoparticle encapsulated and predetermines the stabilization of SARS-CoV-2 spike (S) glycoprotein in its prefusion conformation. This S glycoprotein is essential for viral entry as it mediates host cell attachment.

The randomized, placebo-controlled trial is being conducted in the USA, testing *mRNA-1273* where participants will be receiving 1 intramuscular (IM) injection of 100 micrograms (ug) *mRNA-1273* on day 1 and on day 29 in the experimental arm and 0.9% sodium chloride (normal saline) injection in the placebo comparator arm [10]. The primary outcome will be the preclusion of symptomatic COVID-19. Key secondary outcomes include the prevention of severe COVID-19 disease (as defined by the need for hospitalization) and prevention of infection by SARS-CoV-2 regardless of symptomology. Based on the number of participants with symptomatic COVID-19, the primary efficacy investigation during Phase III study will be an event-driven analysis. To ensure the ongoing safety monitoring of the participants in the trial, data will be reviewed by an independent Data and Safety Monitoring Board organized by NIAID throughout the study.

In January 2020, Moderna announced the development of this vaccine (named *mRNA-1273*) against coronavirus infection. The Phase I human study of the vaccine candidate began in March 2020, in partnership with the US National Institute of Allergy and Infectious Diseases. On May 25, 2020, Moderna began a Phase IIa clinical trial recruiting 600 adult participants to assess safety and differences in antibody response to two doses of its candidate vaccine, *mRNA-1273*. On July 14, 2020, Moderna published preliminary results of the dose escalation during the Phase I clinical trial of the vaccine, exhibiting a dose-dependent induction of neutralizing antibodies against S1/S2, 15 days postinjection. Reported adverse reactions were mild to moderate, such as fever, fatigue, headache, muscle ache, and pain at the injection site in all dose groups, as well as the association of severity with increased dosage. Comparatively low doses were deemed safe and effective in order to proceed with a Phase III clinical trial utilizing two 100-μg doses administered with 29 days interval. A detailed study plan for the clinical trial was published by Moderna in September 2020. On September 30, CEO Stéphane Bancel said that, if the trial is successful, the vaccine might be available to the public as early as late March or early April 2021.

6. *Novavax, USA*

Novavax, Inc., USA, initiated the Phase III clinical trial of its SARS-CoV-2 recombinant spike protein nanoparticle vaccine (SARS-CoV-2 rS) also known as NVX-CoV2373 in September 2020. This interventional study is officially named as "A Phase 3, Randomized, Observer-Blinded, Placebo-Controlled Trial to Evaluate the Efficacy and Safety of a SARS-CoV-2 Recombinant Spike Protein Nanoparticle Vaccine (SARS-CoV-2 rS) with Matrix-M1™ Adjuvant in Adult Participants 18–84 Years of Age in the United Kingdom." NVX-CoV2373 consists of Matrix-M1 adjuvant which improves immunogenicity and a recombinant SARS-CoV-2 nanoparticle vaccine, constructed from the SARS-CoV-2 spike glycoprotein, which plays a role in human angiotensin-converting enzyme 2 (hACE2) receptor binding of the virus. In this latest phase of the trial, this vaccine will be assessed for its efficacy, safety, and immunogenicity against coronavirus infection in approximately 10,000 participants, aged 18–84 years [11]. The trial is aimed to register at least 25 percent of participants older than 65 years of age and will also prioritize groups that are most affected by coronavirus infection, including racial and ethnic minorities.

Participants in the experimental arm of the clinical trial will receive two doses of an intramuscular vaccine comprising 5 μg of protein antigen with 50 μg Matrix-M1™ adjuvant, administered with 21 days interval, while participants in the control arm of the trial will receive a placebo. The trial has two primary outcomes, including the first incidence of PCR-confirmed symptomatic coronavirus infection with onset at least 7 days after the second dose in participants who have not been previously infected with SARS-CoV-2 as the first primary outcome and the first incidence of PCR-confirmed symptomatic moderate or severe coronavirus infection with onset at least 7 days after the second dose in participants who have not been previously infected with SARS-CoV-2 as the second primary outcome. The primary efficacy results will be based on the number of participants with symptomatic or moderate/severe coronavirus infection by an event-driven analysis. After reaching 67% of the desired number of these cases, an interim analysis will be carried out.

This ongoing Phase III trial is based on the Novavax's NVXCoV2373 candidate, engineered from the genetic sequence of SARSCoV2, the virus that causes coronavirus infection. This candidate vaccine was developed by Novavax in January 2020 using its recombinant nanoparticle technology to produce antigen derived from the coronavirus spike (S) protein and also consists of Novavax's patented saponin-based Matrix-M™ adjuvant. It is intended to boost the immune response and stimulate the production of neutralizing antibodies. In May 2020, the first human safety trial of the vaccine was started in Australia. During preclinical trials, NVXCoV2373 demonstrated its critical aspect of efficacy, i.e., to stimulate antibody production that blocks the binding of spike protein to receptors (virus targets). NVXCoV2373 was reported to be well tolerated and elicited robust antibody responses in its Phase I part of its Phase I/II clinical trial. Phase II trials of NVX-CoV2373 which began in South Africa in August and continued in the USA and Australia also evaluated the immunogenicity of the candidate vaccine. Research and development president at Novavax Gregory Glenn expects quick enrollment of phase III clinical trial and hopes to provide the near-term view efficacy data from the study soon. According to him, early data showing promising results are expected to support regulatory submissions for licensure in many countries across the globe including the UK, EU, and USA. The vaccine has a favorable product profile that will allow handling in an unfrozen, liquid formulation that can be stored at 2–8 °C, allowing for distribution using standard vaccine channels. Novavax hopes to manufacture up to 2 billion annualized doses, once all capacity has been brought online by mid-2021.

7. *BioNTech SE and Pfizer Inc. USA*

BioNTech SE in collaboration with Pfizer Inc. has developed a vaccine BNT162b2 which is a nucleoside-modified messenger RNA (modRNA) that expresses the SARS-CoV-2 spike glycoprotein, for strong antiviral effects against SARS-CoV-2 infection. The candidate vaccine is at the investigational stage, and the study is officially named as "A Phase 1/2/3, Placebo-Controlled, Randomized, Observer-Blind, Dose-Finding Study to Evaluate the Safety, Tolerability, Immunogenicity, and Efficacy of SARS-CoV-2 RNA Vaccine Candidates Against COVID-19 in Healthy Individuals" [12]. The study was started on April 29, 2020, and expected to be complete by December 11, 2022. It comprises of two phases: the first identifies the preferred vaccine candidate (by assessing the safety, tolerability, and immunogenicity) and its dosage, and the second will be an expanded cohort to evaluate its efficacy.

After Phase I evaluation of a two-dose schedule (21 days apart) in specified age groups, the candidate vaccine BNT162b2 is being further assessed for safety and efficacy in Phase II/III. The vaccinated participants exhibited an observable quantity of epitopes that are specific to the SARS-CoV-2 spike antigen, recognized in T cell responses. The stimulation of high levels of CD4+ and CD8+ T cell-mediated immunity against the receptor-binding domain (RBD) and the remainder of the spike glycoprotein was also observed. It was well tolerated with mild to moderate fever in less than 20% of participants in all populations. This data supports the decision of BNT162b2 proceeding for Phase II/III, as a two-dose regimen of 30 µg dose level in approximately 30,000 participants that started in July 2020.

Kathrin U. Jansen, senior vice president and head of vaccine research and development, Pfizer, reported these encouraging results after the Phase I safety and immunogenicity data, and her confidence about the potential of the candidate vaccine *BNT162b2* to prevent many millions of SARS-CoV-2 infection cases was high. Ugur Sahin, M.D., CEO, and cofounder of BioNTech, also reported *BNT162b2* dosing completion of more than 11,000 participants and findings of previous trial exhibiting its safety profile and breadth of T cell responses and the progression toward emerging of a safe and effective vaccine candidate. Anticipating clinical success, Pfizer and BioNTech are on way to seek regulatory review of the vaccine candidate BNT162b2 as early as October 2020, and after regulatory approval they plan to provide approximately 100 million doses by the end of 2020 and approximately 1.3 billion doses by the end of 2021 worldwide.

8. *Beijing Institute of Biological Products/Sinopharm, China*

This vaccine is a joint venture of Beijing Institute of Biological Products Co., Ltd., China National Biotec Group Company Limited, and the Huesped Foundation and also sponsored by the Laboratorio Elea Phoenix S.A. They are investigating an inactivated virus vaccine in a Phase III clinical trial in Argentinian population. An inactivated virus vaccine consists of a virus that is obtained by culture and then inactivating it by chemical methods or by simply heating it. Such a virus retains its antigenic potential and can stimulate human immune system but loses its ability to cause the disease in human body. This vaccine (called Vero cells) is an inactivated SARS-COV-2 vaccine which was obtained by infecting African green monkey kidney cells with SARS-COV-2 strain HB02. These cells were then propagated through culturing and then harvested to be inactivated. After that they were concentrated and purified and an adjuvant (aluminum hydroxide) added to it. Volunteers who are administered this vaccine are expected to produce an immune response against SARS-COV-2. This vaccine has already showed in the phase I/II trials that it can produce high levels of antibodies against SARS-COV-2 and is safe.

The clinical trial to test this vaccine is an interventional study aimed to recruit approximately 3000 volunteers in a randomized control trial in which the interventional model is parallel assignment of candidate vaccine and placebo to two separate groups of volunteers. It is a double-blinded study in which both the participant and the physician will be unaware of whether they are in vaccine or placebo group. The official title of the study is "A Randomized, Double Blind, Placebo Parallel-Controlled Phase III Clinical Trial to Evaluate the Efficacy, Immunogenicity and Safety of the Inactivated SARS-COV-2 Vaccine (Vero Cell) in Argentine Healthy Population Aged Between 18 and 85 Years" [13]. The study had started recruiting volunteers on September 16, 2020, and is expected to be complete by December 2021.

9. *Gamaleya Research Institute, Russia*

The Gamaleya Research Institute of Epidemiology and Microbiology, Health Ministry of the Russian Federation, is sponsoring this vaccine called Gam-COVID-Vac in collaboration with the government of the city of Moscow and a contact research organization, Crocus Medical BV. This vaccine is currently being investigated for its immunogenicity, safety, and efficacy in a randomized controlled, Phase III trial in which both the physician and the study participants would not know whether a participant is receiving a placebo or test vaccine. The official title of the study is "A Clinical Trial of Efficacy, Safety and Immunogenicity of Combined Vector Vaccine Gam-COVID-Vac in SARS-COV-2 Infection Prophylactic Treatment in Republic of Belarus" [14, 15].

In this study, for every participant in the placebo group, three participants will be included in the vaccine group. This means a randomization of 3 to 1 in which the reference or placebo group will contain about 10, 000 participants and the study group receiving the Gam-COVID-Vac combined vector vaccine against the SARS-CoV-2-induced coronavirus infection will have 30,000 participants. All participants will be above 18 years of age, and the total volunteers intended to be recruited for the study is 40,000. However, study participants will be further divided into five age groups: 18–30, 31–40, 41–50, 51–60, and 60+ years.

Each study participant will have one screening visit and five on-site visits to study physicians during a period of 180 ± 14 days period after the first dose of the placebo/vaccine. The participants will be administered placebo/vaccine intramuscularly during vaccination visits 1 and 2 which will be day 0 and day 21 ± 2, respectively. Participants will be asked to visit on days 28 ± 2, 42 ± 2, and 180 ± 14 for subsequent visits 3, 4, and 5, respectively. During these visits, participants' condition and well-being will be examined and recorded. Adverse effects, if any, will also be recorded. If physical visits were not possible due to unavoidable reasons, observations will be made through telemedicine consultation. Additional telemedicine consultation will be provided to all participants throughout the trial period.

In a recent press release, the head of the Gamaleya Research Institute of Epidemiology and Microbiology, Alexander Gintsburg, said that none of approximately 2000 volunteers inoculated so far with both portions of Russia's coronavirus vaccine (Gam-COVID-Vac) have contracted the disease. More data will come as more participants are recruited to the clinical trial. However, there has been global concern after Russia pushed ahead with mass vaccinations alongside randomized Phase III clinical trials of the vaccine. Some scientists have also criticized this move and insisted on prioritizing safety and efficacy of the vaccine as evidenced by the solid science rather than national prestige. However, Mr. Ginstburg defended the approach and contended that it is a war-like situation where people are dying every day like during a war and that they are not cutting corners as suggested by some media comp. He said their fast-paced approach seems alien but it is based on solid science.

Conclusions

It takes long years to develop a vaccine. The reasons it takes so long are myriad, but the first step is to investigate its safety and efficacy in animals. It takes about 6 months of strictly following animal care and handling guidelines and stringent laboratory protocols to complete preclinical studies. Once its safety and efficacy are established in animals, then the vaccine is investigated further in human clinical trials. The vaccine is required to be made at industrial scale for large-scale clinical trials. Although Phases I and II are relatively smaller, Phase III are much larger human trials often requiring tens of thousands of participants in each investigative group of the study. Safety of the vaccine is evaluated in Phase I trials, while efficacy is first established in Phase II trials. The safety and efficacy are evaluated on much larger scale in Phase III clinical trials, and it is the result of this phase which usually convinces the regulatory authority to approve or disapprove the vaccine. Although this entire process may be fast-tracked owing to the severity of the COVID-19 pandemic, safety and efficacy must not be compromised at any stage. Even after regulations are tweaked to fast-track the vaccine approval, it will be still unrealistic to expect a safe and effective vaccine within 6 months after the start of clinical trials. Additional challenges may be faced when a vaccine is approved for general use, especially for mass-producing it to meet the global demand. This would be more likely if scaling up the production involves new technologies that have not been tested previously.

Our experience with coronavirus vaccines has taught us that a safe and effective vaccine against SAR-SCOV-2 would be extremely challenging to develop. Some of the vaccines which have shown to enhance survival and decrease mortality in animals could not prevent infections in humans. Others may cause major complications such as lung damage. Therefore, thorough and long-term safety studies are necessary. Another challenge would be to ensure long-term protection against the virus. Establishing long-term immunity against the virus is relevant to COVID-19 as a significant fraction of the infected individuals are reinfected, albeit with milder symptoms. Therefore, immunity offered by the vaccine should last for many months and preferably for years. Finally, the vaccine should be effective in elderly population who are at increased risk of COVID-19 infection and whose immune system usually responds less well to vaccines than the younger population. We should ensure that the ideal vaccine must provide protection to this most vulnerable section of our population.

Acknowledgments The authors wish to thank Mr. Azhar Majeed for his help in preparing the graphical abstract for this manuscript.

Competing Interests The authors declare no competing interest.

FundingFunding for this work was provided by the Shifa Tameer-e-Millat University, Islamabad, Pakistan.

Bibliography

1. Clinical trial of efficacy and safety of Sinovac's adsorbed COVID-19 (inactivated) vaccine in healthcare professionals - full text view. ClinicalTrials.gov. https://clinicaltrials.gov/ct2/show/NCT04456595. Accessed 14 Oct 2020
2. Chinese Clinical Trial Register (ChiCTR) - The World Health Organization international clinical trials registered organization registered platform. http://www.chictr.org.cn/showprojen.aspx?proj=56651. Accessed 14 Oct 2020
3. Clinical trial of recombinant novel coronavirus vaccine (adenovirus type 5 vector) against COVID-19 - full text view - ClinicalTrials.gov. https://clinicaltrials.gov/ct2/show/NCT04540419. Accessed 14 Oct 2020
4. Phase III trial of A COVID-19 vaccine of adenovirus vector in adults 18 years old and above - full text view - ClinicalTrials.gov. https://clinicaltrials.gov/ct2/show/NCT04526990. Accessed 14 Oct 2020
5. CTRI. http://ctri.nic.in/Clinicaltrials/showallp.php?mid1=46186&EncHid=&userName=covid-19%20vaccine. Accessed 14 Oct 2020
6. AZD1222 vaccine for the prevention of COVID-19 - full text view - ClinicalTrials.gov. https://clinicaltrials.gov/ct2/show/NCT04540393. Accessed 14 Oct 2020
7. Phase III double-blind, placebo-controlled Study of AZD1222 for the prevention of COVID-19 in adults - full text view. ClinicalTrials.gov. https://clinicaltrials.gov/ct2/show/NCT04516746. Accessed 14 Oct 2020
8. ISRCTN - ISRCTN89951424: a phase III study to investigate a vaccine against COVID-19. https://doi.org/10.1186/ISRCTN89951424
9. Janssen Vaccines & Prevention B.V. (2020) A randomized, double-blind, placebo-controlled phase 3 study to assess the efficacy and safety of Ad26.COV2.S for the prevention of SARS-CoV-2-mediated COVID-19 in adults aged 18 years and older. clinicaltrials.gov
10. A study to evaluate efficacy, safety, and immunogenicity of mRNA-1273 vaccine in adults aged 18 years and older to prevent COVID-19 - full text view. ClinicalTrials.gov. https://clinicaltrials.gov/ct2/show/NCT04470427. Accessed 14 Oct 2020
11. Clinical trials register - Search for 2020–004123-16. https://www.clinicaltrialsregister.eu/ctr-search/search?query=2020-004123-16. Accessed 14 Oct 2020
12. Study to describe the safety, tolerability, immunogenicity, and efficacy of RNA vaccine candidates against COVID-19 in healthy individuals - full text view - ClinicalTrials.gov. https://clinicaltrials.gov/ct2/show/NCT04368728. Accessed 14 Oct 2020
13. Clinical trial to evaluate the efficacy, immunogenicity and safety of the inactivated SARS-CoV-2 vaccine (COVID-19) - full text view - ClinicalTrials.gov. https://clinicaltrials.gov/ct2/show/NCT04560881. Accessed 14 Oct 2020
14. Clinical trial of efficacy, safety, and immunogenicity of Gam-COVID-Vac vaccine against COVID-19 - full text view - ClinicalTrials.gov. https://clinicaltrials.gov/ct2/show/NCT04530396. Accessed 14 Oct 2020
15. Clinical trial of efficacy, safety, and immunogenicity of Gam-COVID-Vac vaccine against COVID-19 in Belarus - full text view - ClinicalTrials.gov. https://clinicaltrials.gov/ct2/show/NCT04564716. Accessed 14 Oct 2020

Chapter 9
COVID-19: Recent Developments in Therapeutic Approaches

Umar Farooq Gohar, Irfana Iqbal, Zinnia Shah, Hamid Mukhtar, and Muhammad Zia-Ul-Haq

Introduction

Coronaviruses (CoVs) are either pleomorphic or spherical particles which envelop a single positive-stranded RNA genome [1] which is the largest of all known RNA viral genomes [2]. The earliest known coronavirus infection was a severe respiratory infection (SRI) of chicken, reported in the mid-1930s, and was named infectious bronchitis virus (IBV). Human coronaviruses (HCoV) were discovered in and after the mid-1960s, when in 1965 Tyrrell and Bynoe described the presence of an enveloped virion with the feature morphology of previously defined IBV. They named it B814 and proceeded to grow it on tissue cultures at which they could not succeed at that time [3]. Two years later, in 1967, Almeida and Tyrrell established the similarities in morphology of B814 and IBV through electron microscopy of fluid from inoculated organ cultures. They found the particles to be enveloped within a membrane coating, to be pleomorphic in shape, and to have multiple crown-like structures attached to their envelope's surface projections [1]. These crown-like structures are the envelope glycoproteins. Subsequent studies reported more viruses (e.g., 229E, OC43) with a similar morphology to IBV and B814, and in 1968 these viruses were grouped under a new name of "coronavirus" (corona means crown-like), reflecting their characteristic halo- or crown-like surface projections. Before the 2000s, HCoVs were long considered as inconsequential pathogens [4]. Only two HCoVs, 229E [5, 6] and OC43 [7], were known since the 1960s to

U. F. Gohar (✉) · Z. Shah · H. Mukhtar
Institute of Industrial Biotechnology, Government College University,
Lahore, Pakistan
e-mail: dr.mufgohar@gcu.edu.pk

I. Iqbal · M. Zia-Ul-Haq
Office of Research, Innovation and Commercialization, Lahore College for Women University, Lahore, Pakistan

© The Author(s), under exclusive license to Springer Nature Switzerland AG 2021
M. Zia-Ul-Haq et al. (eds.), *Alternative Medicine Interventions for COVID-19*,
https://doi.org/10.1007/978-3-030-67989-7_9

occasionally ever cause severe illness in healthy human subjects [2]. Causing only the "common cold," these viruses never surfaced important enough to be explored. In the twenty-first century, however, two highly pathogenic HCoVs, responsible for severe acute respiratory syndrome (SARS) and Middle East respiratory syndrome (MERS), were identified for causing global pandemics in 2002 and 2012, respectively. These pandemics questioned the unexplored coronaviruses and their potential for causing future outbreaks. Genomic sequencing helped identify SARS-CoV [8] and MERS-CoV [9] as new but highly pathogenic human coronaviruses. Recently, a new strain of HCoV emerged as a local outbreak from a cluster of patients initially diagnosed with pneumonia of an unknown etiology. The infection now called COVID-19 steadily spread from a sea food market located in Wuhan, China, in December 2019 and rapidly transitioned into a global outbreak [10]. This is the third highly pathogenic human coronavirus discovered to date and is referred to as 2019-nCoV or SARS-CoV-2 (SC2). The name SARS-CoV-2 reflects its phylogenetic similarity with that of SARS-CoV [11].

Coronaviruses are taxonomically classified under the order *Nidovirales*, family *Coronaviridae*, by the ICTV: International Committee on Taxonomy of Viruses (Table 9.1). The coronavirus family is split into four genera: α-, β-, γ-, and δ-CoVs [2, 13] (Fig. 9.1). The genera α- and β- are found to be infectious toward mammals, while the other two are known to predominantly cause avian infections, but some γ- and δ-CoVs may also infect mammals [1]. There are seven known HCoVs, including SC2. Four of these are also called common HCoVs as people around the world get infected by these very often. The common HCoVs possess low pathogenicity and are responsible for mild infections – as that of "common cold." SARS-CoV, MERS-CoV, and SC2, unlike the common four, are highly pathogenic HCoVs with high mortality rates. *Betacoronaviruses* further have four lineages and SC2 belongs to lineage b of *Betacoronaviruses* – Fig. 9.2 [12]. SC2 is an alarmingly infectious virus which is capable of human-to-human transmission. Despite having a low mortality rate, strict follow-up of WHO safety guidelines is key to preventing its spread and, in worst-case scenario, its evolution into another novel coronavirus.

The evidence of interspecies transmissibility of coronaviruses was not clear to researchers until when a database of coronavirus gene sequences was made available. Table 9.2 shows lists of some coronaviruses and their hosts. Gene sequence

Table 9.1 Taxonomy of SARS-CoV-2

Lineage	b
Genus	*Betacoronavirus*
Subfamily	*Coronavirinae*
Family	*Coronaviridae*
Order	*Nidovirales*
Class	*Pisonivirecetes*
Phylum	*Pisuviricota*
Kingdom	*Orthornavirae*
Realm	*Ribovaria*

9 COVID-19: Recent Developments in Therapeutic Approaches

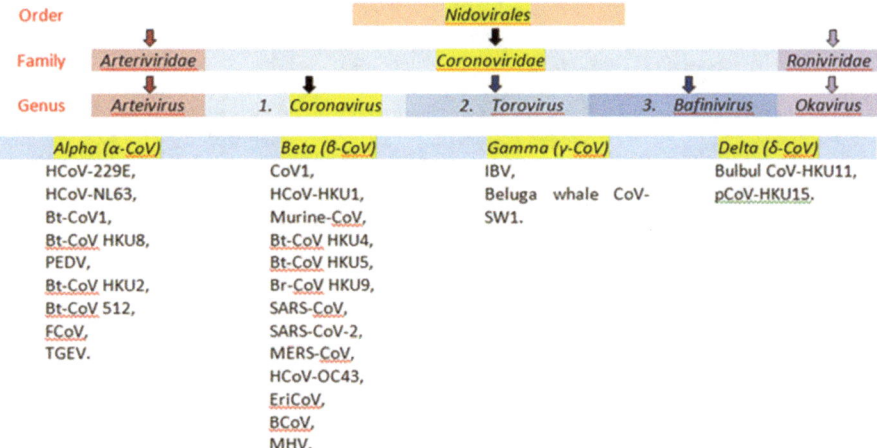

Fig. 9.1 Coronavirus classification. The families *Coronaviridae*, *Arteriviridae*, and *Roniviridae* fall under the order *Nidovirales*. *Coronavirus*, along with the genus *Torovirus*, and a new tentative genus *Bafinivirus* are established under *Coronaviridae* [2]. The genus *Coronavirus* consists of four genera: *alpha, beta, gamma,* and *delta* coronaviruses. The following coronaviruses for each coronavirus genus (α-, β-, γ-, and δ-CoVs) are shown [1, 12]: human coronaviruses (HCoV) 229E, NL63, HKU1, and OC43; *Miniopterus bat coronavirus 1* (Bt-CoV1); *Miniopterus bat coronavirus HKU8*; *Porcine epidemic diarrhea virus* (PEDV); *Rhinolophus bat coronavirus HKU2* (Bt-CoV HKU2); *Scotophilus bat coronavirus 512* (Bt-CoV 512); Feline coronavirus (FCoV); transmissible gastroenteritis virus (TGEV); *Betacoronavirus 1* (CoV 1); *Murine coronavirus* (Murine-CoV); *Tylonycteris bat coronavirus HKU4* (Bt-CoV HKU4); *Pipistrellus bat coronavirus HKU5* (Bt-CoV HKU5); *Rousettus bat coronavirus HKU9* (Bt-CoV HKU9); severe acute respiratory syndrome-related coronavirus (SARS-CoV); severe acute respiratory syndrome coronavirus 2 (SARS-CoV-2); Middle East respiratory syndrome-related coronavirus (MERS-CoV); *Hedgehog coronavirus 1* (ERiCoV); bovine coronavirus (BCoV); mouse hepatitis virus (MHV); infectious bronchitis virus (IBV); *Beluga whale coronavirus SW1* (beluga whale CoV-SW1); *Bulbul coronavirus HKU11* (bulbul-CoV HKU11); and *Porcine coronavirus HKU15* (pCov-HKU15). As seen from the figure, the seven known human coronaviruses are either from genus *Alpha-* or *Betacoronavirus*. Three out of the seven HCoVs are highly pathogenic; these include SARS-CoV, SARS-CoV-2, and MERS-CoV, all belonging to beta-genera

analysis thus revealed that animals can transmit the virus to humans [1]; especially domestic animals can act as intermediate hosts to carry the virus from reservoir animals and pass it to humans. The ability of interspecies jumping has introduced highly pathogenic CoVs in human populations (Fig. 9.3). SARS-CoV was found to have gotten transmitted from bats (reservoir host) to civets (intermediate hosts) and then to humans. MERS-CoV was also found to have originated from bats and got transmitted to humans through camels [14]. Phylogenetic studies of SC2 also revealed bats as the reservoir animal, and Malayan pangolins are being identified as intermediary hosts [1]. These studies further emphasize the potential spillover of new pathogenic coronaviruses into the human populations in the coming future [11]. The alarmingly high diversity of coronaviruses being detected in bats also highlights the importance of using bat coronavirus models for designing strategies against any such future state of viral pandemic.

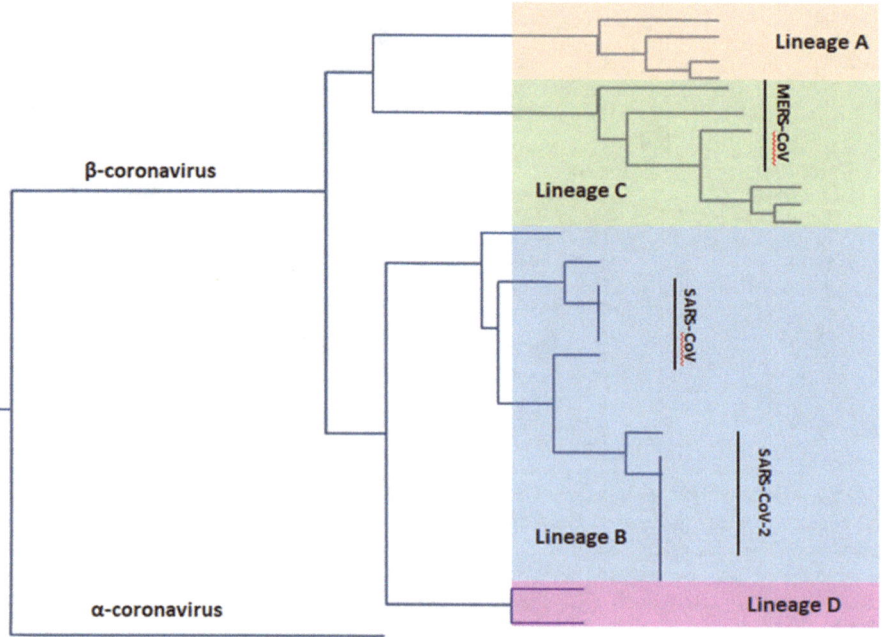

Fig. 9.2 SARS-CoV-2's relatedness with other coronaviruses as evident from protein sequencing studies

Table 9.2 Coronaviruses and their natural hosts

Host animal	Coronavirus
Humans	Human coronavirus 229E (HCoV-229E)
	Human coronavirus OC43 (HCoV-OC43)
	Human coronavirus HKU1 (HCoV-HKU1)
	Human coronavirus NL63 (HCoV-NL63)
	Severe acute respiratory syndrome-related coronavirus (SARS-CoV)
	Middle East respiratory syndrome-related coronavirus (MERS-CoV)
	Severe acute respiratory syndrome 2 coronavirus (SARS-CoV-2)
Cats	Feline infectious peritonitis virus (FIPV)
	Feline coronavirus (FCoV)
Dogs	Canine enteric coronavirus (CCoV)
Chicken	Infectious bronchitis virus (IBV)
Cattle	Bovine coronavirus (BCoV)

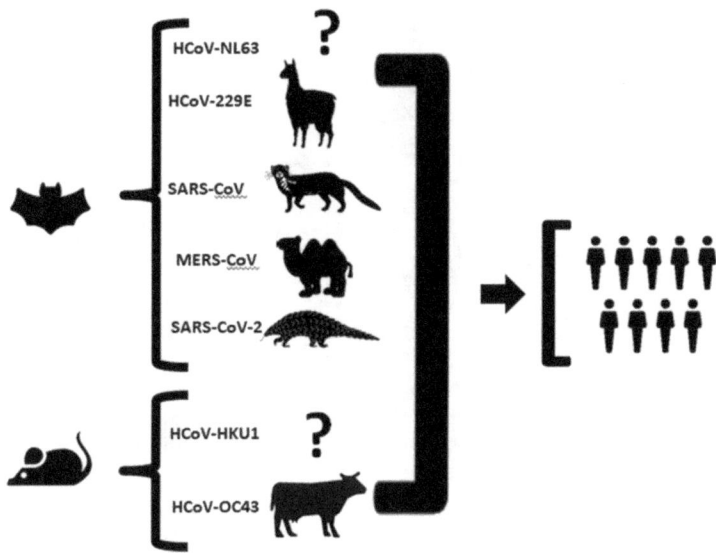

Fig. 9.3 Schematic representation of interspecies transmission route for coronaviruses

Morphology of SARS-CoV-2

COVID-19 has pushed humans in a new struggling battlefield against viral infections. Since December 2019, coronavirology has experienced an expeditious progress revealing morphological and genomic aspects and transmission and replication mechanisms of SC2, the study of which is crucial to decipher its effective therapeutic and preventive strategies.

All coronaviruses share a pronounced morphological and structural resemblance. Corona-virions are pleomorphic to spherical enveloped particles (Fig. 9.4). The diameter of a single corona-virion ranges from 120 nm to 160 nm; the electron microscopy of SC2 revealed a similar range for its diameter, i.e., 60–140 nm [15]. The high similarity shared among SC2 and SARS-CoV's genomic sequence (79%) suggests that the structural characteristics of SC2 would also be quite the same as those of its previously described relative, Table 9.3. The core of coronavirus particle consists of a non-segmented, 3′ polyadenylated, and 5′ capped ssRNA genome of positive polarity [16]. The coronavirus genome is about 27–32 kb which makes it the largest among all known viral RNA genomes [1, 16]. This genome is responsible for the expression of at least four major structural proteins: (N) nucleotide protein, (M) transmembrane protein, (E) envelope protein, and (S) spike proteins [17], all of which are located at the 3′ end of the genome [18]. The RNA genome is found coiled within a helical nucleocapsid which has a diameter of 9–11 nm [16]. The helical nucleocapsid is formed of N-proteins which are known to interact with the C-terminus of the surrounding M-proteins. N-protein is composed of two domains, NTD and CTD, N-terminal domain and C-terminal domain, respectively. Both

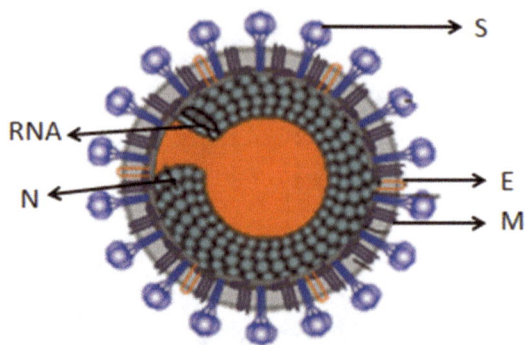

Fig. 9.4 A schematic representation of *Coronavirus* is shown. S spike protein – forms trimmers which are involved in receptor binding, M membrane protein, E envelop protein, N nucleocapsid protein, in all nidoviruses, a single N-protein forms the nucleocapsid which interacts with the genomic RNA as well as the membrane proteins

Table 9.3 Morphological and genomic characteristics of SARS-CoV, MERS-CoV, and SARS-CoV-2

Characteristic	SARS-CoV	MERS-CoV	SARS-CoV-2
Genomic length (nt)	29,727	30,119	29,891
ORFs	11	11	12
Structural proteins	4	4	4
Length of S-protein (aa)	1255	1353	1273
NSPs	5	16	16
Accessory proteins	8	5	6

domains are known to interact with the gRNA but through different mechanisms. The protein itself is found heavily phosphorylated; this phosphorylation is believed to enhance the affinity for viral RNA during viral assembly [17, 18]. The functions of N-proteins [19] have been enlisted in Table 9.4 along with the functions of other common coronavirus proteins [20–25]. The encapsulating membrane is a lipid bilayer derived from the host membrane. All surface proteins including S-, M-, and E-proteins are embedded in this host-derived lipid bilayer [26].

Coronavirus S-proteins are amazing molecules, weighing around 150 kDa [18]. They alone mediate the receptor binding and membrane fusion of viral cells with host cells [20]. S-proteins are heavily glycosylated and [18] exist in two conformations: a pre-fusion and a post-fusion conformation; the understanding of these conformations and factors which trigger its topology has helped identify potential structural targets for therapeutic purposes [57]. Changes in S-glycoproteins (20-nm-long club-shaped protrusions) are largely responsible for the variety in coronavirus tropism.

9 COVID-19: Recent Developments in Therapeutic Approaches

Table 9.4 Coronavirus proteins and their functions

Protein	Function		[Ref]
Spike (S)	Receptor recognition Membrane fusion		[20]
Envelope (E)	Mediate viral assembly Mediate viral release from host cell Involved in virus' pathogenicity		[22, 23]
Transmembrane (M)	Mediate viral release from host cell		
Nucleocapsid	Viral life cycle	Viral core formation Viral assembly Virus budding/envelop formation mRNA replication	[21]
	Cellular response	Chaperone activity Cell cycle regulation Host translational shutoff Immune system interference Signal transduction	
Accessory proteins	NSP1	Cellular mRNA degradation INF inhibition	[24, 27]
	NSP3	Polypeptide cleavage Blocks host innate response Promotes cytokine expression	[28, 29]
	NSP4	DMV formation	[30, 31]
	NSP5	Polypeptide cleaving IFN inhibition $3CL^{pro}$ M^{Pro}	[32–34]
	NSP6	DVM formation Restricts autophagosome expansion	[35, 36]
	NSP7	Cofactor	[37, 38]
	NSP8	Cofactor Primase	[37–39]
	NSP9	Dimerization and RNA binding	[40, 41]
	NSP10	Scaffold protein	[42–45]
	NSP12	Primer-dependent RdRp	[37, 39, 46]
	NSP13	RNA helicase 5′ triphosphatase	[47–49]
	NSP14	Exoribonuclease N7-MTase	[50–53]
	NSP15	Evades dsRNA sensors Endoribonuclease	[54–56]
	NSP16	2′-O-MTase Negative regulation of innate immunity	[25, 43, 44]

Abbreviations: *$3CL^{pro}$* chymotrypsin-like protease, *DVM* double-membrane vesicle, *dsRNA* double-stranded RNA virus, *IFN* interferon, *mRNA* messenger RNA, *M^{pro}* main protease

Of all the nidoviruses, only coronavirus and arterivirus particles possess conserved (E) envelope proteins [2]. The deletion of E-protein in SARS-CoV results in a dramatic reduction of virus infectivity. Though it is the smallest of all viral proteins (8–12 kDa) [18] which SC2 gRNA expresses, yet it is also the most mysterious one. During cell replication cycle, E-proteins are expressed in a huge amount by the viral genome, but only a small amount of these proteins get incorporated into the new viral envelopes [22]. Studies propose three possible roles which E-proteins serve: (1) it mediates viral assembly, (2) its hydrophobic transmembrane domain is important for viral release from the host cells, and (3) it contributes to virus' pathogenicity [22, 23, 58].

M-proteins play a central role at the junction where viral and host factors meet to produce new virus particles. It is a small protein of up to 25–30 kDa [18]; reverse genetic and VLP (virus-like protein) assembly studies suggest the role of M-protein to exhort viral assembly by interacting with peplomers and viral ribonucleoproteins (vRNP) at the budding site [59]. It exists in the form of a dimer inside the virion and may transit between two different conformations which allow it to bind to the N-proteins or promote membrane curvature during budding stage [18].

Genome Organization of SARS-CoV-2

The study of genomic organization is a definite prerequisite for the functional investigation of any virus' replication mechanism, viral protein expression, host-virus interactions, and the viral pathogenicity.

SC2, like all coronaviruses, carries a large genomic RNA (gRNA) comprised of about 27,000–32,000 nucleotides. In general, a coronavirus gRNA contains seven conserved genes which are common to coronaviruses. Two-thirds of this genome is encompassed by (open reading frames) ORF1a and ORF1b; the rest of the genome harbors ORF3 and genes for all the structural proteins [21]. The conserved order of these genes is illustrated in Fig. 9.5. Genomic sequencing has revealed that SARS-CoV-2 genome holds 96% identity to the genome of bat CoV RaTG13 and 79.5% identity with SARS-CoV [60]. SC2 genome encodes several ORFs. Figure 9.6 shows a schematic presentation of SC2 genomic organization [61]. ORF1a/b are translated to produce nonstructural proteins (NSPs). Polypeptide 1a (pp1a, 440–500 kDa) is encoded by ORF1a, which when expressed is cleaved into 11 further NSPs. ORF1b is translated to produce a larger polypeptide, pp1ab (740–810 kDa). Pp1ab is further cleaved into 15 smaller NSPs. The viral genome

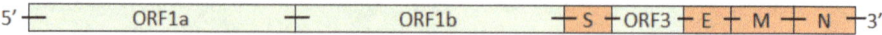

Fig. 9.5 A schematic illustration of coronavirus RNA genome. The genome contains seven genes which are common to all coronaviruses shown in their conserved order. ORF1a and ORF1b make up two-thirds of the entire genome; the rest consists of ORF3 and genes for essential structural proteins; S spike protein, E envelop protein, M transmembrane protein, N nucleocapsid protein

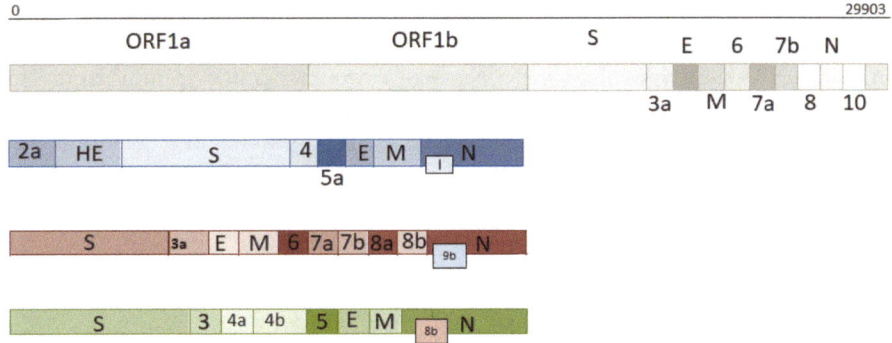

Fig. 9.6 Schematic representation of SARS-CoV-2 genome organization (gray). A 29,903-nt-long mRNA is translated from the entire SC2 gRNA. ORF3a, E, M, ORF6, ORF7a, ORF7b, ORF8, N, and ORF10 are nine major sgRNAs produced by the virus in addition to its gRNA [59, 61]. Illustration of the genomic organization of β-CoVs; MHV, SARS-CoV, and MERS-CoV are shown in blue, maroon, and green, respectively. SC2 SARS-CoV-2, gRNA genomic RNA, sgRNA subgenomic RNA, ORF open reading frame, CoV coronavirus, SARS-CoV severe acute respiratory syndrome-related coronavirus, MERS-CoV Middle East respiratory syndrome-related coronavirus

also serves to encode viral proteases (NSP3/5) which mediate the proteolytic cleavage of pp1a and pp1ab. Nine subgenomic gRNAs (sgRNA) encode for conserved structural proteins (S, E, M, and N) and at least six accessory proteins (3a, 6, 7a, 7b, 8, and 10) according to GenBank: NC_045512.2. CoVs elicit frequent gene recombination which allows them to generate a huge number of variants which may and may also contribute to their better survival and immune evasion activity [62]. Similarly, SC2 gRNA also undergoes frequent events of recombination during its replication cycle, which will be discussed under the next subtopic. The study of its genomic organization opens new directions for investigating SC2's pathogenicity and future pandemic threats.

Replication of SARS-CoV-2

To replicate, the virus cell must first enter into its host target cell. This entry would depend on the host receptor specificity of viral spike glycoproteins (S-glycoP). As for SARS-CoV-2, the S-glycoPs bind to the same ACE-2 receptors on human target cells as SARS-CoV [63]. The S-protein has two subunits: S1 and S2. Subunit 1 consists of an RBD which recognizes ACE-2 receptors specifically on host cell membranes; being the head of S-protein's structure, it is responsible to find and bind to the ACE-2 receptors [64]. Once the binding is done, transmembrane protease serine 2 (TMPRRS2) or cathepsin proteolytically cleaves the S-protein from S2 domains. This cleavage initiates the process of viral-host membrane fusion [18]. This fusion allows the release of SC2 gRNA into host cell's cytosol where pp1a and pp1ab start getting expressed. These large polypeptides are then cleaved into smaller

NSPs by NSP3 and NSP5. Most of these NSPs come together to form the viral replicase-transcriptase complex (vRTC). vRTC is then responsible for encoding the set of sgRNAs which involves the process of ribosomal frame shifting [65]. The viral structure proteins are transported to an ER-Golgi intermediate compartment (ERGIC). Here the gRNA and structure proteins are assembled into new virus particles. After assemblage, the new mature virions are then sent toward the host cell membrane packed in vesicles. As these vesicles reach the budding site, viral M-proteins mediate virus release from the host cell.

Transmission and Pathogenesis of SARS-CoV-2

Viruses also use the known routes of transmission which any other pathogen is known to take. These include:

1. Direct contact transmission – the viral particle gets physically transferred through direct body contact and enters the body through any opening, e.g., eyes, mouth, or a wound.
2. Airborne transmission – this encompasses viral transfer via small droplets or suspended particles which may be inhaled by a host.
3. Ingestion transmission – the virus can be ingested by ingesting contaminated food items.
4. Fomite transmission – the virus gets transmitted via contaminated inanimate objects such as environmental or medical surfaces.
5. Vector-borne transmission – this happens when animals act as carriers/vectors to transfer pathogens; the vectors are mostly arthropods or rodents.
6. Zoonotic transmission – this is spread of diseases from animal populations to human populations and can take any of the five routes of transmission briefed above [66].

Potential for interspecies jumping increases the probability of future zoonotic spillovers (Fig. 9.3); any non-identified intermediate host is one reason that can make it difficult to break the transmission cycle. Figure 9.7 briefly discusses the cross-species transmission mechanisms from host to nonnatural hosts (other humans and animals) and vice versa [1]. Virus transmission from animal to animal relies on the fecal-oral route, whereas the transmission of virus from animal to human may vary and can be depicted through three major stages: (1) viral prevalence and dispersal from animal host; (2) chances of viral exposure, route of entry in host human, and dose of viral particles entering; and (3) genetic, immunological, and physiological state of host human [67]. Transmission from human to human is found to occur due to unprotected and prolonged in-person interaction with the infected individual. Such an exposure builds a constant pathogen pressure on the exposed person, thus leading to the development of infection and possibility of transmission [68]. The major mode of person-to-person transmission is via airborne droplet which can transmit the viral particles in a zone of about 6 ft from the infected individual [69]. The other mode in case of human-to-human transmission is via fomites [70].

9 COVID-19: Recent Developments in Therapeutic Approaches

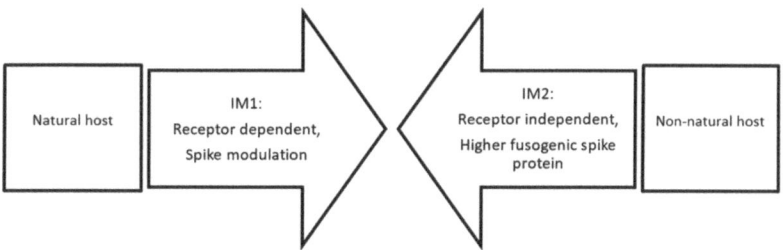

Fig. 9.7 Infection modes (IM1/2) depicting the potential of coronavirus' interspecies transmission. The transmission from natural to nonnatural host is receptor dependent and relies on spike protein modulation (IM1), whereas the transmission in reverse direction is often receptor independent as the spike protein may bypass receptor binding and elicit higher affinity for nonspecific host cell fusion when moving back to a previously known host (IM2)

The airborne droplets can settle on different surfaces and survive for long periods of time. These can then be acquired by healthy individuals through contact with the contaminated surfaces [1].

The pathogenicity of SC2 is thought to be linked with its structural and nonstructural proteins. Although the role of many NSPs haven't been described yet, we do know the crucial role of envelope protein in promoting viral assembly and release [18] and that the NSPs can shut off host immune response. S-proteins, for instance, prominently encourage SC2's pathogenicity by showing high affinity to ACE-2. ACE-2 receptors are widely distributed in the body, but the expression varies between tissues and individuals as well. This also contributes toward the variation in COVID-19's clinical manifestations [63, 71–74] – discussed in "clinical characterization of SARS-CoV-2." The pathogenic mechanism through which SC2 causes pneumonia seems complex. The available data indicates SC2's capability of producing a hyperactive immune response within the host body leading to the development of cytokine storms. That being so, the pathogenic cascade of SC2 implicates several cytokines, such as IL-1β, IL-12, IL-8, TNF-α, IL-6, monocyte chemoattractant protein 1, and macrophage inflammatory protein. IL-6 serves as the prime mover for it can act on numerous types of cells in the body and is mainly involved in the pathogenesis of cytokine release syndrome – which is also the case for COVID-19 – characterized by multiple organ dysfunction and fever. Studies have also proven that the binding of SC2 to TLRs (Toll-like receptors) actuates the release of pro-IL-1β which when cleaved into its active IL-1β form mediates lung inflammation and consequential fibrosis [75].

Clinical Characterization of SARS-CoV-2

The clinical pathology of COVID-19 resembles that of SARS-CoV and MERS-CoV. Once the infection develops in a healthy individual, mild symptoms appear including nonproductive cough, fatigue, low-grade intermittent fever, sore throat, and dyspnea. The disease can then worsen to its severity in 4–5 days from symptom onset, on average [76–78]. COVID-19 patients may therefore present mild, severe, or critical illness (Table 9.5). Gastrointestinal symptoms such as diarrhea and vomiting also develop but are relatively uncommon. As the disease worsens, the infected individuals develop symptoms of ARDS (acute respiratory distress syndrome) and require mechanical ventilation to survive. Figure 9.8 shows a schematic timeline for onset of symptoms in COVID-19 patients.

Table 9.5 Clinical diagnostic features corresponding to the severity of COVID-19 infection in patients [52]

Severity of illness of COVID-19 patients	Diagnostic clinical conditions
Mild	Mild fever
	Unproductive cough
	Sore throat
	Nasal congestion
	Malaise
	Diarrhea
	Headache
	Fatigue
	Sudden loss of taste or smell
	Vomiting
Severe	Severe dyspnea
	Hypoxia
Critical	Acute respiratory distress syndrome (ARDS)
	>50% lung infiltrates
	Extrapulmonary manifestations
	Multiple organ failure

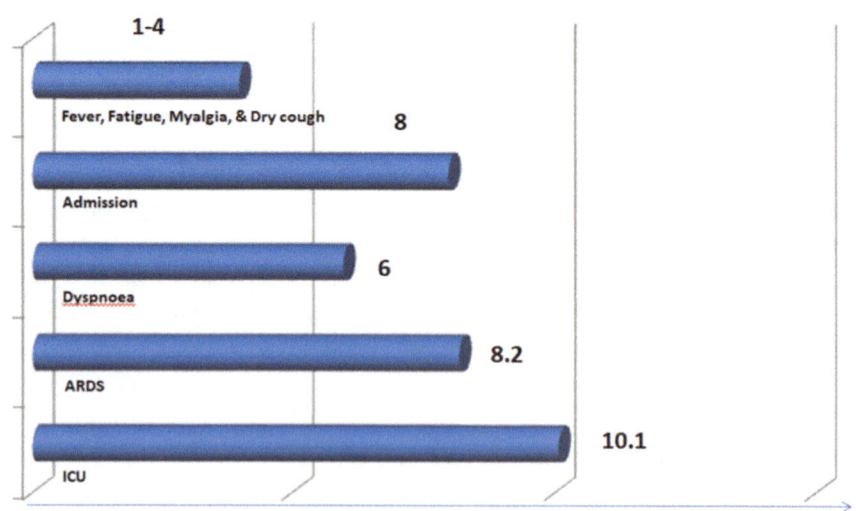

Fig. 9.8 Figure: A timeline of onset of symptoms in COVID-19 patients after development of infection

Underlying clinical complications were found more prevalent in severe COVID-19 cases, the development of which was detected through chest X-rays, T-cell counts (Table 9.4), and tissue biopsy of different organs of the infected person. Rapid progression of pneumonia is often observed in COVID patients' chest X-rays; a reduction in T-cell (CD4 and CD8) counts and hyperactivity of T-cells which triggers a severe cytokine storm [79] are other complications which impede effective treatment. Virus-induced cytopathic effect was observed in intra-alveolar spaces, and moderate microvesicular steatosis and mild lobular and portal activity were observed in liver tissues, while a few inflammatory infiltrates were observed in case of heart tissue. Coronavirus is also suggested to be associated with neurologic manifestations, such as ischemic stroke, axonopathic polyneuropathy, and myopathy [1, 80]. Underlying comorbidities in COVID-19 patients also complicate the course of treatment by increasing the severity of illness and thus the chances of death. The severity of the disease prolongs a patient's stay in ICU and the use of mechanical ventilation, which also adds to the complexity of the disease. Some major underlying comorbidities reported in COVID-19 patients are cerebrovascular disease, hepatitis B, chronic obstructive airway disease, cancer, coronary heart disease, chronic kidney disease, diabetes, hypertension, and immunodeficiency. Table 9.6 enlists the major comorbidities reported in various cohort studies [73, 81, 82].

Table 9.6 Some major underlying comorbidities reported in COVID-19 patients	
	ARDS
	Acute kidney injury
	Acute cardiac injury
	Cancer
	Cerebrovascular disease
	Chronic obstructive pulmonary disease
	Coagulopathy
	Coronary heart disease
	Diabetes
	ECMO
	Hepatitis B
	Hypertension
	Invasive mechanical ventilation
	Pneumonia
	Rhabdomyolysis

Clinical and Laboratory Diagnosis of SARS-CoV-2

A basic clinical diagnosis of SC2 depends on the symptoms described in the previous section "Clinical Characterization of SARS-CoV-2," which are also similar for other virus-induced respiratory infections. COVID-19 was initially misdiagnosed as pneumonia before it emerged as a pandemic. Hence, complimentary diagnosis using serology testing, RT-PCR analysis of blood samples, or chest imaging of the patients can help in differential diagnosis of the infection.

Once the symptoms such as intermittent fever, cough, and fatigue are investigated, serological tests and chest imaging should be considered to establish the presence of SC2 infection. COVID-19 provides somewhat similar serological findings as for other viral respiratory diseases; these include alterations in the levels of ALT, D-dimers, lymphocytes, CRP, thrombocytes, etc. – see Tables 9.7 and 9.8 [73, 81, 82]. Chest radiographs show the presence peripheral or rounded opacities with patchy bilateral GGOs (ground-glass opacity) and mixed attenuations. Lesions in the peripheral region or lower lobes are also predominant findings [83]. These structures, however, may not always be due to COVID-19. The radiologist must compare unique features such as reverse halo or atoll signs to confirm the cause of infection. CT scans are more sensitive than radiographs, but confirmed cases can also have normal chest CTs. As these findings might not always be specific for COVID-19, exposure or travel history of the patient can help avoid misdiagnosis [84].

Differential diagnosis is thus strongly recommended for COVID-19 diagnosis, as it might get misdiagnosed as another form of pneumonia due to the high similarity of serological and radiographic findings with those of other virus-induced respiratory infections (caused by adenovirus, influenza, human metapneumovirus, rhinovirus, parainfluenza, and respiratory syncytial virus). A general flowchart shows how a seldom differential clinical diagnosis of COVID-19 can be carried out – Fig. 9.9.

Table 9.7 Laboratory findings in COVID-19 patients [73, 81, 82]

Variable	Variation observed in patients	Approx. values recorded in patient cohort
ALT	Elevated	>40 U/L
D-dimer	Elevated	>1μg/L
CRP	Elevated	–
Creatinine	Elevated	–
LDH	Elevated	>245 U/L
IL-6	Elevated	>7.4 pg/mL
Ferritin	Elevated	>300μg/L
Creatinine kinase	Elevated	>185 U/L
Procalcitonin	Elevated	>= 0.5 ng/mL
Lymphocytes	Lowered	$<0.8 \times 10^9$/L
Thrombocytes	Lowered	$<100 \times 10^9$/L

Table 9.8 List of few drugs with known mechanism of actions (MOA) against SARS-CoV and MERS-CoV [1]

Drug	MOA
Chloroquine	Chloroquine is attributed to a deficit in glycosylation of ACE-2 receptors (SARS) and target type II transmembrane serine proteases (MERS)
IFN-alpha INF-beta-1a NF-gamma IFN-alpha-2b	Inhibits viral replication (SARS, MERS)
Imatinib Dasatinib	Inhibit viral entry (SARS, MERS).
Selumetinib Trametinib	Inhibit viral entry as well as replication of viral particles (SARS, MERS) via ERK/MAPK signaling pathway
Sirolimus	Targets mTOR signaling pathway to reduce viral infectivity (MERS)
Chlorphenoxamine	Inhibits viral entry
Lopinavir	Targets M^{pro}(SARS)
Camostat mesylate	Inhibits TMPRRS-mediated glycoprotein activation (SARS, MERS)
K11777 E-64-D	Inhibits viral attachment to host cells (SARS, MERS)

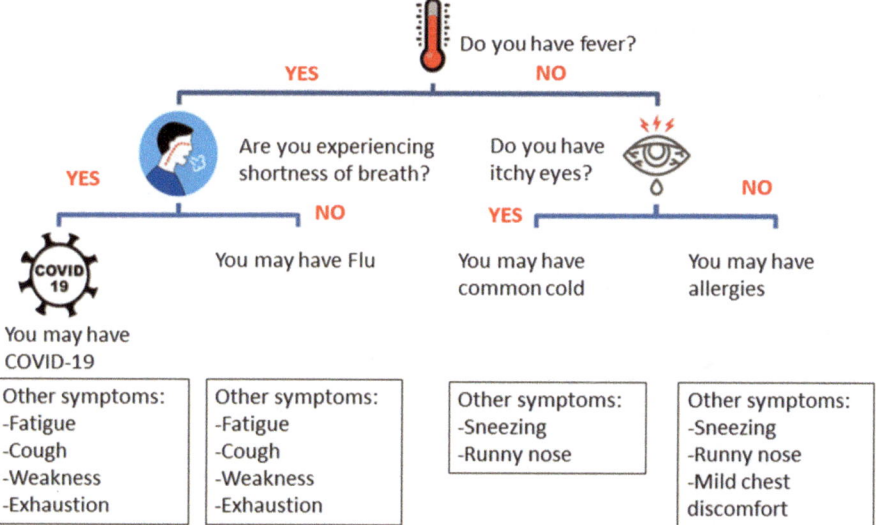

Fig. 9.9 A flowchart to aid in differential clinical diagnosis of COVID-19 and other mimicking conditions

The availability of genome sequence information of SC2 has made it possible for clinicians and biotechnologists to establish an RT-PCR – diagnostic assay. It qualitatively detects the viral nucleic acids in nasopharyngeal or oropharyngeal sputum/swab/aspirates or lavage of the suspected individual. SC2 gRNA is generally detectable during acute phase of infection. Positive results must be correlated with other

Therapeutic Options for the Treatment of COVID-19

Therapeutic options for COVID-19 are meek; a number of FDA-approved drugs were randomly tested on SC2-infected patients under strict supervision of MEURI (Monitored Emergency Use of Unregistered and Investigational Interventions System) [85]. Other than drugs, therapeutic options include vaccine development, convalescent plasma therapy, corticosteroids, and CRISPR/Cas-mediated therapy. Any specific antiviral treatment for COVID-19 has not yet been approved, and the patients must receive supportive care in order to relieve symptoms [86].

Antivirals or Immunomodulatory Drugs for COVID-19

As SARS-CoV-2 has emerged from a similar pool of viruses as SARS-CoV and MERS-CoV, repurposing of drugs (Table 9.8) with well-established pharmacokinetic profiles against previously known highly pathogenic coronaviruses is an approach that readily got considered. The formerly known potent anti-coronavirus drugs either manifest a virus-targeted strategy or a host-targeted strategy – Fig. 9.10.

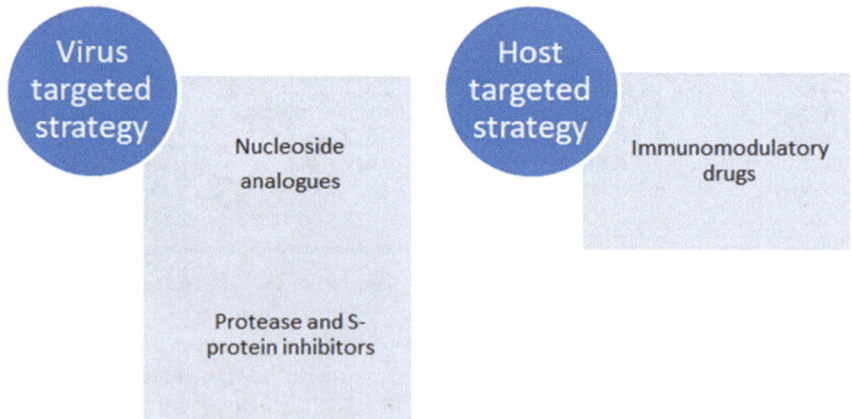

Fig. 9.10 Drug treatment approaches for COVID-19. Two strategies can be considered: (1) Virus-targeted strategy includes any drug which hampers viral development and survival, e.g., nucleoside analogues and protease or S-protein inhibitors. (2) Host-targeted strategy includes treatment with drugs which upregulate host immune response toward viral infection

Nucleoside analogues (NAs) and protease inhibitors (PIs), for instance, demonstrate convincing activity against the progression of SC2 in patients. Fapilavir was the first NA that got approved by NMPA (National Medical Products Administration of China) for the treatment of SC2; ribavirin and favipiravir were also later on approved for the administration in COVID-19 patients [87]. However, the highly frequent ability to mutate provides SC2 with the ability to resist the activity of these NAs. This led to the use of second approach: anti-CoV-NA cocktail therapy which showed enhanced efficacy, especially with coadministration of an additive. The resisting ability of SC2 toward NAs can also be reduced by administering a combination of drugs with different mechanisms of action. Several researchers, however, speculate the use of single broad-spectrum antivirals till a novel anti-SC2 agent is designed/identified [88]. Remdesivir and galidesivir, two experimental NAs, are broad-spectrum antivirals which were also evaluated against SC2 in cell cultures and mouse models [89, 90]. Like typical nucleoside analogues, these drugs inhibit SC2 by causing premature termination of the vRNA chains [87] (Fig. 9.11). Remdesivir is presently in clinical trials' phase for SC2 infection. In animal models, remdesivir was reported to improve pulmonary function and lower viral load (SARS-CoV and MERS-CoV) [90]. Wang et al. studied the effectivity of remdesivir in vitro using human cell line; the study supports the use of intravenous remdesivir as a viable drug treatment for COVID-19 [87].

Protease inhibitors, namely, lopinavir, disulfiram, and ritonavir, possess anti-HCoV activity. Disulfiram is known to inhibit the papain-like protease of SARS-CoV and MERS-CoV in cell-line models [87]. Ritonavir and lopinavir are known to inhibit 3-chymotrypsin-like proteases of SARS-CoV and MERS-CoV. However, the mechanism through which protease inhibitors inhibit proteases still rests in controversy. Clinical trials of lopinavir and ritonavir failed to lower mortality rates in COVID-19 patients and thus have not been considered for furtherance [91]. Other candidate PIs include nafamostat and griffithsin which demonstrate activity against S-proteins, thus hampering viral entry into host cells [92]. A recent study based on activity profiling of SC2 NSP3 (viral papain-like cysteine protease, PL^{pro}) determined crystal structures of two of its proposed potential inhibitors. The study hence

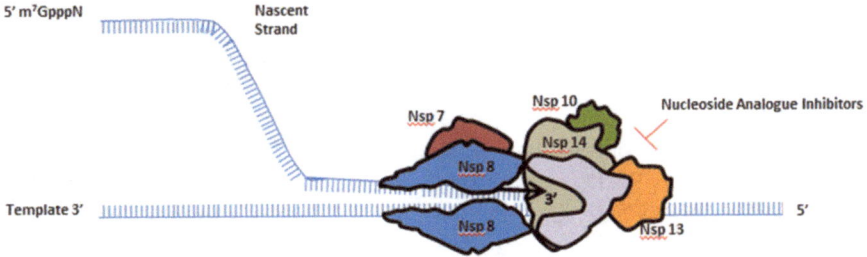

Fig. 9.11 Nucleoside analogues (NAs) are structural mimics of naturally occurring nucleosides. This allows the viral transcriptional machinery to mistakenly incorporate them during RNA translation. NAs either mispair with natural nucleosides to result in lethal mutations or cause premature termination of the viral RNA chain

propounds a framework for the development of protein inhibiting antivirals with promising therapeutic value as an anti-SC2 drug [93].

Chloroquine and hydroxychloroquine (antimalarial drugs) were also repurposed against SC2. 500 mg of chloroquine and 200 mg of hydroxychloroquine every 12 hours were proposed [52]. Gautret et al. [94] reported significant reduction in viral load until complete disappearance associated with the use of hydroxychloroquine. This effect was also documented to have enhanced by the administration of azithromycin with hydroxychloroquine. Like PIs, these drugs were also discontinued due to their inability to lower COVID-19 mortality rate [52]. Howbeit, further studies are recommended to evaluate the use of azithromycin against viral infections despite their established potential to mitigate and modulate immune system in in vivo and in vitro systems.

IFNs are innate components of the human body which naturally respond in defense to viral infections. These have been previously used against hepatitis C virus and in chemotherapy against numerous malignancies [52] (Table 9.8) and thus can be speculated for anti-SC2 activity. However, the data presently available is not enough to decide in favor or against their use for COVID-19 [95].

Vaccine for COVID-19

More than 100 potential vaccines for coronaviruses are being evaluated; these include the following types:

1. Whole virus (live/attenuated)
2. Antibody-based
3. Small subunit-based
4. Vector-based
5. Nucleic acid-based

Live attenuated vaccines help build a long-term immune response toward a specific virus; however, it may sometimes develop complications in the recipient. Antibody-based vaccines are based on (mAbs) monoclonal antibodies. The mAbs are strain specific but only provide a limited protection against the subjected virus. Small subunit-based vaccines are the most safe to use, are simple to produce, and are broad-spectrum as well. Nucleic acid-based vaccines are also referred to as DNA or RNA vaccines. These are also safe to use and provide long-term protection.

Some of the most promising candidates for SC2 as reviewed by the global COVID-19 vaccine R&D landscape were PiCoVacc, INO-4800, Ad5-nCoV, mRNA-1273, LV-SMENP-DC, and aAPC [96, 97] and are currently under clinical trials [52]. S-protein-based vaccines are a type of small-subunit-based vaccines; having demonstrated to be the most effective against both SARS-CoV and MERS-CoV [1], they are now being readily considered as safe, simple, and stable [98] option against SC2 but are presently underway clinical trials. In Wuhan (China), the very first in-human trials for a recombinant adenovirus type 5 (Ad5) vectored SC2 vaccine were carried out between 16th and 27th of March 2020. This vaccine is also an S-protein-based vaccine, and no serious adverse effects were reported within 28 days of its administration. The specific Ab response to SC2 peaked after day 28. This vaccine is currently undergoing phase II trial to confirm its safety and immunogenicity for mass use [99].

Plasma/Serotherapy for COVID-19

The use of convalescent plasma (CP) lowered the mortality rate in SARS- and MERS-infected patients. Likewise, passive immunization with CP in SC2 patients exhibited improved outcomes. CP is speculated to contain therapeutic levels of anti-SC2 neutralizing antibodies (Abs). Studies report the efficacy of convalescent plasma as a treatment without any adverse side/posttreatment effects [100]; however, some known complications otherwise associated with serotherapy include circulatory overload, anaphylaxis, and transfusion-related acute lung injury [101], and so further study is advised. Various treatment protocols mention serotherapy as treatment of last resort; fortunately, no serious unfavorable events have been reported so far. Instead, successful cases of COVID-19 treatment are surfacing [56]. A clinical trial to investigate a cocktail of Abs purified from CP in COVID-19 patients was launched in June 2020 [52]. Anyhow, the maximum benefits or associated complications might be realized through routine administration of these therapies.

Alternative Therapies

Anticoagulant Therapy

Anticoagulation therapy is also being suggested as SC2-infected patients have a higher incidence of venous thromboembolism. The therapy has been associated with reduced ICU deaths in COVID patients. Furthermore, in case of thrombosis or thrombophilia, 1 mg/kg dose of enoxaparin – twice a day for full-intensity anticoagulation – is indicated [102].

Glucocorticoids (GC)

Systemic GCs have been investigated for SARS-CoV back in 2003, but these studies are few and do not provide any conclusive evidence regarding their efficacy [103]. Corticosteroids have been studied in animal models and been extensively reported as potential therapeutic agents which can help reduce inflammation, lung injury and improve patient survival [104]. These however, were not recommended as a treatment option for critical cases of viral pneumonia or ARDS [52], but as a subset of population infected with COVID-19 may develop cytokine storm syndrome [79, 105] and multiple organ failure, the therapeutic role of glucocorticoids was re-hypothesized for these patients in this current dramatic emergency. A recent study demonstrates dexamethasone to have reduced deaths by 33% among critically ill SC2-infected patients [106]. However, other studies claim that there is no convincing evidence available to prove the efficacy of corticosteroids in decreasing the mortality rate for COVID-19 [107]. The debate on the use of glucocorticosteroids has hence been reignited and needs further experimental evidence to comprehend any benefits it might have for treating COVID-19.

Stem Cell-Based Therapy

A study [108] introduced mesenchymal stem cell transplantation (MSCT) as a promising alternative to antiviral drugs for COVID-19. They report that MSCT can help regulate immune homeostasis and induce specific immune tolerance to help reduce inflammation Based on these findings, the study proposes it as one viable therapeutic approach toward COVID-19 as MSCT were also reported safe and effective for use especially for critically ill COVID-19 patients.

ACE-2-Mediated Therapies

ACE-2 (angiotensin-converting enzyme 2) is a peptidase widely expressed in organs including lungs, kidneys, and the GIT. One major similarity between SC2 and SARS is their affinity for ACE-2 receptors. The entry of SC2 via ACE-2 is followed by the accumulation of angiotensin-II which may mediate acute lung injury. ACE-2 blockers and administration of soluble ACE2 are potential therapeutic approaches for COVID-19 [109]. ARBs (angiotensin receptor blockers) propose another possible treatment option [110]. These therapies are convenient and can be applied.

CRISPR/Cas System

The diversity of HCoVs requires a flexible antiviral technique to pace up with its mutative frequency. CRISPR/Cas9 system has already been successfully used to enhance immunity by reprogramming B- and T-cells as a chemotherapeutic strategy. It can be used as a promising antiviral technique as well, as it can be used to manipulate immune system against the pathogen or can destroy viral cells directly [111].

Conclusion

SARS-CoV-2 is a highly pathogenic coronavirus which is found to have originated in bats and carried over to humans by Malayan pangolins. COVID-19 emerged as a local outbreak of pneumonia from a seafood market in Wuhan (China) and rapidly transitioned into a global pandemic. SC2 is highly infectious and possesses an aggressive ability of person-to-person transmission, yet the mortality rate in COVID patients is less than that in SARS-CoV and MERS-CoV. COVID patients have been reported with flu-like symptoms in mild to ARDS, cytokine storm, and MOD in critical cases. The most common detection method for SC2 is RT-PCR diagnostic assay, but because RT-PCR is prone to produce false results and other diagnostic findings might share similarity with those of other virus-induced respiratory complications, differential diagnosis must be performed to confirm COVID illness. Various therapeutic options have been explored; however, any potential anti-SC2 agent is currently undergoing clinical trials, and so there is no approved treatment yet. The patients thus rely on supportive treatment to relieve symptoms. Till a treatment is approved, prevention methods including the use of face masks, proper hygiene, and social distancing should be strictly followed, and in case of any symptoms, early detection and treatment must be considered.

References

1. Shailendra KS (2020) Coronavirus disease 2019 (COVID-19): epidemiology, pathogenesis, diagnosis, and therapeutics. 1st ed. Singapore: Springer.
2. Enjuanes L, Gorbalenya AE, deGroot RJ, Cowley JA, Zeibuhr J, Snidjer EJ (2008) Nidovirales. Encyclopedia of virology. Elsevier.
3. Tyrrell DA, Bynoe ML (1966) Cultivation of viruses from a high proportion of patients with colds. Lancet 1(7428):76–77
4. Paules CI, Marston HD (2020) Coronavirus infections – more than just the common cold. J Am Med Assoc 323(8):707–708
5. Hamre D, Procknow JJ (1966) A new virus isolated from the human respiratory tract. Proc Soc Exp Biol Med 121(1):190–193
6. McIntosh K, Dees JH (1967) Recovery in tracheal organ cultures of novel viruses from patients with respiratory disease. Proc Natl Acad Sci U S A 57(4):933–940
7. Lau SKP, Lee P, Tsang AKL, Yip CCY, Tse H, Lee RA et al (2011) Molecular epidemiology of human coronavirus OC43 reveals evolution of different genotypes over time and recent emergence of a novel genotype due to natural recombination. Virology 85:11325–11337
8. Drosten C, Gunther S (2003) Identification of a novel coronavirus in patients with severe acute respiratory syndrome. N Engl J Med 348:1967–1976
9. Ramadan N, Shaib H (2019) Middle east respiratory coronavirus (MERS-CoV): a review. Germs 9(1):35–45

10. https://www.who.int/emergencies/diseases/novel-coronavirus-2019/situation-reports
11. Lu R, Zhao X, Li J, Niu P, Yang B, Wu H et al (2020) Genomic characterisation and epidemiology of 2019 novel coronavirus: implications for virus origins and receptor binding. Lancet 395(10224):565–574
12. Jaimes JA, Andre NM (2020) Phylogenetic analysis and structural modeling of SARS-CoV-2 spike protein reveals an evolutionary distinct and proteolytically sensitive activation loop. J Mol Biol 432(10):3309–3325
13. Woo PC, Lay SKP (2009) Coronavirus diversity, phylogeny and interspecies jumping. Exp Biol Med (Maywood) 234(10):1117–1127
14. Song Z, Xu Y, Bao L, Zhang L, Yu P, Qu Y et al (2019) From SARS to MERS, thrusting coronaviruses into the spotlight. Viruses 11(1):pii:E59
15. Zhu N, Zhang D, Wang W, Li X, Yang B, Song J et al (2020) A novel coronavirus from patients with pneumonia in China, 2019. N Engl J Med 382(8):727–733
16. Pellet PE, Mitra S, Holland TC (2014) Basics of virology. Handb Clin Neurol 123:45–66
17. Hasoksuz M, Sarac F (2020) Coronaviruses and SARS-CoV-2. Turk J Med Sci 50. https://www.researchgate.net/deref/http%3A%2F%2Fdx.doi.org%2F10.3906%2Fsag-2004-127
18. Fehr AR, Perlman S (2015) Coronaviruses: an overview of their replication and pathogenesis. Methods Mol Biol. New York, NY, USA 1282:1–23
19. McBride R, Zyl MV (2014) The coronavirus nucleocapsid is a multifunctional protein. Viruses 6(8):2991–3081
20. Fang L (2016) Structure, function, and evolution of coronavirus spike proteins. Annu Rev Virol 3:237–261
21. Pyrc K, Berkhout B (2007) The novel human coronaviruses NL63 and HKU1. J Virol 81:3051–3057
22. Schoeman D, Fielding BC (2019) Coronavirus envelope protein: current knowledge. Virol J 16(69):2–22
23. Satija N, Lal SK (2007) The molecular biology of SARS coronavirus. Ann N Y Acad Sci 1102(1):26–38
24. Huang C, Lokugamage KG, Rozovics JM, Narayanan K, Semler BL et al (2011) SARS coronavirus nsp1 protein induces template- dependent endonucleolytic cleavage of mRNAs: viral mRNAs are resistant to nsp1-induced RNA cleavage. PLoS Pathog 7(12):e1002433
25. Shi P, Su Y (2019) PEDV nsp16 negatively regulates innate immunity to promote viral proliferation. Virus Res 265:57–66
26. Finlay BB, See RH (2004) Rapid response research to emerging infectious diseases: lessons from SARS. Nat Rev Microbiol 2(7):602–607
27. Tanaka T, Kamitani W (2012) Severe acute respiratory syndrome coronavirus nsp1 facilitates efficient propagation in cells through a specific translational shutoff of host mRNA. J Virol 86(20):11128–11137
28. Lei J, Kusov Y (2018) Nsp3 of coronaviruses: structures and functions of a large multi-domain protein. Antiviral Res 149:58–74
29. Serrano P, Johnson MA, Chatterjee A, Neuman BW, Joseph JS, Buchmeier MJ et al (2009) Nuclear magnetic resonance structure of the nucleic acid-binding domain of severe acute respiratory syndrome coronavirus nonstructural protein 3. J Virol 83(24):12998–13008
30. Beachboard DC, Anderson-Daniels JM (2015) Mutations across murine hepatitis virus nsp4 alter virus fitness and membrane modifications. J Virol 89(4):2080–2089
31. Gadlage MJ, Sparks JS, Beachboard DC, Cox RG, Doyle JD, Stobart CC et al (2010) Murine hepatitis virus nonstructural protein 4 regulates virus-induced membrane modifications and replication complex function. J Virol 84(1):280–290
32. Stobart CC, Sexton NR, Munjal H, Lu X, Molland KL, Tomar S et al (2013) Chimeric exchange of coronavirus nsp5 proteases (3CLpro) identifies common and divergent regulatory determinants of protease activity. J Virol 87(23):12611–12618
33. Zhu X, Fang L, Wang D, Yang Y, Chen J, Ye X et al (2017) Porcine deltacoronavirus nsp5 inhibits interferon-beta production through the cleavage of NEMO. Virology 502:33–38

34. Zhu X, Wang D, Zhou J, Pan T, Chen J, Yang Y et al (2017) Porcine deltacoronavirus nsp5 antagonizes type I interferon signaling by cleaving STAT2. J Virol 91(10):pii:e00003-17
35. Angelini MM, Akhlaghpour M (2013) Severe acute respiratory syndrome coronavirus nonstructural proteins 3, 4, and 6 induce double-membrane vesicles. MBio 4(4):pii:e00524-13
36. Cottam EM, Whelband MC (2014) Coronavirus NSP6 restricts autophagosome expansion. Autophagy 10(8):1426–1441
37. Kirchdoerfer RN, Ward AB (2019) Structure of the SARS-CoV nsp12 polymerase bound to nsp7 and nsp8 co-factors. Nat Commun 10:2342
38. Zhai Y, Sun F, Li X, Pang H, Xu X, Bartlam M et al (2005) Insights into SARS-CoV transcription and replication from the structure of the nsp7-nsp8 hexadecamer. Nat Struct Mol Biol 12(11):980–986
39. te Velthuis AJ, van den Worm SH (2012) The SARS-coronavirus nsp7+nsp8 complex is a unique multimeric RNA polymerase capable of both de novo initiation and primer extension. Nucleic Acids Res 40(4):1737–1747
40. Egloff MP, Ferron F, Campanacci V, Longhi S, Rancurel C, Dutartre H et al (2004) The severe acute respiratory syndrome-coronavirus replicative protein nsp9 is a single-stranded RNA-binding subunit unique in the RNA virus world. Proc Natl Acad Sci U S A 101(11):3792–3796
41. Zeng Z, Deng F, Shi K, Ye G, Wang G, Fang L et al (2018) Dimerization of coronavirus nsp9 with diverse modes enhances its nucleic acid binding affinity. J Virol 92(17):e00692–e00618
42. Bouvet M, Lugari A, Posthuma CC, Zevenhoven JC, Bernard S, Betzi S et al (2014) Coronavirus Nsp10, a critical co-factor for activation of multiple replicative enzymes. J Biol Chem 289(37):25783–25796
43. Chen Y, Su C, Ke M, Jin X, Xu L, Zhang Z et al (2011) Biochemical and structural insights into the mechanisms of SARS coronavirus RNA ribose 2′-O-methylation by nsp16/nsp10 protein complex. PLoS Pathog 7(10):e1002294
44. Decroly E, Debarnot C, Ferron F, Bouvet M, Coutard B, Imbert I et al (2011) Crystal structure and functional analysis of the SARS-coronavirus RNA cap 2′-O-methyltransferase nsp10/nsp16 complex. PLoS Pathog 7(5):e1002059
45. Ma Y, Wu L, Shaw N, Gao Y, Wang J, Sun Y et al (2015) Structural basis and functional analysis of the SARS coronavirus nsp14-nsp10 complex. Proc Natl Acad Sci U S A 112(30):9436–9441
46. Ahn DG, Choi JK (2012) Biochemical characterization of a recombinant SARS coronavirus nsp12 RNA-dependent RNA polymerase capable of copying viral RNA templates. Arch Virol 157(11):2095–2104
47. Adedeji AO, Lazarus H (2016) Biochemical characterization of Middle East respiratory syndrome coronavirus helicase. mSphere 1(5):e00235–e00216
48. Hao W, Wojdyla JA, Zhao R, Han R, Das R, Zlatev I et al (2017) Crystal structure of Middle East respiratory syndrome coronavirus helicase. PLoS Pathog 13(6):e1006474
49. Jia Z, Yan L, Ren Z, Wu L, Wang J, Guo J et al (2019) Delicate structural coordination of the severe acute respiratory syndrome coronavirus Nsp13 upon ATP hydrolysis. Nucleic Acids Res 47(12):6538–6550
50. Eckerle LD, Becker MM, Halpin RA, Li K, Venter E, Lu X et al (2010) Infidelity of SARS-CoV Nsp14-exonuclease mutant virus replication is revealed by complete genome sequencing. PLoS Pathog 6(5):e1000896
51. Bouvet M, Imbert I (2012) RNA 3′- end mismatch excision by the severe acute respiratory syndrome coronavirus nonstructural protein nsp10/nsp14 exoribonuclease complex. Proc Natl Acad Sci U S A 109(24):9372–9377
52. Cascella M, Rajnik M, Cuomo A et al (2020) Features, evaluation, and treatment of coronavirus (COVID-19). StatPearls Publishing. https://www.ncbi.nlm.nih.gov/books/NBK554776/
53. Minskaia E, Hertzig T, Gorbalenya AE, Campanacci V, Chambillau C et al (2006) Discovery of an RNA virus 3′→5′ exoribonuclease that is critically involved in coronavirus RNA synthesis. Proc Natl Acad Sci U S A 103(13):5108–5113
54. Bhardwaj K, Sun J (2006) RNA recognition and cleavage by the SARS coronavirus endoribonuclease. J Mol Biol 361(2):243–256

55. Deng X, Hackbart M, Mettelman R-C, O'Brien A, Mielech AM, Yi G et al (2017) Coronavirus nonstructural protein 15 mediates evasion of dsRNA sensors and limits apoptosis in macrophages. Proc Natl Acad Sci U S A 114(21):E4251–E4E60
56. Zhang L, Li L, Yan L, Ming Z, Jia Z, Lou Z et al (2018) Structural and biochemical characterization of endoribonuclease Nsp15 encoded by middle east respiratory syndrome coronavirus. J Virol 92(22):e00893–e00818
57. Yuan M, Wu NC, Zhu X, Lee DCC, So RTY, Lv H et al (2020) A highly conserved cryptic epitope in the receptor binding domains of SARS-CoV-2 and SARS-CoV. Science 368(6491):630–633
58. Nieto-Torres JL, Dediego ML, Verdia-Baguena C, Jimenez-Guardeno JM, Regla-Nava JA, Fernandez-Delgado R et al (2014) Severe acute respiratory syndrome coronavirus envelope protein ion channel activity promotes virus fitness and pathogenesis. PLoS Pathog 10:e1004077
59. Khailany RA, Safdar M, Ozasalan M (2020) Genomic characterization of a novel SARS-CoV-2. Gene Rep 19:100682.
60. Mehmood Z, Alrefai H, Hetta HF, Kader HA, Munawar N, Rahman SA et al (2020) Identify potential targets for developing COVID-19 treatment and prevention strategies. Vaccine 8(3):443
61. Kim D, Lee JY (2020) The architecture of SARS-CoV-2 transcriptome. Cell 181:914–921
62. Kariko K, Buckstein M (2005) Suppression of RNA recognition by toll-like receptors: the impact of nucleoside modification and the evolutionary origin of RNA. Immunity 23(2):165–175
63. Zhou P, Yang XL, Wang XG, Hu B, Zhang L, Zhang W et al (2020) A pneumonia outbreak associated with a new coronavirus of probable bat origin. Nature 579:270–273
64. Wrobel AG, Benton DJ, Xu P, Roustan C, Martin SR, Rosenthal PB et al (2020) SARS-CoV-2 and bat RaTG13 spike glycoprotein structures inform on virus evolution and furin cleavage effects. Nat Struct Mol Biol 27:763–767
65. Hoffman M, Kleine-Weber H, Schroeder S, Krüger N, Herrler T, Erichsen S et al (2020) SARS-CoV-2 cell entry depends on ACE2 and TMPRSS2 and is blocked by a clinically proven protease inhibitor. Cell 181(2):271–280
66. https://www.aaha.org/aaha-guidelines/infection-control-configuration/routes-of-transmission/
67. Plowright RK, Parrish CR (2017) Pathways to zoonotic spillover. Nat Rev Microbiol 15(8):502–510
68. Ghinai I, McPherson TD (2020) First known person-to-person transmission of severe acute respiratory syndrome coronavirus 2 (SARS-CoV-2) in the USA. Lancet 395(10230):1137–1144
69. del Rio C, Malani PN (2020) COVID-19—new insights on a rapidly changing epidemic. JAMA 323:1339–1340
70. Galbadage T, Peterson BM (2020) Does COVID-19 spread through droplets alone? Front Public Health 8:163
71. Raj VS, Mou H, Smits SL, Dekkers DHW, Müller MA, Dijkman R et al (2013) Dipeptidyl peptidase 4 is a functional receptor for the emerging human coronavirus-EMC. Nature 495(7440):251–254
72. Li W, Moore MJ, Vasilieva N, Sui J, Wong SK, Berne MA et al (2003) Angiotensin-converting enzyme 2 is a functional receptor for the SARS coronavirus. Nature 426(6965):450–454
73. Zhou F, Yu T, Du R, Fan G, Liu Y, Liu Z et al (2020) Clinical course and risk factors for mortality of adult inpatients with COVID-19 in Wuhan, China: a retrospective cohort study. Lancet 395(10229):1054–1062
74. Andersen KG, Rambaut A (2020) The proximal origin of SARS-CoV-2. Nat Med 26:450–452
75. Conti P, Ronconi G, Caraffa A, Gallenga CE, Ross R, Frydas I et al (2020) Induction of pro-inflammatory cytokines (IL-1 and IL-6) and lung inflammation by Coronavirus-19 (COVI-19 or SARS-CoV-2): anti-inflammatory strategies. J Biol Regul Homeost Agents 34(2):327–331

76. Harvala H, Robb M, Watkins N, Ijaz S, Dicks S, Patel M et al (2020) Convalescent plasma therapy for the treatment of patients with COVID-19: assessment of methods available for antibody detection and their correlation with neutralising antibody levels. MedRxiv
77. Oberfeld B, Achanta A, Carpenter K, Chen P, Gilette NM, Langat P et al (2020) SnapShot: COVID-19. Cell 181:954
78. Xu Z, Shi L, Wang Y, Zhang J, Huang L, Zhang C et al (2020) Pathological findings of COVID-19 associated with acute respiratory distress syndrome. Lancet Respir Med 8(4):420–422
79. Tang Y, Liu J (2020) Cytokine storm in COVID-19: the current evidence and treatment strategies. Front Immunol 11:1708
80. Tian S, Hu W (2020) Pulmonary pathology of early-phase 2019 novel coronavirus (COVID-19) pneumonia in two patients with lung cancer. J Thorac Oncol 15(5):700–704
81. Huang C, Wang Y (2020) Clinical features of patients infected with 2019 novel coronavirus in Wuhan, China. Lancet 395:497–506
82. Guan WJ, Ni ZY, Hu Y, Liang WH, Ou CQ, He JX et al (2020) Clinical characteristics of coronavirus disease 2019 in China. N Engl J Med 382(18):1708–1720
83. Shi H, Han X, Jiang N, Cao Y, Alwalid O, Gu J et al (2020) Radiological findings from 81 patients with COVID-19 pneumonia in Wuhan, China: a descriptive study. Lancet Infect Dis 20(4):425–434
84. Kong W, Agarwal PP (2020) Chest imaging appearance of COVID-19 infection. Radiol Cardiothorac Imaging 2(1):e200028
85. Wang M, Wu Q, Xu W, Qiao B, Wang J, Zheng H et al (2020) Clinical diagnosis of 8274 samples with 2019-novel coronavirus in Wuhan. MedRxiv
86. https://www.cdc.gov/coronavirus/2019-ncov/hcp/therapeutic-options.html
87. Wang M, Cao R, Zhang L, Yang X, Liu J, Xu M et al (2020) Remdesivir and chloroquine effectively inhibit the recently emerged novel coronavirus (2019-nCoV) in vitro. Cell Res 30(3):269–271
88. Corman VM, Landt O, Kaiser M, Molenkamp R, Meijer A, Chu DKW et al (2020) Detection of 2019 novel coronavirus (2019-nCoV) by real-time RT-PCR. Euro Surveill 25(3):pii=2000045
89. Agostini ML, Andres EL, Sims AC, Graham RL, Sheahan TP, Lu X et al (2018) Coronavirus susceptibility to the antiviral remdesivir (GS-5734) is mediated by the viral polymerase and the proofreading exoribonuclease. MBio 9(2):pii: e00221-18
90. Sheahan TP, Sims AC, Leist SR, Schäfer A, Won J, Brown AJ et al (2020) Comparative therapeutic efficacy of remdesivir and combination lopinavir, ritonavir, and interferon beta against MERS-CoV. Nat Commun 11(1):222
91. Cao B, Wang Y, Wen D, Liu W, Wang J, Fan G et al (2020) A trial of Lopinavir-Ritonavir in adults hospitalized with severe Covid-19. N Engl J Med 382:1787–1799
92. Barton C, Kouokam JC, Lasnik AB, Foreman O, Cambon A, Brock G et al (2014) Activity of and effect of subcutaneous treatment with the broad-spectrum antiviral lectin griffithsin in two laboratory rodent models. Antimicrob Agents Chemother 58(1):120–127
93. Rut W, Lv Z, Zmudzinski M, Patchett S, Nayak D, Snipas SJ et al (2020) Activity profiling and crystal structures of inhibitor-bound SARS-CoV-2 papain-like protease: a framework for anti-COVID-19 drug design. Sci Adv 6(42):eabd4596
94. Gautret P, Lagier JC, Parola P, Hoang VT, Meddeb L, Mailhe M et al (2020) Hydroxychloroquine and azithromycin as a treatment of COVID-19: results of an open-label non-randomized clinical trial. Int J Antimicrob Agents 56(1):105949
95. https://www.covid19treatmentguidelines.nih.gov/immune-based-therapy/immunomodulators/interferons/#:~:text=Interferons%20are%0a%20family%20ofaivap
96. Le TT, Andreadakis Z, Kumar A, Roman RG, Tollefsen S, Saville M et al (2020) The COVID-19 vaccine development landscape. Nat Rev Drug Discov 19:305–306
97. Organization, WHO (2020) Draft landscape of COVID-19 candidate vaccines

98. Kim TW, Lee JH, Hung CF, Peng S, Roden R, Wang MC et al (2004) Generation and characterization of DNA vaccines targeting the nucleocapsid protein of severe acute respiratory syndrome coronavirus. J Virol 78(9):4638–4645
99. Zhu FC, Li YH, Guan XH, Hou LH, Wang WJ, Li JX et al (2020) Safety, tolerability, and immunogenicity of a recombinant adenovirus type-5 vectored COVID-19 vaccine: a dose-escalation, open-label, non-randomised, first-in-human trial. Lancet 395(10240):1845–1854
100. Mair-Jenkins J, Saavedra-Campos M, Baillie JK, Cleary P, Khaw FM, Lim WS et al (2015) The effectiveness of convalescent plasma and hyperimmune immunoglobulin for the treatment of severe acute respiratory infections of viral etiology: a systematic review and exploratory meta-analysis. J Infect Dis 211(1):80–90
101. Pandey S, Vyas GN (2012) Adverse effects of plasma transfusion. Transfusion 52(Suppl 1):65S–79S
102. Kollias A, Kyriakoulis KG (2020) Thromboembolic risk and anticoagulant therapy in COVID-19 patients: emerging evidence and call for action. Br J Hematol 189(5):846–847
103. Stockman LJ, Bellamy R (2006) SARS: systematic review of treatment effects. PLoS Med 3:e343
104. Meduri GU, Bridges L (2016) Prolonged glucocorticoid treatment is associated with improved ARDS outcomes: analysis of individual patients' data from four randomized trials and trial-level meta-analysis of the updated literature. Intensive Care Med 42(5):829–840
105. Ragab D, Eldin HS (2020) The COVID-19 cytokine storm; what we know so far. Front Immunol 11:1446
106. Ledford H (2020) Coronavirus breakthrough: dexamethasone is first drug shown to save lives. Nature 582(7813):469
107. Veronese N, Demurtas J, Yang L, Tonelli R, Barbagallo M, Lopalco P et al (2020) Use of corticosteroids in coronavirus disease 2019 pneumonia: a systematic review of literature. Front Med 7:170
108. Zhao RC (2020) Stem cell-based therapy for coronavirus disease 2019. Stem Cells Dev 29(11):679–681
109. Zhang H, Penninger JM (2020) Angiotensin-converting enzyme 2 (ACE2) as a SARS-CoV-2 receptor: molecular mechanisms and potential therapeutic target. Intensive Care Med 46(4):586–590
110. Vaduganathan M, Vardeny O (2020) Renin-angiotensin-aldosterone system inhibitors in patients with COVID-19. N Engl J Med 382:1653–1659
111. Liu D, Chen M, Mendoza B, Cheng H, Hu R, Li L et al (2019) CRISPR/Cas9-mediated targeted mutagenesis for functional genomics research of crassulacean acid metabolism plants. J Exp Bot 70(22):6621–6629

Index

A
Abyssinone II, 128
Acute respiratory distress syndrome (ARDS), 178, 261
Acute respiratory syndrome, 65
Adenosine triphosphate (ATP), 172
Agrobacterium tumefaciens, 215
Airborne transmission, 258
Alcea digitata (Boiss.) Alef., 37
Alkaloids, 36, 118–120
Allicin, 46
Allium Sativum (Garlic), 46
Alnus japonica (Thunb.) Steud., 42
Alphacoronavirus, 7
Alternative drug therapies, 34
Amentoflavone, 128
Amino acid swap, 15
Andrographis paniculata, 36
Angiotensin-converting enzyme (ACE2), 36, 86, 87, 146, 157, 199–201, 268
    binding sites, 14
    glycosylation, 14
    protein, 14
    receptor, 8, 36, 257, 259
Animal models
    ACE2 receptors, 9, 10
    age and health factors, 11
    airborne coronavirus transmission, 10
    airborne nature, 11
    coronavirus studies, 9, 11
    digestive/respiratory tracts, 11
    disease transmission, 9
    epidemiological studies, 11
    HACE2, 10
    influenza, 11
    primates, 9
    rhesus monkey, 9
    SARS-CoV study, 10
    shed virus, 10
    traditional models, 10
    transgenic animals, 10
    transgenic knockout mice, 10
    types, 10
    vaccine formation, 12
    vulnerability, 11
Animal reservoir, 19
Animal welfare, 12, 13
Anthemis hyalina, 36
Anthraquinone, 146
Anticoagulation therapy, 267
Anti-coronavirus drugs, 265
Anti-CoV-NA cocktail therapy, 265
Anti-inflammation, 89
Anti-inflammatory properties, 156
Anti-inflammatory responses, 162
Anti-SARS-CoV characteristics, 35
Anti-SARS-CoV effect, 38
Anti-SC2 activity, 266
Anti-SC2 agent, 265
Anti-SC2 drug, 266
Antiviral screening, 112
Antiviral therapeutic mechanisms, 51
Antiviral therapy, 19
Antivirals
    SC2, 263–266
Apigenin, 128
Apoptosis, 92, 93
Arachidonic acid, 95, 96
Artemisia annua L., 37
Artogomezianone, 127

Asteraceae, 41
Astragalus membranaceus, 163
Atherosclerosis, 46
Autophagosomes, 92
Autophagy, 92
Azadirachta indica (Neem), 47

B
B-and T-cell response and immunity, 18
Belgium scientist group, 8
Berbamine, 120
Berberidaceae, 41
Berberis integerrima, 41
Betacoronavirus, 3, 7, 15
Betacoronaviruses, 250, 252
Bioactive compound curcumin, 47
Bioactive molecules, 158
Black cumin seeds, 49
BLT/LOM supported Zika virus, 15
Bovine corona, 5
Brassicaceae family, 142
Bupleurum species
   A. annua L., 37
   A. digitata, 37
   A. japonica, 42
   B. integerrima, 41
   C. aurantium, 40
   C. avium, 40
   C. microphylla, 41
   H. cordata, 39
   herbal medicine, 36
   I. indigotica, 37
   L. aggregata, 39
   L. radiata, 37
   O. acanthium, 41
   P. lingua, 38
   P. multiflorum, 40
   P. tomentosa, 42
   Q. tinctoria, 41
   R. palmatum, 39
   R. tinctorum, 40
   T. nucifera, 39
   T. terrestris, 43

C
Caesalpinia sappan L., 35
Calophyllum inophyllum, 131
Calorimetric detection, 9
Cancer patients, 177
Cardiovascular disorders, 52
Carriers, 4

Case casualty rates (CFR), 207
Cathelicidins, 169
Cathepsin proteolytically, 257
Cedrela sinensis, 36
Cell culture systems, 217, 218
Cell signaling pathways, 168
Cellular immunity, 167
Cepharanthine, 119
Cerasus avium (L.) Moench., 40
Chalcones, 122, 123
Chest computed tomography, 7
Chinese animal welfare organization, 12
Chinese Health Commission, 7
Chitin, 146
Chitosan, 146
Chlorophyll, 145
Chloroquine, 87, 196, 197, 266
Chromone-based polyphenolic
        phytochemicals, 121
Chromones, 142, 143
Chronic conditions, 168
Chronic diseases, 159, 161
Chronic human diseases, 111
Chronic inflammation, 156
Chronic inflammatory diseases, 158
Chrysophanic, 146
Citrus aurantium L., 40
Cleansing herb, 50
Clerodendrum inerme, 36
Clinical characterization
   SC2, 260–262
Clinical symptom studies, 19
CMV pathogenesis, 15
Coalition for Epidemic Preparedness
        Innovations (CEPI), 212, 226, 229
Combined booster strain vaccine, 16
Compositional profiling, 48
Convalescent plasma (CP), 267
Coriandrum sativum, 142
Corona-infected intestinal system, 5
Coronaviridae, 15, 33, 206
Coronavirus (COVID-19), 20, 249
   cell-mediated immune reactions, 229
   classification, 250, 251
   clinical and immunological
        endpoints, 217–220
   HBD, 78, 82
   herbs, TCM formulations, 69–72
   HZ capsule, 83
   immunity, 229
   interspecies transmission route, 251, 253
   JQ, 83
   LQ, 84

Index 277

monoclonal antibodies, 230
MXSGD, 85
and natural hosts, 250, 252
nucleic acid, 229
plasma/serotherapy, 267
QFPD, 75, 76
RdI, 84
SFJC, 84
S-proteins, 254
therapeutic options, 264
treatment, 264
utilization frequency, 69, 72, 74, 75
vaccines, 266, 267
viral shedding, 229
virus neutralizing antibodies, 229
XbI, 84
XFBD, 83
YPFS, 85
YQS, 85
Coronavirus-infected laboratory animals, 4
Coumarins, 142, 143
COVID-19 pandemic
   animal host and carriers, 4
   animal models, 9–13
   animal welfare, 12–13
   antiviral drugs, 34
   emergence and epidemiology, 5–7
   human-animal interaction, 7–9
   infection categories, 7
   influence types, 1
   jumping phenomena, 3
   molecular differences, 13–16
   origin, 2–5
   phylogenetic analysis, 3
   social and economic losses, 1
   social distancing, 1
   subfamily genera, 3
   symptoms, 5
   transmission hypothesis, 3
   zoonotic link, 1, 2
Coxsackievirus B1 (CVB1), 49
Crataegus microphylla C. Koch, 41
CRISPR/Cas-mediated therapy, 264
CRISPR/Cas System, 269
Cucurbitacin B, 160
Curcuma domestica (Turmeric, Haldi), 47
Curcumin, 144
Cures, 224
Cyclooxygenase (COX), 95
Cyclooxygenase-2, 96
Cytochrome P450 (CYP450), 95, 97
Cytokine release syndrome, 259
Cytokine storm, 89, 176, 259
Cytokines, 259
Cytosolic phospholipase A2-α (cPLA2α), 96

D
Death rate, 207
Defensins, 169
Dehydroascorbic acid (DHA), 138
Depression, 175
Dexamethasone, 194–196
Diabetes, 176
Diamine oxidase (DAO), 99
Dianella longifolia, 146
Diet, 154–157
Dietary fibers, 158
Dietary polyphenols, 178
Dihydrochalcones, 123, 126
Dihydroflavonols, 127
Direct contact transmission, 258
Disodium copper chlorin, 145
Disulfiram, 265
DNA vaccines, 16
Double-blind trials, 19
Dried garlic powder, 51

E
Echinacea purpurea L. (Echinacea), 47
Ecosystem, 63
Emetine, 88, 121
Emodin, 146
Encephalomyocarditis virus (EMCV), 144
Endocrine resistance, 98
Energy-dense foods, 161
Enterovirus 71 (EV71), 49
Envelop protein, 256
Epidemiological studies, 6
Epidemiology phases, 6
E-proteins, 256
ER-Golgi intermediate compartment (ERGIC), 258
Essential oils oleanane triterpenes, 34
Extracts, 120, 121, 123, 124, 128, 131, 136, 142, 143, 146

F
Fagaceae family, 41
Fangchinoline, 119
Fapilavir, 265
Feline calicivirus (FCV), 51
Fennel seed essential oil, 47
Fiber-rich plants, 161

Flavones, 124
Flavonoids, 36, 121
Flavonol, 127, 128
Flavonones, 124, 126
Foeniculum vulgare Mill. (Fennel), 47
Fomite transmission, 258
Foods, COVID-19
　cancer patients, 177
　cardiovascular patients, 176, 177
　depression, 175
　diabetes, 176
　diet, 154–157
　digestive disorders, 177, 178
　diseases, 153
　diversified food, 153
　fruits, 158–160
　geological regions, 154
　grains, 157, 158
　herbs, 163–165
　human-to-human transmissible disease, 153
　immune system, 173, 174
　legumes, 161, 162
　lowering stress, 175
　maladies, 154
　malnutrition, 179
　meat, 162
　metabolic processes, 154
　milk and milk products, 163
　minerals
　　iron, 170, 171
　　magnesium, 172, 173
　　selenium, 171
　　zinc, 172
　noncommunicable diseases, 153
　nutrients, 154, 174
　nutrition, 154–157, 179, 180
　nutritional intervention, 180
　nuts, 160, 161
　omega-3 PUFAs, 173
　pathogens, 180
　pulmonary disease patients, 178, 179
　resilience, 173, 174
　unhealthy substances, 154
　vegetables, 158–160
　vitamins
　　vitamin A, 166
　　vitamin B, 166, 167
　　vitamin C, 167, 168
　　vitamin D, 168, 169
　　vitamin E, 169
French Food Agency, 8

Friend murine leukemia helper virus (F-MuLV), 146
Fruits, 158–160
Functional foods, 160, 162

G
Galidesivir, 265
Gastrointestinal symptoms, 261
Genomic organization, 256, 257
Genomic RNA (gRNA), 256, 257
Gingerols, 145
Glucocorticoids (GCs), 268
Glycosaminoglycan (GAG), 136
Glycycoumarin, 141
Glycyrrhiza Glabra, 35, 48, 142
Glycyrrhizin, 35
Grains, 157, 158
Griffithsin, 266

H
Hampering, 266
HCoV target proteins, 36
Hdroxychloroquine, 196
Health approach strategies, 13
Health collaborative approach, 12
Healthcare professional, 52
Helichrysum aureonitens, 121
Hepatitis B virus (HBV), 122
Hepatitis C virus, 266
Herbal medication, 34
Herbs, 163–165
Herpes simplex virus (HSV-1), 122, 144
Hhydroxychloroquine, 196
Homoharringtonine, 119
Hong Kong Society, 13
Host-targeted strategy, 264, 265
Houttuynia cordata, 36, 39
Huashi Baidu Decoction (HBD), 78, 82
Human ACE2 similarity, 14
Human angiotensin-converting enzyme 2 (HACE2), 10, 243
Human coronavirus disease 19 (hCoV-19), 205
Human coronaviruses (HCoV), 249, 250
Humanized animals, 15
Humoral cell-mediated immunity, 16
Humoral immune response, 15
Huoxiang Zhengqi (HZ) capsule, 83
Hydroxychloroquine, 162, 266
Hyoscyamus niger, 34, 36
Hyperphagia, 175

Index

Hypertension-and hypokalemic-induced secondary disorders, 51
Hypophagia, 175

I

Immune regulation, 89
Immune-deficient, 15
Immunity, 154, 155, 158, 162, 163, 165, 167–172, 174, 178, 180, 229
Immunocompetence, 155
Immunohistobiochemistry assay, 8
Immunomodulation, 155
Immunomodulatory drugs
    SC2, 263–266
Immunosenescence, 155
Indian herbal plants, 34
Inducible nitric oxide synthase (iNOS), 99
Infection and immune physiology, 15
Infectious bronchitis virus (IBV), 207, 249
Infectious respiratory tract, 34
Inflammatory diseases, 177
Ingestion transmission, 258
Innate immune system, 170
Intercellular adhesion molecule 1 (ICAM1), 95
Interleukins (ILs), 91
International Committee on Taxonomy of Viruses, 5
Iron, 170, 171
Isatis indigotica fortune ex Lindl., 37
Isoflavanones, 131
Isoflavones, 130
Isoflavonoids, 129

J

JAK-STAT signaling pathway
    DPP4, 88
    EGFR, 89
    ILIB, 89
    JUN, 89
    sEH, 88
    VEGF, 88
Jinhua Qinggan (JQ), 83
JNK/MAPK phosphorylation, 38
Juglanin, 129

L

Legumes, 161, 162
Leukotrienes (LTs), 97
Lianhua Qingwen (LQ), 84
Licopyranocoumarin, 141

Lindera aggregata (Sims) Kosterm., 39
Lipopolysaccharide-induced inflammation, 168
Lipoxygenase (LOX), 95
Loop-mediated amplification tool, 9
Lophatherum gracile, 124
Lopinavir, 265, 266
Luteolin, 87
Lycorine, 121
Lycoris radiata, 35, 37
Lymphocyte-mediated adaptive immune system, 92

M

Ma Xing Shi Gan Decoction (MXSGD), 85
MAB management, 19
Macronutrients, 159, 160
Macrophage inflammatory protein, 259
Magnesium, 172, 173
Malnutrition, 179
Meat, 162
Mechanical ventilation, 261
Mechanism of actions (MOA), 263
Medicinal plants
    antiviral potential, 43–46
    basil, 49
    black seeds, 49
    echinacea, 47
    fennel herb, 47
    garlic, 46
    ginger, 51
    lemon balm, 48
    licorice, 48
    neem, 47
    oregano, 49
    peppermint, 49
    rosemary, 50
    sage, 50
    Senna Makki, 50, 51
    turmeric, haldi, 47
Mediterranean diet, 155
Melatonin, 175
Melissa officinalis L. (Lemon Balm), 48
Mentha piperita L. (Peppermint), 49
MERS-CoV, 251, 261
    camels, 18
    inactivated, 17
    mortality rate, 6
    pathology, 7
    receptor, 17
    reproductive number, 7
    vaccines, 9

Mesenchymal stem cell transplantation (MSCT), 268
Meta-analysis, 99, 102, 103
Micronutrients, 159, 160
Middle East respiratory syndrome (MERS), 250
Middle East respiratory syndrome corona (MERS-CoV), 6
Mild symptoms, 261
Milk and milk products, 163
Monitored Emergency Use of Unregistered and Investigational Interventions System (MEURI), 264
Monoclonal antibodies (mAbs), 266
Morphology
   SC2, 253–256
Mpro enzymes, 43
M-proteins, 256
mRNA-based vaccine, 17
Mulberry tree, 34
Multidimensional approach, 175
Mutant temperature-sensitive virus strain, 16
MVA-S vaccine, 17
Myocytes, 176
Myricetin, 127

N
Nafamostat, 266
Nanoparticles, 19
Naringin, 127
Nasal shedding, 18
National Center for Complementary and Integrative Health, 64
National Health Commission, 63
National Institute of Allergy and Infectious Diseases (NIAID), 212
Natural plant-derived compounds, 34
Neoflavonoids, 131, 133
Neurologic manifestations, 261
NF-kB pathway, 34
N-glycosylation, 13
Nicotiana benthamiana, 214
Nidoviruses, 256
Nigella sativa L. (Black Seeds), 49
Nonhuman model, 9
Nonhuman primate animals, 20
Nonhuman primates (NHPs), 242
Nosocomial infection, 6
Novel coronavirus (nCoV), 239
N-proteins, 256
Nucleic acid, 66, 211, 212
Nucleic acid-based vaccines, 19

Nucleocapsid, 66
Nucleocapsid protein, 256
Nucleoside analogues (NAs), 265
Nutrients, 174
Nutrition, 154–157
Nutritional syndromes, 48
Nuts, 160, 161

O
Ocimum bacilicum L. (Basil), 49
Olive oil, 155
Oliverine, 122
Omega-3 polyunsaturated fatty acids (PUFAs), 173
Onopordum acanthium L., 41
Oral vaccines, 18
Organic sulfur compounds (OSCs), 46, 143
Origanum vulgare L. (Oregano), 49
Oxostephanine, 122

P
Pachystaudine, 122
Palivizumab, 230
Papain-like protease (PLpro), 36
Pathogenesis, 10, 17
   SC2, 258, 259
Pathophysiological pathways, 166
Paulownia tomentosa (Thunb.) Steud., 42
Pelargonium sidoides, 163
Pericalline, 118
Pets adoption, 13
Pharmacokinetic profiles, 265
Pharmacological effects, 52
Phospholipase A_2 (PLA$_2$), 96
Phospholipase C (PLC), 96
Phylogenetic analysis, 15
Phytochemicals, 52, 112, 159, 161
Phytochemicals, antiviral agents
   alkaloids, 118–120
   flavonoids, 121
Phytonutrients, 159, 160
Plant metabolites, 65
Plant natural products, COVID-19
   anthraquinone, 146
   antiviral screening, 112
   chalcones, 122, 123
   chitin, 146
   chitosan, 146
   chlorophyll, 145
   chromones, 142, 143
   coumarins, 142, 143

Index

curcumin, 144
dihydrochalcones, 123, 126
dihydroflavonols, 127
flavones, 124
flavonol, 127, 128
flavonones, 124, 126
gingerols, 145
herbal medicines, 112
isoflavanones, 131
isoflavones, 130
isoflavonoids, 129
mechanism, 112–117
neoflavonoids, 131, 133
phytochemicals, 112
procyanidin A2, 130
procyanidin B1, 130
protection of human health, 111
selenium compounds, 144
tannins, 135, 136
terpenoids, 133
viral infections, 111
virucidal/antiviral agents, 111
vitamins, 136
Plant secondary metabolites (PSMs), 65, 66
Plant selection, 112
Plant's potential therapeutic applications
Allium genus, 35
C. sappan L., 35
chalcone compounds, 36
cytopathogenic effects, 35
H. cordata, 36
Morus plant species, 35
T. nucifera L., 36
Plasma therapy, 19
Plasma/serotherapy, 267
Platinum drug resistance, 91
Polyamine metabolism, 98
Polygonum multiflorum Thunb., 40
Polyphenols, 159, 175
Polyproteins, 87
Porcine epidemic diarrhea virus (PEDV), 3
Pretazettine, 121
Primates and non-primates, 18
Pro-inflammatory cytokine, 91
Pro-inflammatory cytokines prostaglandin E2 (PGE2), 37
Protease, 87
Protease inhibitors (PIs), 265
Protein Kinase B (PKB), 93
Public Health Emergency of International Concern, 63
Pyrrosia lingua (Thunb.) Farw., 38

Q
Qingfei Paidu Decoction (QFPD), 75, 76
Quercetin, 127
Quercetin-3-β-galactoside, 35
Quercus tinctoria, 41
Quinine, 87

R
Radix bupleuri, 36
Radix isatidis (dried root), 37
Ranunculus sceleratus, 122
Ranunculus sieboldii, 122
Receptor-binding domain (RBD), 207, 212, 244
Recombinant vaccine scenario, 17
Reduning Injection (RdI), 84
Remdesivir, 197–199, 265
Renin-angiotensin-aldosterone system (RAAS), 90, 200
Respiratory epithelium, 5
Respiratory infection, 16
Respiratory syncytial virus (RSV), 49, 142
Respiratory system, 18
Rheum palmatum L., 39
Rheumatism, 34
Rhinolophus sinicus, 14
Ritonavir, 265, 266
RNA-dependent RNA polymerase (RDRP), 36, 197
RNA positive-sense virus, 15
RNA vaccines, 17
RNA viral genomes, 249
Rosaceae, 41
Rosmarinus officinalis L. (Rosemary), 50
RT-PCR testation, 8
RT-qPCR test, 8
Rubia tinctorum L., 40
Rubiaceae, 40, 142

S
Salvia officinalis L. (Sage), 50
SARS-CoV
animal origin, 3
bovine coronavirus, 3
inflammatory cytokines, 34
natural carrier, 3, 6
origin, 6
reproductive number, 7
vaccine, 17, 19
SARS-CoV-1 S protein, 39

SARS-CoV-2 (SC2), 65, 66, 205
   animal models, 215–217
   cell culture systems, 217, 218
   antivirals/immunomodulatory drugs, 263–266
   Betacoronaviruses, 250, 252
   clinical characterization, 260–262
   comorbidities reported, 261, 262
   electron microscopy, 253
   E-protein, 256
   genomic characteristics, 253, 254
   genomic organization, 256, 257
   MOA, 263
   morphology, 253–256
   M-proteins, 256
   pathogenesis, 258, 259
   proteins and functions, 254, 255
   replication, 257, 258
   structural characteristics, 253
   taxonomy, 250
   transmission, 258, 259
   transmission and replication mechanisms, 253
S-based MERS vaccine, 17
Scotch thistle, 41
Scrophulariaceae, 42
Scutellaria plant genus, 35
Secondary metabolite quercetin, 35
Secondary plant metabolites, 159
Selaginella uncinata, 142
Selenium, 144, 171
Senna alexandrina Mill. (Senna Makki), 50, 51
Senna-induced dermatitis, 52
Serotonin, 175
Severe acute respiratory syndrome (SARS), 6, 146, 205, 250
Severe acute respiratory syndrome coronavirus 2 (SARS-CoV-2), 153
Severe respiratory infection (SRI), 249
S-glycoP, 257
S-glycoproteins, 254
Shufeng Jiedu Capsule (SFJC), 84
Silybum marianum, 124
Single-stranded RNA, 34
Size of coronavirus (HCoVs), 33
Spike protein, 256
Spike protein inhibitors, 201, 202
Stem cell-based therapy, 268
Strange pneumonia, 5
Strobilanthes Cusia, 34
Structure-activity relationship (SAR), 118
Subgenomic gRNAs (sgRNA), 257
Subunit-based immunizations, 212
Sulforaphane, 160
Swabs, 12

T

Tannins, 135, 136
Terpenoids, 133
Tetrandrine, 119
T-helper cells, 167
Therapeutic approaches
   ACE-2, 268
   anticoagulation therapy, 267
   COVID-19, 264
   CRISPR/Cas system, 269
   GCs, 268
   stem cell-based therapy, 268
Thymoquinone, 49
Toll-like receptors (TLRs), 259
Torreya nucifera L., 36, 39
Traditional Chinese medicine (TCM), 34
   Akt, 93
   anti-inflammation, 89
   antiviral activities, 64
   antiviral medicine, 104
   AOC1, 99
   apoptosis, 92, 93
   arachidonic acid, 95, 96
   BCL proteins, 93
   CALM, 97
   CASP, 93
   CCL2, 99
   COVID-19 patients, 90
   cPLA2α, 96
   cyclooxygenase-2, 96
   cytochrome P450, 97
   cytokine-mediated positive and negative immunity, 79
   detoxification category, 64
   endocrine resistance, 98
   formulation, 65
   FOS, 98
   HIF-1 signaling pathways, 98
   ICAM1, 95
   ILs, 91
   immune regulation, 89
   LOX, 97
   LTs, 97
   lymphocyte-mediated adaptive immune system, 92
   MAPK, 90, 91
   mechanism
      ACE2, 86, 87

direct and indirect actions, 86
DNA intercalation, 87, 88
plant secondary metabolites, 86
protease, 87
signaling pathways, 86
meta-analysis, 99, 102, 103
NF-κB pathway, 94
NOS$_2$, 99
plant metabolites, 65
platinum drug resistance, 91
polyamine metabolism, 98
RAAS, 90
RELA, 95
SARS-CoV, 65–69
TNF, 92
TP53, 94, 95
tyrosine kinase resistance, 91
usage, 64
Traditional week and inactive vaccines, 16
Transcription factors, 160
Transgenic animals, 10
Transmembrane protease serine 2 (TMPRRS2), 257
Transmembrane protein, 256
Transmissible gastroenteritis virus (TGEV), 3
Transmission, 6
  SC2, 258, 259
Treatment, COVID-19, 264
  angiotensin-converting enzyme inhibitors, 199–201
  chloroquine, 196, 197
  dexamethasone, 194–196
  pharmacologic and nonpharmacologic interventions, 194
  remdesivir, 197–199
  repurposing of drugs, 193, 194
  spike protein inhibitors, 201, 202
Tribulus terrestris L., 43
Trisodium copper chlorin, 145
T-suppressor cells, 167
Tylophora indica, 119
Tylophorine, 120
Tyrosine kinase resistance, 91

U
Umbelliferae, 142
Upper respiratory tract, 18
Urtica dioica (stinging nettle), 35
US Centers for Disease Control and Prevention, 13
US Food and Drug Administration (FDA), 239
Utrica dioica Agglutinin (UDA), 35

V
Vaccine development
  Beijing Institute of Biological Products/Sinopharm, China, 245
  Beijing Institute of Biotechnology, 240
  BioNTech SE and Pfizer Inc. USA, 244, 245
  CanSino Biological Inc., 240
  Gamaleya Research Institute, Russia, 246
  immunity, 237
  international organizations, 237
  Janssen Pharmaceutica, USA, 241, 242
  Moderna/NIAID, USA, 242, 243
  Novavax, USA, 243, 244
  secondary outcomes, 239
  Sinovac Biotech, China, 238, 239
  University of Oxford/AstraZeneca, UK, 240, 241
  Wuhan Institute of Biological Products Co., LTD, 239
Vaccines, 266, 267
  CFR, 207
  coronaviruses, 206
  developments, 220, 223, 226
  genome, 206
  geographical distribution, 228, 229
  global and equitable distribution, 224, 226–228
  helical nucleocapsid (N), 207
  immunity, 208, 210
  immunization-related disease, 208
  immunopathology, 208
  inactivated/live-attenuated virus vaccines, 213
  infection fatality rate (IFT), 207
  infectious diseases, 208
  nucleic acid, 211, 212
  off-target effects, 214
  phylogenetic assessment, 206
  plant, 214, 215
  protein subunit, 212
  vaccine platforms, 208, 210
  viral vectors, 213
  VN antibodies, 210
Vascular endothelial cells, 176
Vector-borne transmission, 258
Vegetables, 158–160
Vero cells, 239, 245
Viral diseases, 205
Viral infections, 89, 111
Viral pathogens, 34
Viral protein, 16

Viral replicase-transcriptase complex (vRTC), 258
Viral ribonucleoproteins (vRNP), 256
Viral RNA genomes, 253
Viral transmission studies, 20
Viral vectors, 213
Viremia, 18
Virophagy, 93
Virus-host-specific interactions, 52
Virus-induced cytopathic effect, 261
Virus-like protein (VLP), 256
Virus-mediated antibody, 17
Virus neutralization assay (VNA), 242
Virus neutralizing (VN) antibodies, 210
Virus-targeted strategy, 264, 265
Virus transmission, 259
Vitamin A, 166
Vitamin B, 166, 167
Vitamin C, 167, 168
Vitamin D, 168, 169
Vitamin E, 169
Vitamins, 136
Vitex trifolia, 36

W
Web-based dash board, 9
World Combat Zone, 104
World Health Organization (WHO), 13, 33

X
Xenophagy, 92
Xuanfei Baidu Decoction (XFBD), 83
Xuebijing Injection (XbI), 84

Y
Yin Qiao San (YQS), 85
Yupingfeng San (YPFS), 85

Z
Zinc, 172
Zingiber officinale Rosc. (Ginger), 51
Zoonotic diseases, 15
Zoonotic transmission, 258